INSTRUCTOR'S RESOURCE MANUAL
with TEST BANK & BIO-ART™
to accompany

Botany
An Introduction to Plant Biology
SECOND EDITION

Mauseth

INSTRUCTOR'S RESOURCE MANUAL
by **Marshall Sundberg**, *Louisiana State University*

TEST BANK
by **Ann M. Mickle**, *LaSalle University*
and James E. Mickle, *North Carolina State University*

SAUNDERS COLLEGE PUBLISHING
Harcourt Brace College Publishers

Fort Worth • Philadelphia • San Diego • New York • Orlando • Austin
San Antonio • Toronto • Montreal • London • Sydney • Tokyo

Copyright ©1995 by Harcourt Brace & Company
Copyright © 1991 by Saunders College Publishing

All rights reserved. No part of this publication may be reproduced or transmitted in any form or by any means, electronic or mechanical, including photocopy, recording, or any information storage and retrieval system, without permission in writing from the publisher, except that, until further notice, the contents or parts thereof may be reproduced for instructional purposes by users of BOTANY, Second Edition by James D. Mauseth, provided each copy contains a proper copyright notice as follows: © 1995 by Harcourt Brace & Company.

Printed in the United States of America.

Sunberg/Mickle/Mickle: Instructor's Manual, Test Bank, Bio-Art to accompany BOTANY, Second Edition.

ISBN 0-03-005892-9

567 017 987654321

Preface

This resource guide includes the instructor's manual, Test Bank and Bio-Art to accompany the second edition of *Botany* by James D. Mauseth. The instructor's manual, written by Dr. Marshall Sundberg of Louisiana State University, contains chapter outlines, laboratory suggestions, and ideas for fieldwork projects. It also has a wealth of supplemental information that can be included in lectures to provide extra motivation for students. Dr. Ann Mickle and Dr. James Mickle have prepared a test bank of multiple-choice, true/false, matching, and short essay questions. The Test Bank is also available as an *ExaMaster Computerized Test Bank* in IBM 5.25" and 3.5", as well as Macintosh and Windows. Bio-Art consists of 37 illustrations without labels. These can be used as part of a labeling exercise in an exam or can be photocopied and given to students for taking notes. Instructors can add their own labels to customize them to their course.

Table of Contents

1. Instructor's Manual by Marshall Sundburg, Louisiana State University

2. Test Bank by Dr. Ann Mickle, LaSalle University and Dr. James Mickle, North Carolina State University

3. BioArt

INTRODUCTION

The purpose behind this manual is to assist you to help your students learn botany. Botany is an exciting discipline and one in which new discoveries are making a dramatic impact on our quality of life. Your enthusiasm and excitement about the subject will do more than anything else to foster learning in your class. The majority of your students will probably not be majoring in a science and may, if not most, will have some degree of science anxiety at the beginning. Again, the attitude you project towards the material and towards the class will do wonders towards overcoming your students' initial apprehension.

In each chapter I have listed my objectives for that unit when I teach introductory botany. They do not cover every topic listed in the chapter. Similarly I have discussed the concepts which I feel are important to my students and included an example of a concept map for each chapter. For those of your who are not familiar with concept maps, I have made them a bit more elaborate for the beginning chapters to give you an idea of how to set them up. I find them to be a very useful tool for me, to organize my lectures and to check up on students' understanding of concepts. They are also a very useful tool to teach your students to aid them in their studying. In later chapters the maps are much more spartan - these are, in fact, what I use as lecture notes. Again, not every concept covered in the chapter is included in the concept map; I do not discuss every topic covered in each chapter in lecture.

I have also provided some hints and suggestions for presenting each of the topics in lecture. I must admit my bias, however; I much prefer asking questions and generating discussion to traditional lecturing. Over the years I have discovered that a significant number of my students do not follow even my most logically and meticulously crafted lectures - a fact I never knew at the end of such a session. It was only when I started asking questions, and WAITING for the student called upon to answer, that I began to realize the many unjustified assumptions I was making about my students' background and understanding. Wait time is critical! Rephrase the question if you have to, but wait at least 30 seconds to allow students to repond. It may take this long at first. Students are generally not used to be called upon to think and answer questions in class. Once students realize you are serious, however, they begin to respond much more rapidly and true interaction begins to take place. You cannot possibly give too many examples! Bring the "real thing" into class whenever you can!

Each chapter is written as a unit, complete with some suggested laboratory exercises. In most cases there is enough material in each unit for two or three days of lecture and laboratory. Your must pick-and-choose from this buffet for what you want to present in your class. The laboratory exercises are not meant to be complete in themselves - only to give you ideas and suggestions. In most cases I have specifically listed some references where you can obtain detailed procedures.

I hope you will find botany to be as rewarding to teach as I do. I also hope this manual will be useful to you in designing and implementing your course. I encourage you to let me know how you handle difficult concepts and other class-related problems. Finally, I want to thank Dr. Michael Dini for his stimulating discussions and useful suggestions, and my family for their patience.

CONTENTS

CHAPTER 1 INTRODUCTION TO PLANTS	1
CHAPTER 2 INTRODUCTION TO THE PRINCIPLES OF CHEMISTRY	10
CHAPTER 3 CELL STRUCTURE	16
CHAPTER 4 CELL DIVISION	24
CHAPTER 5 TISSUES AND PRIMARY STEM GROWTH	32
CHAPTER 6 LEAVES	40
CHAPTER 7 ROOTS	46
CHAPTER 8 WOODY PLANTS	53
CHAPTER 9 FLOWERS	61
CHAPTER 10 PHOTOSYNTHESIS	69
CHAPTER 11 RESPIRATION	79
CHAPTER 12 TRANSPORT	86
CHAPTER 13 SOILS AND MINERALS	94
CHAPTER 14 MORPHOGENESIS	98
CHAPTER 15 GENES	106
CHAPTER 16 GENETICS	115
CHAPTER 17 POPULATION GENETICS AND EVOLUTON	122
CHAPTER 18 CLASSIFICATION	130
CHAPTER 19 KINGDOM MONERA	139
CHAPTER 20 FUNGI	147
CHAPTER 21 ALGAE	156
CHAPTER 22 NON-VASCULAR PLANTS	163
CHAPTER 23 SEEDLESS VASCULAR PLANTS	169
CHAPTER 24 GYMNOSPERMS	174
CHAPTER 25 ANGIOSPERMS	180
CHAPTER 26 POPULATIONS AND ECOSYSTEMS	187
CHAPTER 27 BIOMES	196
APPENDIX A TREE KEYS	201
APPENDIX B FIELD TRIPS	210

CHAPTER 1: INTRODUCTION TO PLANTS

The first day's lecture is the time to do general course housekeeping, go over the syllabus, explain grading policy, outline your general objectives, etc. It is also the time to set the tone for the course. One of the great advantages of studying plants, as compared to studying animals, is that plants are so much more easily observable. Plants do not run away as you approach them and they are much easier to care for in the house (or dormitory room) than are animals. They are also much more amenable to experimentation; few people are concerned with "plant rights" and will object to dissection or surgical experimentation. Take advantage of these opportunities and encourage your students to follow Mauseth's advice - "You can figure out a great deal by observing a plant and thinking about it... you will be surprised at how much you already know about plants..."

OBJECTIVES

1. Generate students' interest in plants and the study of plants.
2. Describe the roles of hypothesis, experimentation and verification (or rejection) in the scientific method.
3. Describe two areas in which the scientific method is inappropriate to provide an answer to a question.
4. Explain the roles of mutation and natural selection in the process of evolution.

CONCEPTS

This chapter introduces several key concepts: plants, scientific method, evolution, and diversity of living things. There are many interrelationships between these concepts, two of which are mapped on the next page. The main concepts, in boxes, are arranged in a hierarchy with the more specific concepts placed beneath the more inclusive ones. Lines, labelled with linking words, show relationships between concepts.

It is important for students to realize that there is no single "right" concept map, just as there is no one "right" way for everyone to teach botany. Each chapter of this manual contains a concept map to help you organize your lecture. Of course, you must decide the flow pattern to use in guiding your class presentation. This chapter has two different concept maps to illustrate how a change in emphasis may alter the way in which the same material is presented. Ideally, you will want to construct your own map to represent your emphasis of the material you will cover. Note that the concept maps do not outline everything discussed in the chapter - neither should you feel compelled to cover everything in each chapter in your lectures. The textbook is a supplement to aid you in presenting the material you feel is important for your class.

Mauseth does a good job of explaining the limits of science and the areas that are inappropriate for scientific investigation. Most students, especially non-science majors, will bring to the course either a misconception that science can provide "all the answers," or worse, a distrust of any scientifically supported theory because of a perceived conflict with religious or moral beliefs. It is important to make a clear distinction between the areas appropriate for each kind of study at the beginning of the course. This is especially true given the evolutionary approach adopted in the text; evolution is the single most misunderstood and, therefore, controversial concept in biological science.

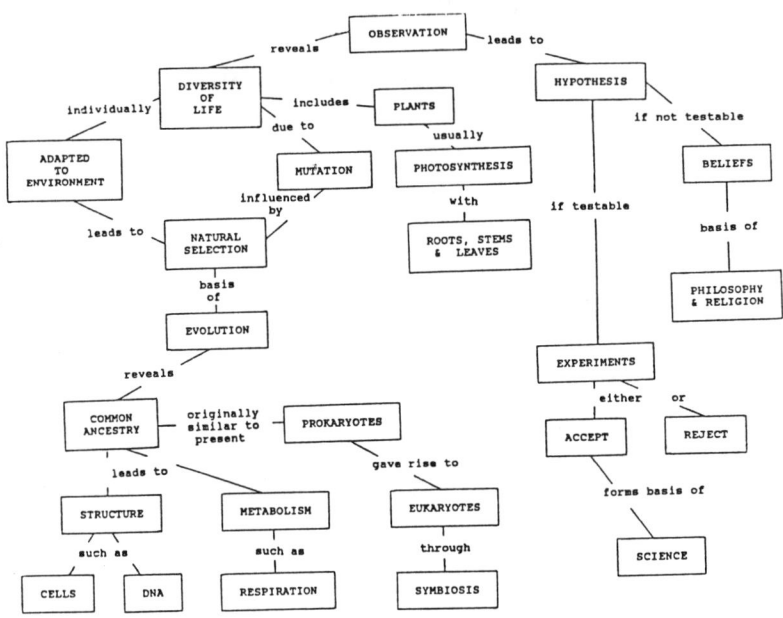

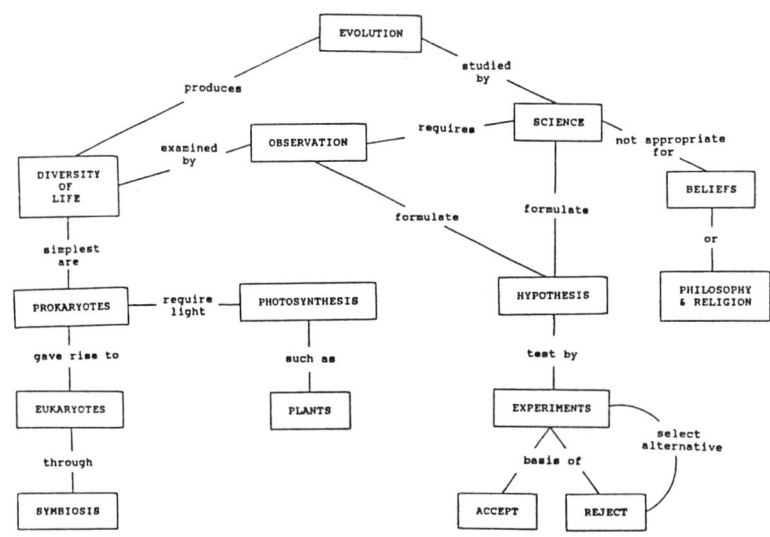

Introduction to Plants

INSTRUCTIONAL AIDS

Lecture - Many students in non-majors' science courses bring science anxiety to class with them. One of our most difficult tasks is to convince these students that science is not more inherently difficult than any other way of knowing. In other words, the scientific method is simply a more formalized and rigorous form of everyday common sense in problem solving. A good way to illustrate this is to start the semester with a hypothetical situation in which the students must use common sense to solve a problem. For example, you may describe the following scenario:

> On the first day of classes you [the observer] arrive for your 8:00 botany lecture a half-hour early to be sure to get a good seat - front row, center. After about ten minutes the first student arrives. This strikingly attractive coed proceeds directly to the far back corner of the lecture hall. A few minutes later a handsome young man enters, hesitates as he quickly scans around the room, then moves to the back and sits in the seat next to the coed.

Now ask your class to guess what, if any, relationship exists between these two students. Write some of their guesses on the board. Responses may be slow in coming at first, but once a few students volunteer an answer you will soon have a list of possibilities from which to choose.

Given the list on the board, ask the students how they would determine which, if any of them, was actually correct. Some students may suggest to drop some of the more capricious guesses, to narrow the list by process of elimination. Others may want to examine one of the more likely alternatives. In either case, they will come up with ways to "check out" one of these guesses in order to accept or reject it. For each test proposed, ask what criteria should be looked for to either accept or reject the hypothesis. Be sure to caution your students that even if the criteria suggest that the hypothesis should be accepted, there might be some other possible explanation that was overlooked.

You now have an example of the scientific method in action, in the context of everyday problem solving. At this point you could formally outline the scientific method, using the examples the class just developed, to illustrate each step. You also could move to a more specific example that forces students to look more closely at plants as living things.

For example, you could pass out some carrots to the class and ask your students to decide if a carrot is alive or not. A technique that works well here, even with several hundred students, is to have a vote - all those in favor of living? Of not alive (everyone must make a commitment and vote one way or the other)? This situation invariably presents an example of the failure of democracy: the majority will vote "not living!" But for you, the teacher, this is a blessing because you now have an hypothesis -- a harvested carrot is not alive -- which is eminently testable and in fact could be the basis of your first laboratory session!

How can your students test if the carrot is alive or not? The first thing they will have to do is agree on some characteristics of living things, some criteria of life. Ask for suggestions to construct another list on the board. Some typical suggestions include: living things are composed of cells [you will have to get some criteria for recognizing cells, especially plant cells]; living things grow [you will have to ask for a definition of growth - is this increase in size due to cell enlargement?, increase in number of cells? increase in complexity of the organism?, use of food or some chemical process to obtain energy?]; living things move [first get a definition of "move." Students usually mean that the entire organism is motile. This of course fits most, but not all, animals and illustrates that when students think about living things, they really think about animals, and more specifically, vertebrate animals. This is a bias that you as a botany teacher will want to be aware of throughout the semester - and consciously use as a source of familiar comparison.

Post-lecture Review - As you might expect, concept mapping is also a good learning tool for students. A useful technique is to take the last five minutes of lecture to make a list of the concepts that you covered that day (your concept map of the lecture should be a good guide). Ask students to generate the concepts while you list them on the board. Given this list of concepts, ask the students to map them in their notes as part of a homework assignment. One or two "volunteers" will put their concept map on the board at the beginning of the next class (a technique that works even in large classes)!

The first day you will have to spend a little more time at the end of class to demonstrate how to construct a concept map. After the students have generated a list of what they consider to be the key concepts covered that day, assisted by your coaching, ask for a vote as to what was the single most inclusive concept (here is the place to emphasize that there is no one right answer, but there are different ways of organizing information). Map the highest vote getter at the top of the diagram; place it on the board as such. The other high vote getters are important subconcepts. Choose one and ask how it is related to the first. Map this on the board with one or a few connecting words to describe the relationship. Continue to map one or two additional concepts to be sure that the class understands what they are supposed to do with the entire list of concepts.

Pre-lecture Review - The following day, begin by asking one or two students to put their concept map for the previous lecture on the board (volunteers can be asked to do this before class begins). You can use these student concept maps to quickly review the previous day's lecture and to identify misconceptions which may have arisen or information you emphasized but was not "picked up" by the student. You can also use concept maps as an evaluation tool, as an alternative to essay questions, or to evaluate student understanding of course material.

Introduction to Plants

LABORATORY - EXERCISES

1. Diversity and Classification

If a fair assortment of plant material is readily available, an interesting and useful first lab is to set out stations of three specimens for students to observe. Two of the three specimens at each station should be closely related, the third should be different. The student's task is to observe carefully the specimens at each station, record observations, and decide which specimens are related most closely and which is the "outlier."

The emphasis for the students is observation. In some cases it may be useful to have hand lenses or dissecting microscopes available at a station for students to observe minute details. A brief demonstration of proper use and care of these tools is sufficient at this point. Encourage students to make rough sketches of significant features, including descriptive labels as appropriate. Many students will complain that they are "not artists." Reassure them that artistic ability is not required to make a useful record of an observation.

Some of the stations should be straightforward, e.g., branches of two gymnosperms and one angiosperm or flowering shoots of two monocots and one dicot, but others should be difficult, eg. a foliose lichen, a liverwort, and a moss or two gymnosperm seedlings and a lycopod ("ground pine"). Use many familiar examples, but also specimens, or parts of specimens, frequently overlooked. The larger and more showy the collection you can provide, the more positive the initial impression will be on your students.

After each member of the class has had an opportunity to observe all stations, begin to review, station by station, the specimens on display. Ask a volunteer (or get a class vote) to identify the outlier at the first station. Then justify this hypothesis by presenting their supporting observations. Here is your opportunity to give a first introduction to the variety of living things traditionally studied in botany courses, morphological variations and adaptations, the concepts of homology and analogy, etc.

Some additional examples of comparisons: shoots of Juniper, Pine, and *Casuarina*; cactus, succulent euphorb, and poinsettia; bracket fungus, mushroom, and morel; *Psilotum*, unbranched and branched *Equisetum; Salvinia, Azolla,* and duckweed; two grasses and a sedge; leafy shoots of maple (simple leaves), boxelder (compound leaves), and buckeye (compound leaves); willow catkins, flowering shoot, and cluster of male pine cones.

2. Hypothesis-testing laboratory: "Is a carrot alive?"

Relist characteristics of living things from lecture. Lead students to design simple experiments that will allow them to check for the presence or absence of some of these characteristics. The original hypothesis was probably that the carrot is not alive. A logical

Chapter 1

approach is to deduce that if the carrot is not alive, then it will not exhibit any of the characteristics of living things. Let groups of students design several experiments, under your guidance, then divide responsibilities for carrying them out. At the end of the laboratory period, each group reports on the results of its experiment and the class can reevaluate their original hypothesis based on the research results. For example:

A. Hypothesis: Cellular structure is a characteristic of living things; therefore, if the carrot is not alive, it will not have cells.

At this point, it will be necessary to discuss briefly the nature of cells - how can students recognize cells. Many students will remember something about cell walls, a nucleus and, perhaps, other organelles from their high-school biology. You may briefly review these concepts with the aid of an outline sketch on the board. How can students test this hypothesis? With the microscope. You may want to do a formal introduction to the use and care of the microscope at this point, although it is not necessary in order to complete this experiment. At a minimum, you will want to describe the light path and relate this to the need to have thin specimens. You also will have to identify the low- and medium-power objectives, high power will be unnecessary, and the coarse- and fine-focus knobs.

The most critical aspect will be to explain and demonstrate how to make a hand section of the carrot. Emphasize that the section should be as thin as possible, but it need not be uniformly thin. Wedge-shaped sections are easier to cut and the thin edge is usually adequate for examination. Be sure to caution students about safety when handling glass slides and razor blades, and point out disposal receptacles and the location of a first-aid kit.

Tell the students doing this experiment to sketch a group of 5 or 6 cells from their preparation, large enough to fill at least half a side of paper. The sketch should be in pencil and include all of the structures observed, labeled with descriptions. Students will observe cell walls and orange chromoplasts and occasionally they will see a nucleus or starch granules.

B. Hypothesis: living things obtain energy for growth from respiration; therefore, if the carrot is not alive, it will not undergo respiration.

Most students will associate respiration with breathing; use this as a starting point for a brief discussion of respiration. Exchange of gases, in particular release of CO_2, provides a useful index of respiration. As a demonstration, have a student blow through a straw into a half-filled beaker of phenol red solution. Phenol red is an acid-base indicator that changes from red to yellow as pH decreases in the range of from 8.4 to 6.8. In solution, CO_2 reacts with water to form carbonic acid, thus lowering the pH of the solution. As students blow bubbles into the red solution, it will gradually change to orange, then yellow. Given this phenomenon, ask students how they could use phenol red to test the above hypothesis?

Most students will realize that a reasonable experiment to test this hypothesis will be to place some pieces of carrot into a tube of phenol red solution. Some will realize that a control tube, phenol red without carrot tissue, also should be set up. If the hypothesis is correct, there should be no color change in any of the tubes. A change in color indicates that respiration is probably occurring; therefore, material in the tube may be living. The phenol red is reusable and yellow solution will gradually shift back to red under atmospheric conditions.

C. Hypothesis: living things reproduce; therefore, if the carrot is not alive, it will not be able to reproduce.

To test this hypothesis students must use a considerably more elaborate experiment than those outlined above - carrot tissue culture. Carrots were one of the first materials from which entire plants were grown from tissue culture explants. Most manuals of tissue culture include at least one experiment involving carrot culture, and kits are available from major biological supply houses. The key is to maintain sterility! Ideally the procedures should be done in a hood, but this is not essential. I have had students construct reasonably effective "hoods" from foil-lined cardboard boxes with a Saran wrap window. The procedures may even be done, with some success, on an open bench top if students keep movement around the laboratory to a minimum. In the latter situation, I can usually expect to find two or three plates uncontaminated plates from a class of 24 non-majors. These eventually produce callus. Even the "failures" are useful, however, because the contaminants exhibit a variety of growth forms, many of which will be examined later in the course. Contaminants can also be used to emphasize the ubiquity of bacterial and fungal spores, members of two widespread and extremely important kingdoms of living things that are usually overlooked or ignored by students.

There are several special safety precautions that must be considered before beginning this experiment. First, alcohol and a flame are used to sterilize instruments. The alcohol container should be kept away from the flame and a cover for the container should be easy to apply to extinguish a possible fire. There should be a fire extinguisher readily available and you must caution students about the potential danger of fire. Students will also be using either razor blades or a scalpel so, again, there is the possibility of a student being cut.

Finally, any contaminants that turn up in a plate should be considered potentially harmful. Students should seal their plates with tape or Parafilm™ when they are first set. Do not reopen contaminated plates unless the plates are first flooded with disinfectant or autoclaved.

LABORATORY - MATERIALS, EQUIPMENT, AND PREPARATION

1. Diversity and Classification.

This exercise will initially require a good deal of effort to come up with a suitable number of comparison stations. The effort required will vary depending on time of year the

Chapter 1

course is offered (what is available locally to be brought into the laboratory) and the greenhouse facilities and landscape plants on your campus. In some cases you can use preserved specimens, but it is always better to use living material if at all possible. Set up for subsequent years is simply a matter of collecting appropriate material from sources already identified.

-hand lenses and/or dissecting microscopes (demonstrate proper use of these instruments at the beginning of the laboratory period)

2. Hypothesis testing; "Is a carrot alive?"

A. Cellular Structure.
-four carrots - for a class of up to 24 students
-large kitchen knife
-slides, cover slips, razor blades (the key to hand sectioning is a sharp razor blade - the sharper the blade, the easier it is to obtain a usable sections. The very best blade is the Gilette Blue-blade®, followed by other double edge blades and finally single edge blades. Double-edged blades may be cut in half with a scissors, then a piece of tape doubled over the cut side to form a handle.)
-stain, e.g., 1% Safranin or Toluidine Blue

B. Cellular Respiration.
- two test tubes and a holder (rack or small beaker) per group (additional tubes will be needed for replications and/or additional conditions, e.g., light vs. dark, as dictated by student experimental designs.
- large kitchen knife
- container of ca. 50 ml Phenol Red per group (add enough Phenol Red indicator dye to water to produce a noticeable color - if red it is ready to use, if it is yellow, add enough NaOH to shift to the reddish form)
- 10% commercial liquid bleach with a drop of surfactant, such as Tween 20 or liquid detergent if students wish to surface sterilize their tissue
- sterile water (two changes should follow surface sterilization to remove excess bleach
- prepare ahead of time and have at room temperature).

C. Tissue Culture
per student group
- two petri dishes of growth medium (usually a variation of Murashigi/Skoog)
- five sterile petri dishes (one for surface sterilizing, three for rinsing tissue samples, and one to use as a sterile dissection field)
- liquid bleach and sterile water, as above
- forceps and scalpel (individually wrapped and autoclaved)

- capped tube of alcohol and alcohol lamp or burner (to resterilize instruments during use).
- Parafilm, or tape (to seal cultures)

REFERENCES
General:

Attenborough, David. 1979. *Life on Earth.* Little, Brown and Company. Boston. 319 pp. Although concentrating on animal life, this book, and the video tape set of the same name, provides an excellent introduction to the diversity of living things.

Darwin, Charles. 1962. *The Voyage of the Beagle.* Doubleday and Company, Inc. Garden city, New York. 524 pp. Darwin's own account of the voyage which provided the inspiration to develop his theory of evolution by means of natural selection.

Futuyma, Douglas. 1983. *Science on Trial: The Case for Evolution.* Pantheon Books, New York. 251 pp. One of several books dealing with the science/creation controversy; this one does a good job of making clear the limits of science and religion.

Gould, Stephen Jay. 1977. *Ever Since Darwin: Reflections in Natural History.* W.W. Norton and Company, New York. 285 pp. This is the first of his collections of essays from Natural History - any one of them provides interesting perspective on the Theory of Evolution.

Heinrich, Bernd. 1984. *In a Patch of Fireweed.* Harvard University Press, Cambridge, Mass. 194 pp. The science in this book is insect physiology, but it provides an excellent commentary on why to become a scientist and how scientists think.

Resources:

Dodds, John and Lorin Roberts. 1982. *Experiments in Plant Tissue Culture.* Cambridge University Press, Cambridge. 178 pp. Several tissue culture technique manuals are available - this is one of the more inclusive small ones suitable for undergraduates.

Morholt, Evelyn, Paul Brandwein, and Joseph Alexander. 1966. *A Sourcebook for the Biological Sciences, 2nd ed.* Harcourt, Brace and World, Inc. 795 pp. This is an invaluable reference for materials and techniques. A new edition is available.

Scagel, R.F., R.J. Bandoni, J.R. Maze, G.E. Rouse, W.B. Scofield, and J. R. Stein. 1984. *Plants: An Evolutionary Survey.* Wadsworth Publishing Co., Belmont, California. 757pp. This text provides a wealth of information concerning the diversity of living things traditionally covered in Botany courses - monerans, fungi, algal protists, and non-vascular and vascular plants.

CHAPTER 2: INTRODUCTION TO THE PRINCIPLES OF CHEMISTRY

Many of your students will be taking botany because it fulfills a science requirement at the college or university, and anything dealing with biology is preferable to physics or chemistry. As a result, students are generally not too excited about the prospect of having to learn some chemistry at the beginning of their botany course. You will probably explain the rationale that much of modern life science is really biochemistry-related, so it is essential that students have a grasp of the basic concepts; unfortunately this is not usually enough of a motivation for most students. Probably the most important thing to do is reassure students that the basic concepts are not really all that difficult and that they should concentrate on understanding these concepts and not worry about getting bogged down in details. It is, then, your responsibility to cover these basic ideas in lecture, one step at a time, and let the text provide the details. Of course this means that when it comes time for testing, you also should concentrate on these basic concepts, with only enough of the details thrown in to separate the A's from the B's.

OBJECTIVES

1. Understand the nature of covalent, ionic, and hydrogen bonds and the relationship between chemical bonds and energy.
2. Describe the biologically important properties of water resulting from its chemical bonds.
3. Understand the significance of polymerization in constructing biological molecules.
4. Recognize the basic types of biological molecules: carbohydrates, lipids, proteins, and nucleic acids.
5. Describe the role of dehydration synthesis and hydrolysis in the formation and breakdown of biological molecules.
6. Understand how co-factors may couple exergonic and endergonic reactions.

CONCEPTS

There are many more concepts covered in this chapter than you will have time to discuss in lecture; however, there are a few key concepts that typically give students the most trouble. If you spend your lecture time giving your students a solid foundation in these concepts, they should be able to pick the rest up from the text. The first, and most basic, of these concepts is that of the chemical bond. The critical thing is that bonds store energy and that it takes energy to form a bond, and that energy is released when bonds are broken. The second major concept is that of polymerization. Biological molecules are composed of relatively few basic building blocks that may then be strung together to make larger molecules (or the larger molecules taken apart into their more basic components). The process by which this occurs, dehydration synthesis, (removing water to join blocks together) or hydrolysis (adding water to split larger molecules apart) is basically the same for all biological molecules. Of course, polymerization or depolymerization require the formation or breakdown of chemical bonds that ties back to the first main concept. Ask students, where the energy comes from to drive energy requiring reactions? This leads to the third key concept, that of co-factors (particularly ATP) transferring energy from one reaction to another.

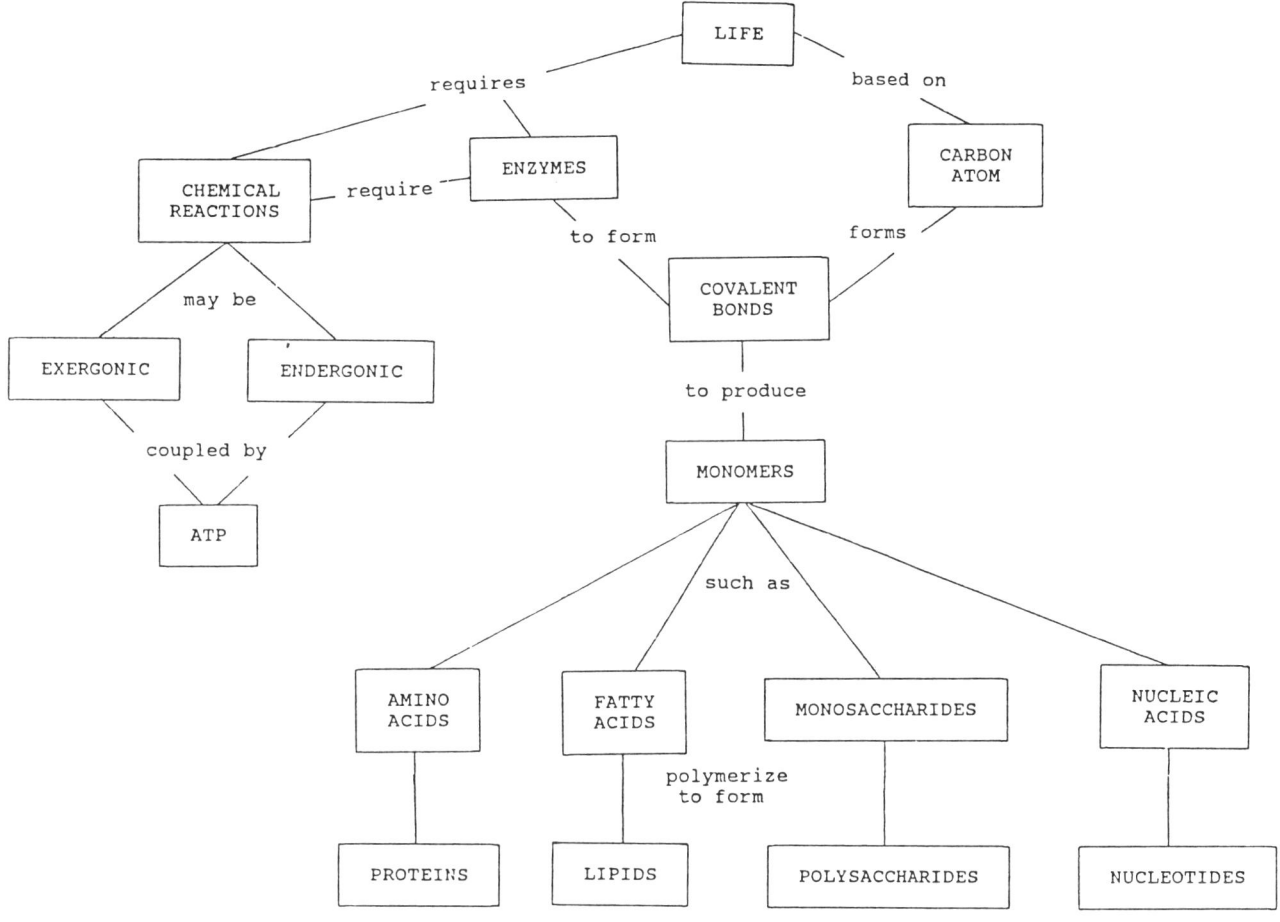

INSTRUCTIONAL AIDS

Lecture - Before beginning today's lecture, don't forget to have a student put her/his concept map from the last class period on the board to use as a quick review. Many of the characteristics of living plants, such as metabolism and the construction of cells themselves, depend directly on the chemical nature of cells. This is your tie-in to today's material.

To understand chemistry, you must first have a feel for the concept of an atom. Students can read about protons and neutrons, atomic weight and atomic mass, etc.; concentrate on the role of electrons in forming bonds. A set of magnetic stir bars, projected through an overhead projector, can be a useful demonstration of attraction of oppositely charged particles, whether ions or polar molecules such as water.

You can use the same set of stir bars to illustrate the concept of polymerization. Each bar represents a monomer, of which there can be many different types (eg. different-length stir

Chapter 2

bars). The formation of chemical bonds between individual monomers will join them together into larger molecules of various types. Perhaps only two units join, but they may be two of the same or two different monomers. Perhaps many units have polymerized. The nature of the polymer depends, to a large degree, on which monomer units are used and the order in which they are linked. The point to stress is that no matter how many individual units are joined together, or what their individual order may be, the same basic process holds one to the next. Biochemically, we are demonstrating the principle of dehydration synthesis reactions.

You should also stress the principles of dehydration synthesis and hydrolysis when covering the structure of basic biomolecules. Again students can read the details, the names of particular molecules and functional groups. Lecture time is better spent on recognizing the characteristic structure of a class of molecules, eg. $(CH_2O)n$ for a carbohydrate, and how the monomers of that class are linked to form polymers (dehydration synthesis) and polymers split back down to the basic units (hydrolysis).

So far, we've been dealing with structural features of chemistry: recognizing the structure of molecules and how individual units may be joined or split off from one another. The main purpose for covering basic chemistry in a botany course, though, is to gain a better understanding of cellular energetics.

Mauseth has a nice discussion of exergonic and endergonic reactions and their relation, or lack of relation, to activation energy. Examples of these processes usually involve heat, and rather large amounts, but some of these reactions proceed at room temperature as well and it is frequently useful to be able to demonstrate this. For instance, you can have two beakers of water at the front of the room - have a student read the temperature. To one beaker, add a pellet or two of NaOH, add a few ml of acidified DMP to the other [1 drop of conc. acetic acid / 100 ml 2,2-dimethoxypropanol (DMP)]. After a minute or two, read the temperatures of the respective solutions. Heat is released as NaOH dissociates, heat is absorbed as DMP hydrolyses into acetone and methanol.

This chapter provides some good opportunities to reinforce evolution as a common theme in plant science. For instance, it is not coincidental that all life is based on carbon chemistry or that the same basic processes of dehydration synthesis / hydrolysis are used in polymerization and depolymerization of all types of biomolecules. The process of polymerization itself mirrors the evolution of organisms. Polymers are built of simpler pre-existing units, and of the multitude of possible configurations which could be constructed, a relatively few are characteristic of living things; non-useful or harmful molecules must be selected against. Finally, the same molecule, DNA, directs development in all living things. Analysis of this molecule is now one of the strongest tools used by botanists to investigate evolutionary relationships between different organisms.

Chemistry

Post-lecture Review - Don't forget to take the last five minutes of lecture to have your students map the concepts covered today. You may argue that this is a waste of valuable class time and that you could tell your students to do this as a homework assignment (to have ready to put on the board at the beginning of the next class). Unless you demonstrate that the exercise is valuable enough to devote some class time to, most students will not take it seriously (until the first exam comes along!)

LABORATORY - EXERCISES

Each class of biologically important molecules has one or more characteristic functional groups that may be used to identify molecules in that class. A chemical test, specific for a particular functional group, will be positive only if that group is present. For instance:

1. Carbohydrates
A. Monosaccharides and some disaccharides have a free aldehyde group and are called "reducing sugars". When heated in the presence of an excess of **Benedict's Reagent**, the color of the solution changes from blue to green to orange, depending on the amount of reducing sugar present.

Add 1 ml of each solution to be tested to a separate, labeled test tube. Then add 2 ml of Benedict's Reagent to each tube and heat in a boiling water bath for 3-5 min.

B. Starch is a polymer with a specific three-dimensional structure that interacts with iodine-potassium iodide (I_2KI) to produce a blue-black color.
Add I_2KI, drop by drop, to the unknown and look for color change

Lipids
A. Lipids are hydrophobic, thus insoluble in water. This in itself is a useful indicator - a mixture of lipid in water will separate into two distinct layers.
(test solutions of lipids will have to be prepared with 95% alcohol, not water). Chemical tests for lipids depend on the multiple hydrocarbon groups of the fatty acids.

The stains **Sudan III** and **Sudan IV** are red dyes which stain lipids. Place a drop of each unknown solution on a piece of filter paper (be sure to label each drop if more than one is placed on each paper). Allow the drops to dry - use a hair drier to speed up this process. Soak the paper for 2-3 minutes in Sudan stain, then rinse the paper in water for a minute. The intensity of orange staining indicates the amount of lipid present.

Proteins
A. Amino acids are characterized by an amine group which joins to a carboxyl group to form a peptide bond. The bound amine group in this bond reacts with copper ions in **Biuret**

Chapter 2

Reagent causing a color change from blue to violet. Add equal amounts of unknown solution and Biuret Reagent.

Working in groups, students should try each of the tests, along with a control, to see what a positive test looks like. Then the fun begins. Have a number of unknowns available. Each group must determine the chemical composition of the unknowns assigned to them. Run as a contest for speed and accuracy, this is a good way to develop group cooperation, spirit and class rapport.

LABORATORY - MATERIALS, EQUIPMENT, AND SUPPLIES

- test tubes and test tube racks
- hotplate and beaker (boiling water bath)
- test tube holders
- hair dryer
- large finger bowls or beaker (for sudan stain and water rinse)
- filter paper
- squeeze bottles or dropping bottles to dispense solutions
- Benedict's Reagent. Mix 173 g of sodium (or potassium) citrate and 200 g of Na_2CO_3 in 850 ml of distilled H_2O; heat and stir to dissolve, then cool. If a precipitate forms upon cooling, filter this solution. Then dissolve 17.3 g copper sulfate ($CuSO_4$) in 100 ml distilled H_2O and slowly stir this solution into the first. Bring the total volume to 1 liter with distilled H_2O.
- I_2KI. Dissolve 3 g of potassium iodide (KI) in 25 ml distilled H_2O. Add 0.6g iodine crystals (I_2) and stir until dissolved. Bring volume to 200 ml with distilled H_2O and store in a brown bottle or covered with aluminum foil.
- Biuret Reagent. Make a stock solution of 3% copper sulfate (3 g $CuSO_4$ / 100 ml H_2O). Add 25 ml of this stock solution to one liter of 10% potassium (or sodium) hydroxide (100 g KOH or NaOH / liter H_2O).
- some possible test materials.
    1% gelatin - add 1 g gelatin to 100 ml H_2O, heat and stir until clear.
    1% glucose (1 g / 100 ml H_2O)
    1% sucrose (1 g / 100 ml H_2O)
    1% soluble starch (1 g / 100 ml H_2O) First make a paste by mixing a small amount of the water with the starch. Add the rest of the water while stirring, then heat and stir until clear.
    1% cornstarch (1 g / 100 ml H_2O)
    1% salt (1 g / 100 ml H_2O)
    1% maple syrup (1 ml / 100 ml H_2O)
    1% corn syrup (1 ml / 100 ml H_2O)
    1% honey (1 ml / 100 ml H_2O)

- juiced fresh food (eg. potato, onion, carrot, etc.) Cut into chunks and add 60 g and 500 ml H_2O to a blender and blend thoroughly. Filter through several layers of cheesecloth.

REFERENCES

General:

Breed, Allen, Thomas Rodella and Ronald Basmajian. 1975. *Through the Molecular Maze.* William Kaufman, Los Altos, Calif. 72 pp. This little book is a useful supplement for students intimidated by basic chemistry.

Stephenson, William K. 1967. *Concepts in Biochemistry: A Programmed Text.* John Wiley and Sons, Inc. New York. 222p. A self-paced learning guide for basic biological chemistry.

Resource:

Baker, J.J.W. and G.E. Allen. 1981. *Matter, Energy and Life: An Introduction to Chemical Concepts.* 4th ed. Addison-Wesley Publishing Co., Reading, Mass. 241 p. Well written introduction the basics of biological chemistry.

Windholz, M. et al., eds. *The Merck Index, 9th ed.* 1976. Merck and Co., Rahway, New Jersey. This is an invaluable reference for preparing chemical solutions.

CHAPTER 3: CELL STRUCTURE

Cell structure is one of the more straightforward topics covered in introductory life science courses. Students can pick up much of the information in this chapter simply by reading the text on their own. You will want to spend most of your lecture time discussing the general concept of biological membranes, both structurally and functionally. Throughout the chapters in Part 1 of the text, Mauseth stresses how a particular structure facilitates a particular metabolic process. He also suggests what the selective advantages of a particular structure might be. Metabolism is a frightening word for most students. They will appreciate any additional explanation you can provide about particular organelles as you lecture Now is also a good time to begin clarifying the difference between selection pressure causing mutations , as opposed to selection pressure acting on already present mutations. The role of natural selection (and mutation) in the evolutionary process is probably the single most misunderstood concept in evolutionary biology. Mauseth does an excellent job of stressing the role of selection in the evolutionary process. Continually challenge your students to explain correctly the examples Mauseth provides.

OBJECTIVES

1. Become familiar with the scale of cells and the units we use to measurethem; be able to quantify dimensions of cells and organelles.
2. Describe the general composition and properties of biological membranes.
3. Describe the basic characteristics of cells and explain the differences between prokaryotic and eukaryotic cells.
4. Recognize the cellular organelles visible with the light microscope; be able to describe their metabolic functions and the selective advantage of having these functions sequestered in a specialized organelle.

CONCEPTS

Students frequently have difficulty visualizing what we are trying to explain in cell lectures simply because we are describing structures at a scale much smaller than they are used to seeing. The concept of size, and with it metric units, requires careful explanation at the beginning of a cell lecture. Related to the simple matter of orientation to scale is the problem of visualizing three- dimensional geometry, particularly the relationship of surface area and volume to size. Virtually all the images of cells a student sees, for example, photomicrographs, line drawings, sketches or images seen through a microscope, are two-dimensional representations of a three-dimensional object. Mauseth provides a good summary of these concepts in the box at the end of this chapter, but you shouldn't rely on your students simply to read through this material and be able to apply it. They will, on the other hand, be able to read and apply most of the information in this chapter on their own - if they understand the fundamental properties of size and the relationship of size to function.

Cell Structure

A second basic concept that tends to cause problems for students is that of biological membranes. The fluid mosaic model is fairly straightforward in principle, but students frequently get the impression, presumably from photomicrographs and diagrams, that the membrane is a static barrier. In fact, membranes are dynamic structures in which the component molecules are constantly in movement. You will again have to spend a considerable amount of lecture time not only explaining the physical structure of membranes, but also why the chemical properties of the component molecules dictate this structure. You also should plan to spend time explaining how membranes differ from each other, how this facilitates cell function and is selectively advantageous.

Most students will have some familiarity with the basic structural components of typical cells, and this material is covered well in the text. One frequent misconception students have, however, relates to chloroplasts and mitochondria. Students know that chloroplasts, for photosynthesis, are found only in plant cells, but they tend to believe that mitochondria, for respiration, occur only in animal cells (we must do a good job somewhere along the line, of convincing our students that the overall reactions of photosynthesis and respiration are exactly the reverse of each other, therefore, if only plants are capable of one process, then animals must be solely responsible for the other)! With this misconception cleared up, students can learn the general structural characteristics and functions of the other cellular organelles from the book.

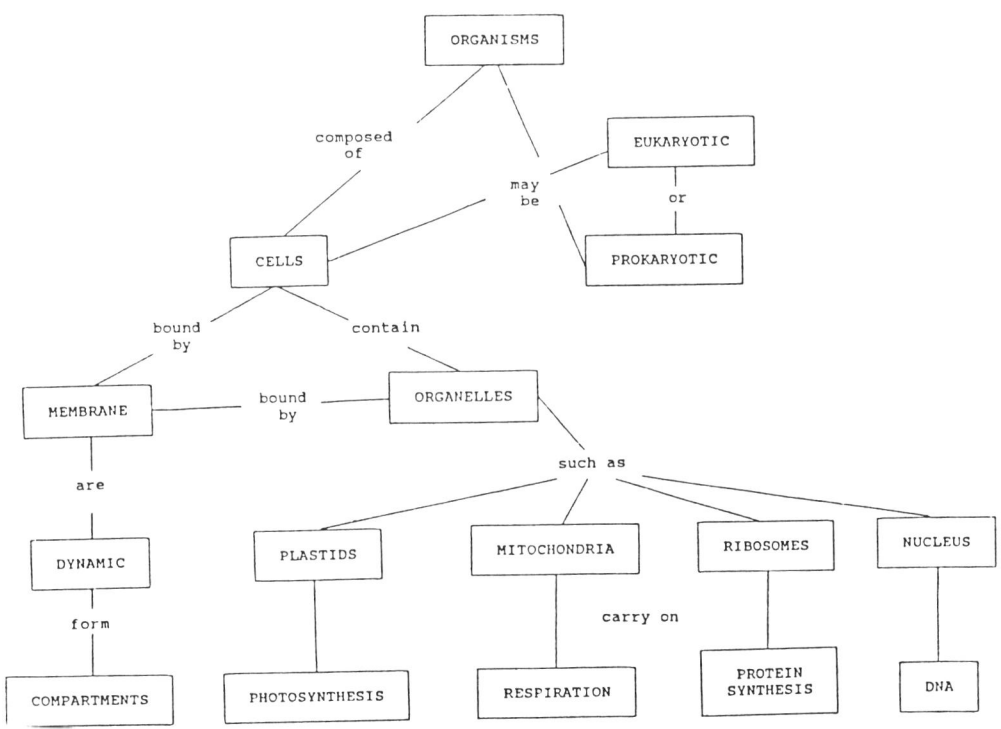

Chapter 3

INSTRUCTIONAL AIDS

Lecture - Before beginning a litany of the structure and ultrastructure of cells you will want to orient your students to the microscopic scale of cells.

Begin by reviewing metric units of measurement, concentrating on microscopic units smaller than millimeters. This can effectively be done by placing the relative size of a unit into a more familiar frame of reference. For instance, a millimeter is approximately 1/16 of an inch. If you represent a millimeter on the street outside as a line 0.6 miles long (about half a mile), you could draw a micrometer on the board proportionally as a line one meter long; a nanometer drawn to the same scale is a segment one millimeter long. The students, sitting in their seats, will not be able to resolve one mm on your board diagram, but they might, be able to visualize it as about 1/16 of an inch. Finally, an angstrom is 1/10 of that. At this scale you could represent a typical dividing cell, about 12 μm on a side, by a volume the size of a small lecture room. The students themselves would approximate the size of a small mitochondrion. This example also gives you an opportunity to describe the three-dimensional nature of cells and some of the problems this poses for our study of cell structure. For instance, if your classroom represented a single cell and the instructor's desk at the front of the cell was a solitary organelle such as the nucleus, what would be the probability of a thin slice through that cell (microscope slide section) including the nucleus? Would the nucleus necessarily be visible in every cell on a microscope slide? If you don't see it, does that mean that it isn't there?

Another factor related to the size of cells and their organelles is the relationship between the surface area of a membrane and the volume it encloses. Bring your business students into the discussion by presenting this as a problem of supply and demand. Both surface area and volume increase exponentially as size increases, but they do so at different rates. The concept of exponential increase is difficult for many students so you will want to illustrate this on the board with a simple example. Sketch a square on the board and label the length of the sides 1 unit. Now make a table with columns for length (units), surface area (units^2), and volume (units^3). Fill in the table for length of one unit, then repeat for two and three unit lengths. Graph the results to illustrate exponential increase at different rates and the relationship between the square and cube functions. How does this relate to the size of a cell? The living cytoplasm occupies the volume within the cell membrane and has certain demands for raw materials (e.g., nutrients and gases) as well as waste removal. These materials must move into or out of the cell across the cell membrane; therefore, supply is a function of surface area. As a cell increases in size, both the demand and supply increase exponential, but demand increases as a cube function while supply increases as a square. It does not take long for supply to limit demand.

If supply/demand were the only factor influencing cell size, all cells would be about the same size. This is not the case; there are several modifications that have a selective advantage in allowing cells to attain greater size. Two widespread adaptations of plant cells are to have a lower metabolic rate (relative to animal cells) and to fill a substantial portion of the cell volume with non-living material, a vacuole. In both cases the "demand" of a cell is considerably lessened without decreasing the "supply" capability. Other factors include altering the shape of the cell (spheres and cylinders are more supply/demand efficient than are cubes) and elaborating the surface area with folds.

The surfaces across which "supply" occurs to individual organelles and the cell as a whole, are the cellular membranes. Most students know that oil and water do not mix, a phenomenon that you can demonstrate nicely with an overhead projector and a few ml of oil in a petri dish of dilute aqueous stain. This hydrophobic property of lipids is exploited in forming a lipid boundary between different aqueous components of the cell. The special properties of phospholipids, which also contain a hydrophilic head, causes individual sheets of lipid to line up "tail to tail" forming the lipid bilayer characteristic of biomembranes. This concept is not too hard for students to visualize. What is difficult to conceptualize is that this boundary is fluid and dynamic. A useful analogy is the surface of water at the air/water interface. This surface forms a boundary, surface tension, that is strong enough to permit many insects to "walk on water." Yet the individual water molecules are fluid and can move about within the surface layer. Individual phospholipid molecules can move about laterally in their membrane layer in the same way. Mauseth makes the point that the intrinsic proteins of membranes are also capable of moving about laterally in the membrane. This is easier for students to visualize because it is analogous to corks floating on the surface of water - a fluid mosaic.

Biological membranes are dynamic structures that are alike in their basic fluid-mosaic composition, yet they serve many different functions in different compartments of a cell. What makes a chloroplast different from a mitochondrion and both of these different from the cytoplasm as a whole? In large part, these differences are due to differences in their constituent membranes. Slightly different molecular compositions may be responsible for greatly different permeability properties and the complement of intrinsic and extrinsic proteins will result in different metabolic pathways associated with their respective membranes.

A thorough explanation of the above concepts can easily consume an entire lecture period, but it is well worth the students' time. With this background, they can study the remainder of the material on cell structure as more than an exercise in vocabulary enhancement and memorization. If the students are given a fairly explicit guide as to what your expectations are, they should be able to pick up the rest of what you want them to know about cell structure and ultrastructure simply from reading the text.

Pre-lecture Review - Understanding the properties of membranes is critical to understanding the structure and function of cells and organelles. The basic components of membranes are phospholipids and proteins. Briefly review the characteristics of these important groups, in preparation for your discussion of cell membranes.

Post-lecture Review - A review of students' concept maps from the previous lecture is always a good review technique, but information from this chapter lends itself especially well to the use

Chapter 3

of overhead transparencies or projection slides to "quiz" students on structure/function. Is this cell prokaryotic or eukaryotic? What is this organelle? What is its function? What might be an advantage of having this shape?, etc. Finally, emphasize that mitochondria are characteristic of all eukaryotic cells, including plants, therefore plants also undergo respiration!

LABORATORY -EXERCISES

The basic technique for studying cells is microscopy. Two alternatives are available. First, you can use commercially prepared slides This has the aesthetic advantage of uniformly thin sections stained in multiple colors. The second alternative is for students to produce their own slides by making hand sections or epidermal peels and water mounts. The advantage here is that the students may orient themselves to exactly where the section was taken from in the intact plant and there is much greater flexibility in what students can examine. My own preference is for the latter - here are a few hints. The sharper the blade, the easier it is to section; the sharpest blades are double-edge Gillette Blue Blades®. Cut these in half with a scissors prior to use. Attach a strip of masking tape, doubled back over the cut edge, to provide a handle. Stainless and chromium- plated double-edge blades are sharper than single edge blades, but not as good as the old "blues." Students frequently get frustrated trying to cut the perfect hand section. Encourage them to cut several sections fairly rapidly, placing them in a small dish of water. Later they can go back and choose their best section(s) to observe - as long as a section is thin at one edge, it is usable. A simple stain to use is Toluidine Blue for a minute or two, followed by a water rinse. It is not necessary to fix the cells in alcohol prior to staining.

1. Microscopy.

Every freshman biology or botany laboratory manual will have an exercise on use and care of the microscope. What most do not have, however, is an explanation of some simple modifications students can make on their microscopes to enhance their capabilities. Two techniques that have multiple uses in botany are polarization and dark field microscopy, both of which you can introduce after students have mastered the basics of brightfield microscopy.

A. Polarization. Polarization microscopy provides spectacular images of plant material, both of prepared, stained slides and fresh material. The technique takes advantage of crystalline material or regular layers of molecules, such as cellulose, that are birefringent, i.e., they are capable of diffracting plane polarized light. Place a polarizing filter, the polarizer, between the light source and the specimen to produce plane polarized light. Place a second polarizing filter, the analyzer, between the specimen and the observer's eye - usually in or on the ocular lens. As the analyzer is rotated, the field of view will lighten or darken to extinction depending on the relative orientation of the polarizer and analyzer. You can demonstrate this with two sheets of polarizing material on an overhead projector. With the filters rotated to extinction, the field will be dark unless the specimen between the two filters contains some birefringent material.

In this case, the birefringence will appear bright on a dark background. A piece of acetate placed between crossed sheets of polarizing material on the overhead illustrates this principle. Birefringent materials frequently found in plant cells include cellulose, crystals, and starch grains. Students can use polarization for simple diagnostic tests such as for the presence of starch.

B. Darkfield. Darkfield provides brilliant images of solitary cells against a dark background but depends on reflection of light by structural elements of the cell, not birefringence of the material. Darkfield will not be effective for examining thick tissues. It is very useful when examining the unicellular organisms covered in the diversity chapters. To obtain darkfield, place an opaque circle, about the size of a dime, in the filter holder under the condenser (the optimum size is determined by the diameter of the iris when the microscope is adjusted for Köhler illumination). This opaque **field stop** blocks all the light that normally passes directly through the specimen and into the objective lens. By opening the iris diaphragm to its maximum position, additional light will pass through the specimen, but at an angle too wide to enter the objective, thus the specimen will not be seen - unless material in the cell itself reflects some light into the lens. In this case it will appear that the specimen itself is generating the light as it "glows" against a dark background. If your students look for starch grains or crystals in sections of fresh material, they will see individual structures floating in the water on the slide. These were released from cut cells and are easily seen with darkfield.

C. Measuring with a Microscope. Many laboratory manuals include an exercise on calibrating the diameter of the microscope's field of view using a plastic ruler on the stage. Measure low and medium power magnifications directly to the nearest fraction of a millimeter. The field diameter under high power must usually be calculated by setting up a proportion with a measured field (two magnifications and one diameter are known, you solve for the diameter of the higher magnification). Students can then use their microscope to estimate the sizes of cells and tissues they observe. The ability to do such quantification is useful when studying tissues and organs, so it is well worth learning. An extension of this is to practice calculating the surface areas and volumes of various cells, for instance those listed in Table 3.1a. Students may be asked to use the information in the box on geometry of cells to complete the table as below:

Cell Type	Surface Area (μm^2)	Volume (μm^3)	Surface Area/Volume
dividing cell	864	1728	1/2
epidermis	18,510	96,525	1/5.2
photosynthetic cell	1,365	2,365	1/1.7
vessel	305,363	12,882,493	1/42.2
fiber	1.3×10^{10}	1.8×10^7	1/0.01

Chapter 3

A final note on microscopy; have students record their own observations by sketching what they see, in pencil, on a sheet of notepaper. The sketch should be LARGE enough so that the image of the field of view at least half fills the sheet. It does not require great artistic ability to make meaningful drawings, but it does require careful observation and patience. Not only will students tend to learn the material better, but you will be able to check their work quickly to ensure that they are seeing, and comprehending, what you want them to see.

LABORATORY - MATERIALS, EQUIPMENT, AND PREPARATION

1.A. Polarization.
- standard student microscope for each student.
- slides, cover slips, dropper bottle of water, and double edge razor blades (if living material will be examined).
- prepared slides (optional, *Ranunculus* roots with starch grains and woody material are particularly impressive)
- 2 disks of polarizing material (cut with a scissors from square sheets of polarizing filter available from scientific supply houses) for each microscope. If your microscopes have a filter holder under the condenser, one disk should be cut to fit this. If not, the filter may be taped onto the bottom of the condenser. The second disk can be hand held on top of the ocular.

B. Darkfield.
- standard student microscope
- clear plastic filter with opaqued center spot about 2mm diameter (india ink provides a good opaque spot).
- slides, cover slips, and droppers.
- suitable only for solitary material, e.g., pond water or suspensions.

C. Measuring.
- standard student microscope
- plastic ruler
- calculator

REFERENCES

General:
Headstrom, Richard. 1977. *Adventures with a Microscope*. Dover, New York. A compendium of project suggestions using a microscope.

Paturi, Felix R. 1976. *Nature, Mother of Invention: The Engineering of Plant Life*. Harper and Row, New York. 208 p. This interesting translation of a German work describes, in layman's terms, the mechanical advantages of a variety of plant structures.

Resources:

Eberhard, Carolyn. 1988. *Experiments in Biology*. Saunders College Publishing, Philadelphia. 432pp. This is a very useful general biology laboratory manual.

O'Brien, T.P. and Margaret E. McCully. 1969. *Plant Structure and Development: A Pictorial and Physiological Approach*. Macmillan, London. 114 p. This is a wonderful photographic atlas of plant structure and diversity.

Sundberg, Marshall. 1984. Special Microscopy Using a Standard Student Microscope. *Amer. Biol. Teach.* 46:113-115. A brief introduction to polarization, darkfield, and fluorescence microscopy using a standard student microscope.

CHAPTER 4: CELL DIVISION

Cell division is an essential component of the growth of plants; it is a method of increasing the number of cells in the plant body and/or producing gametes for sexual reproduction. Mauseth begins his discussion with the cell cycle. His treatment will help you to emphasize the continuous nature of both the growth and division phases of the cell cycle. Most students will have heard the terms mitosis and meiosis before (and probably assume that they already know the processes involved), but at most they will have memorized the names of some stages and perhaps some of the characteristics of each stage. Your biggest challenge will be to help your students conceptualize the processes of mitosis and especially meiosis. A related challenge will be for you to convince some of your students that they really don't have as firm a grasp of these processes as they think they have and that they should study the material seriously, in spite of what they think they already know.

OBJECTIVES

1. Describe the significance of the cell cycle in growth and reproduction of a line of cells.
2. Relate the structure of the eukaryotic chromosome to the cell cycle.
3. Describe the overall process of mitosis and the events that characterize each stage.
4. Relate the events of meiosis to the chromosomal requirements of sexual reproduction: gamete formation and syngamy.
5. Differentiate between cytokinesis and karyokinesis.

CONCEPTS

In order to understand the cell cycle, mitosis, or meiosis, it is first necessary for students to understand the structure of a eukaryotic chromosome. The carrier of genetic information is the double-stranded DNA molecule. Make sure that students understand this is different from the two chromatids that constitute a chromosome following the S-phase of the cell cycle. With an understanding of the structure of a typical chromosome, students generally have little difficulty following the process of mitosis presented in diagrammatic form. One key point to emphasize, however, is that mitosis (and meiosis) is a continuous process. We only subdivide the process into stages to facilitate our learning and study.

Meiosis is more difficult for students to comprehend than mitosis. Particularly troublesome is the concept of homologous chromosomes (another kind of pair that students tend to confuse with the paired chromatids and the double strands of DNA). Students understand that parents each contribute a set of genetic information, chromosomes, to his/her offspring through their gametes. The problem becomes how to reduce the number of chromosomes in the gametes so one of each homologous chromosome will be available. Mauseth introduces the terminology "reduction division" as a synonym for meiosis; it would be good to point this out to the students. The older terminology emphasizes that meiosis actually has two components, first, to reduce the number of chromosomes, second, to divide each of the resulting daughter nuclei in two.

The transition from mitosis to meiosis is a good point to emphasize that evolution works by adding to or modifying existing processes or structures. Thus, the machinery of cell division used in mitosis, e.g., microtubules, centromeres, kinetochores, and the chromosomes themselves, are adapted to the first (reduction) meiotic division; meiosis II (division) is essentially a mitotic division. The major differences between mitosis and meiosis, then, occur during meiosis I.

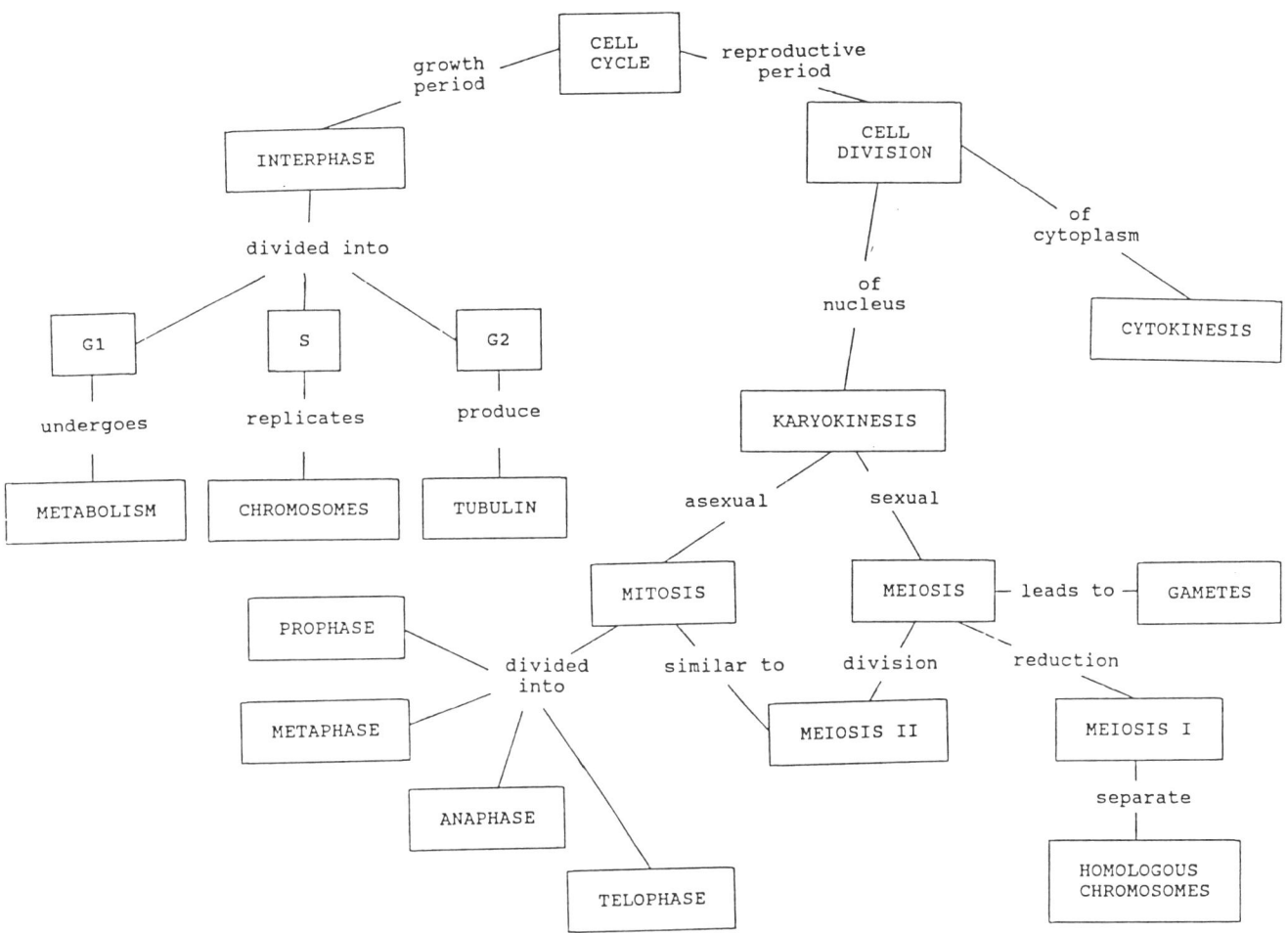

INSTRUCTIONAL AIDS

Lecture - When describing the structure of the chromosome, a useful analogy is a back-up disk of a computer program or a duplicate copy of a favorite audio tape. The duplicate is an exact copy of the original. For two daughter cells to be identical to the parent, there must be two identical copies of the genetic information, the two chromatids. One of each chromatids is donated to each cell. Thus, each daughter cell will receive one of the chromatids from each of the original chromosomes.

Chapter 4

A quick review of the basics of sexual reproduction is helps students understand meiosis. During sexual reproduction, both parents contribute to the chromosomal makeup of their offspring, thus the offspring will have two of every type of chromosome - one from each parent. These homologous pairs must again separate before the next reproductive cycle so that the new gametes formed will have only one of each type of chromosome. When diagramming the process of meiosis, use two different colors to represent the chromosomes originally donated by the two parents. At this point introduce the terms haploid and diploid. Be sure that your students understand the problem that would arise if reduction did not occur - there would be a doubling of the number of chromosomes from one generation to the next until there would literally be no room left in the cell.

If students have a firm conceptual grasp of the difference between mitosis and meiosis, in terms of their functions in a generalized life cycle, they will have a much easier time understanding the mechanistic differences between the two processes. The first key difference is that during Prophase I, when the chromosomes begin to condense, they also pair, homologue to homologue. This pairing of homologous chromosomes, called synapsis, is essential to assure proper separation of homologous pairs. It is not necessary to go into the details of substages, or even to mention crossing over at this point. Concentrate on the basic concept of the cell changing from the diploid to the haploid condition.

The next key difference occurs at Metaphase I when homologous chromosomes, already paired, migrate to the metaphase plate. It is useful to return briefly to the previous diagram of mitotic metaphase (illustrating a cell with the same number of chromosomes) to emphasize that while in mitosis it doesn't matter where one chromosome is relative to any other, because each one will divide in two, in meiosis it is extremely important that homologues be paired so the pairs may be separated. The centromere of each chromosome attaches by a short microtubule to one pole of the spindle. Its homologue, the other chromatid, attaches to the other pole. In Mitosis, on the other hand, each centromere attaches to both poles. It is also useful at this point to illustrate the random nature of the orientation of maternally to paternally derived homologues - the basis of independent assortment. Colored pipe cleaners, or in a large class, colored acetate chromosome cutouts to use on the overhead projector, are useful to illustrate this point.

The final trouble spot is Anaphase I, although if students clearly understand metaphase there will be little trouble here. The key is that because the individual homologous chromosomes attach to opposite poles, intact homologues will separate into different daughter cells thus reducing the total number of chromosomes by one half. More importantly, each daughter cell receives one of each type of chromosome pair, independently assorted.

Explain meiosis II as essentially a mitotic division of each of the products of meiosis I. Some students will recognize, however, that when you discussed mitosis you were dealing with a diploid cell, and now each cell is haploid. They will question if that doesn't make a difference. Here is an excellent opportunity to explain that both haploid and diploid cells (or even triploid or pentaploid cells!) can undergo mitosis. All that is required is for each chromosome to consist of two chromatids. Mitosis will produce two daughter cells identical to the parent cell, regardless of ploidy. If students understand this concept well, plant life cycles will be much easier for them to comprehend.

A thorough understanding of meiosis is essential to understanding life cycles and genetics that otherwise become exercises in meaningless memorization. The best plan is to repeat, from the beginning, each phase of the process as you proceed to the next. Although this is time consuming, and you will not be able to lecture about the details of prophase I, crossing over, cytokinesis, etc., it is time well spent in the long run.

Pre-lecture Review - Have a student place her/his concept map from the previous lecture on the board. Answer any class questions about this material, then ask -- given the relative number of each of the organelles listed, which would present the greatest problem to a cell reproducing (dividing)? Most organelles are present in multiple numbers. The nucleus, however, is both a solitary organelle in most cells and is responsible for cellular control. It is essential for each cell to have one, therefore the nucleus must somehow divide equitably.

Post-lecture Review - Place a summary diagram of meiosis on the overhead. Now ask individual students to explain, from the beginning, the overall process. Rotate several students through this explanation, e.g., let one student begin with prophase, have another pick up with metaphase, etc.

LABORATORY - EXERCISES

Onion root tip slides are the workhorse for laboratory exercises on mitosis. The following exercises are based on these standard slides, however, we employ two little twists. Rather than turning the students loose to find examples of each stage of mitosis, comparing what they see to photomicrographs in a textbook or laboratory manual, lead them through a tutorial on identification of stages. Then ask them how they can use their slides, sectioned at one instant of time, to estimate the duration of each of the recognizable stages of division.

1. Identification of Stages of Mitosis.

This is a rather straightforward exercise similar to that found in any laboratory manual of biology, but it is an essential prerequisite to the next section. For that reason it is generally best to take the class, in lock step, through an identification of each stage. Begin by describing the onion root tip slide and explaining where on the specimen to look for the greatest likelihood of finding mitotic figures - the apical meristem just behind the root cap. Ask one student to describe the general shape of cells in this region - nearly square. Outline this shape on the board, then describe the appearance of cell in prophase - nuclear area with numerous thick

Chapter 4

"dots" or "tangle of loose spaghetti." Challenge the students to find a cell in prophase on their slides, then center that cell under the pointer in their microscope and raise their hand when they're ready to be checked.

When you check a student's work and he/she has an appropriate cell under the microscope, ask him/her to assist you in checking other students. If the cell under the pointer is not in the desired stage, find one that is and move it under the pointer to show the student what this stage looks like - then ask the student to find another example on the slide to show you!

When every student in the class has been checked by you or one of your assistants, move on to the next stage. Make a sketch on the board of the general appearance of a cell in the stage you want, then ask students to find an example of that stage, center it, and again raise their hands to have it checked. Students will not have any trouble finding examples of prophase or metaphase on their slides, but occasionally a section will not have a cell in anaphase or telophase. As students will discover in the next exercise, the latter two phases occur relatively quickly, therefore proportionally fewer cells in a meristem will be in that stage at the instant the tissue was fixed. As a result, there is a greater likelihood that a single section through the meristem will not contain a cell in that stage (It also depends on the time of day the slidemaker fixed the tissue, mitotic cell division in root apical meristems shows a distinct circadian periodicity - a biorhythm!). If students cannot find a stage on their specimen, have them check other sections on their slides - there are usually two or three on a single slide. If you still cannot find an example, check another slide.

2. Determination of Duration of Mitotic Stages.

After completing the above exercise, students should feel confident that they can recognize each stage of mitosis; this is essential to successfully complete this exercise. Have the students center the most mitotically active region of their root tip in their field of view under high power. Once centered, the slide will not be moved. Now obtain an estimate of the total number of cells in the field of view. One may do this by counting all the cells in one segment of the field, e.g., one quadrant, then multiplying this count by the proportion of the total field not counted. For instance, if students count the cells in 1/4 of the field, the total field would contain approximately 4 times this number.

A preferable way to make this estimate is to calculate the average area of a single cell and divide this into the area of the field of view. Not only will this provide a more accurate estimate in most cases, but it reinforces quantitative thinking and basic number-handling skills. Students will already have calculated the diameter of the field of view, thus they can readily calculate the area of the field from the formula for the area of a circle = r^2 where r = 1/2 diameter. Estimate the length or width of a single cell by counting the number of cells across the field of view and dividing this into the field diameter. The average area of a single cell, length times width, divided into the area of the field estimates the number of cells in the field.

Once the number of cells is known, add this number to the table below. It is now a simple matter to count the number of cells in recognizable stages of division and add these tallies to the appropriate column in the table. Estimate the number of interphase cells by subtracting the number of cells in all other stages from the total number of cells in the field.

A few students may question why so many cells appear not to have a nucleus at all. Remind them that cells are three dimensional and that the tissue is cut thin to prepare microscope slides, typically 5-10 um thick on commercial slides. Therefore, it is quite possible that a section through a cell did not include a portion of the nucleus. A related question will be how do we know that all of the cells whose nuclei are not visible were in interphase? They aren't, of course, but the vast majority will be, so this is a reasonable estimate.

STAGE	YOUR COUNT		CLASS COUNT	
	NUMBER	DURATION	NUMBER	DURATION
TOTAL				
PROPHASE				
METAPHASE				
ANAPHASE				
TELOPHASE				
INTERPHASE				

The duration of the cell cycle in a typical onion root tip is about 16 hours; therefore, we can estimate the duration of time spent by an individual cell in each stage of mitosis by the proportion of cells in that stage at the instant the tissue was fixed. Use the following formula to calculate the duration of each stage, in minutes.

$$\text{Duration of stage} = \frac{\text{number of cells at stage}}{\text{total number of cells}} \times \frac{16 \text{ hours}}{\text{cycle}} \times \frac{60 \text{ minutes}}{\text{hour}}$$

Each student should calculate the duration of each stage (min./cell cycle) based on his/her own data and enter this into the table above. It is also useful to get class totals and to generate duration times from the pooled data.

Chapter 4

The increased sample size provides a more reliable estimate - in fact, durations of metaphase, anaphase and telophase usually approximate published values. Prophase is frequently underestimated by about half. There are two reasons for this. First, and most important, students have a difficult time distinguishing early prophase from interphase. They will tally many prophases as interphase. Second, a significant number of cells whose nuclei were not visible will have been in prophase and these will not have been tallied. Published values for Prophase, Metaphase, Anaphase, and Telophase respectively are: 71, 6.5, 2.4, and 3.8 minutes.

3. Meiosis

Microgametogenesis, pollen grain development, is the typical material used to study meiosis in plants. Use the same technique of asking the class to find a particular stage on their microscope, then checking for accuracy. The added difficulty is that because the cells are free in the pollen chamber of the anther at the time the material is fixed and prepared for microtomy, their orientation is random relative to the plane of section. For instance, the metaphase plate may be anywhere from in perfect side view, the plane usually illustrated, to in perfect face view with the chromosomes arranged in a disk shape. A model of a small disk (representing the equatorial plane), held between the spread fingertips of each hand (to represent the spindle fibers) is useful to demonstrate this spatial arrangement. By rotating your hands, the shape of the disk visible to the class appears to change. One other clue is that the cells in an anther (actually all the anthers of a particular flower) divide more or less in synchrony. Therefore, by scanning each of the locules represented on a slide, students may: 1) look for a familiar chromosomal arrangement to determine the stage, and 2) identify various appearances of the chromosomes, due to different orientations of the plane of division, at the same stage of meiosis.

LABORATORY - MATERIALS, EQUIPMENT, AND PREPARATION

Per student:
- compound microscope
- *Allium* root tip slides
- *Lilium* anther, set of various stages of meiosis
- calculator

REFERENCES

General:
- Mazia, Daniel. 1974. The cell cycle. *Sci. Amer.* 54. An older, but very readable, explanation of the cell cycle.

Resources:
- Sundberg, Marshall. 1981. Making the most of onion root tip slides. *Amer. Biol Teach.* 43:386-388. Instructions for obtaining duration of mitotic stages from root tip slides.

- Sundberg, Marshall. 1989. *Biology 1208: Biology for Science Majors, 2nd ed.* Burgess Publishing, Minneapolis, Minn. 168 pp. A laboratory manual emphasizing quantitative approaches to introductory biology.

CHAPTER 5: TISSUES AND PRIMARY STEM GROWTH

Just as cell division is an essential aspect of growth for all plants, differentiation of cells into different cell types is an essential aspect of growth for all multicellular plants. Furthermore, different arrangements of specialized cells allow for specialization of function - tissue differentiation and organ formation. These concepts will be foreign to students' thinking about plants, but will be familiar from animal models. It is useful to take advantage of this as you present information during lecture.

OBJECTIVES

1. Characterize three basic plant cell types: parenchyma, collenchyma, and sclerenchyma.
2. Describe the structure and function of the three primary tissue systems: cortex, epidermis, and vascular.
3. Describe the external morphology of stems.
4. Relate the primary structure of cells and tissues in stems to their functional adaptations.
5. Describe shoot growth and the roles of apical and intercalary meristems in stem elongation.

CONCEPTS

The main concepts in this chapter are adaptation and the structure/function relationship. Students frequently think of adaptation as being an animal characteristic, because of their obvious behaviors, rather than as a characteristic of all living things. Mauseth provides a good introduction to the variety of modifications of stems that have selective advantage under particular environmental conditions. You can extend this concept to the cellular and tissue levels as well.

Two of the tenets of the cell theory are that cells are basically alike and that all cells come from preexisting cells. These suggests that any cell of a plant can potentially give rise to a complete plant in its own right - the concept of totipotency (first proposed by the German plant physiologist Haberlandt in the 1880s). In 1958 Muir et al. first demonstrated that under certain conditions, single cells of some plants can generate entire plants. This tissue culture process now is exploited commercially in the horticulture and biotechnology industries. Not unexpectedly, parenchyma cells are most easily induced to propagate intact plants. These cells are most similar to undifferentiated meristem cells.

Collenchyma and sclerenchyma are more highly differentiated, with thickened walls related to their function in providing strength and support. Sclerenchyma is unique because the cell usually dies BEFORE it becomes functional. Mature, functioning sclerenchyma, then, is composed of dead cells!

Tissues and Stem

The concept of tissue, a group of similar cells having a specific function, is also a familiar one for students when thinking of animal models; the transfer to plants is frequently not made. The closest parallels between plant and animal tissues are in the dermal systems; these provide good examples for comparison.

The concept of stem ties closely to that of leaves. The stam plus its leaves constitute the shoot system. Much of the characteristic external morphology of stems depends on the associated leaves. Be careful not to carelessly interchange the terms "shoot" and "stem" when discussing vascular plant morphology. Students tend to become confused unless the two terms are clearly defined: stem is the above-ground axis, shoot is the stem plus associated leaves.

Xylem and phloem, as specialized tissue types, present two conceptual difficulties. The first is that the cell wall itself does not necessarily provide an impermeable barrier to water flow. This is because the term wall implies "barrier" to many students. A pit, which is a region lacking secondary wall but in which the primary walls of adjacent cells remains intact, does not present a major barrier to water flow.

The second difficulty is that the sieve tube element remains alive as it functions, in spite of its lack of a nucleus and most of its intracellular machinery. Most of the metabolic functions of the sieve tube element is taken over by the companion cell.

Apical growth is relatively easy for students to comprehend. They are used to seeing the tips of woody shoots elongate every spring. Intercalary growth, however, is less intuitive (and is actually responsible for most of the rapid growth associated with bud burst). If you adequately explain leaf initiation, as well as the structural differences between nodes and internodes, the concept of intercalary growth should not be too difficult.

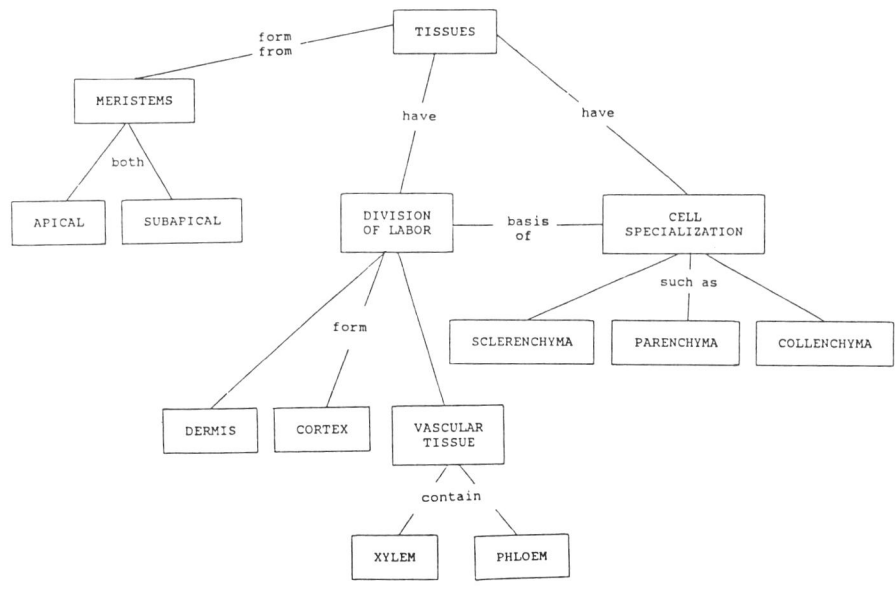

Chapter 5

INSTRUCTIONAL AIDS

Lecture - Students will have a general idea of what constitutes a stem, but will probably not be able to provide a list of distinguishing characteristics that they could use to evaluate unknown organs. In a small or even medium sized class, much of your instruction about stems can effectively be done outside on an impromptu campus "field trip." Be sure to bring your pocket knife or a pruning shears to collect specimens along the way. You can use a clipboard with plain paper and a marking pen to sketch details or features not readily seen in intact specimens. Not only can you point out external morphology from such samples, but you can show general tissue regions with a cleanly cut cross section. For example, the scattered vascular bundles characteristic of monocot stems, will be visible in the cut end of a large grass stem. If you want to provide the detailed structure of a single bundle, sketch it on your clipboard pad and hold it up for your students to see - a miniature chalkboard of sorts. A less satisfactory substitute, for rainy days, is to bring in specimens of shoots to pass out to students. As you discuss various features of stem morphology and anatomy they will have a sample to examine.

The perfect time to discuss the characteristic growth pattern of stems, and the role of intercalary meristems, is when examining the external structure of stems, especially woody stems. Nodes and internodes are easy to distinguish, even on leafless twigs. Ask students to think about what caused the internodes to form, thus separating the nodes? How does this external morphology relate to the terminal buds? Buds, of course, already have the next season's leaves and nodes already formed, but no internodal elongation. The characteristic spurt of stem growth at the beginning of the next growing season, due to the combined activity of intercalary meristems at each of the already formed nodes, is responsible for the rapid "bud break".

Large classes will probably limit you to using projected slides for examples. This is not entirely bad, however, as you can then easily include examples of not-so- characteristic stems, or non-stems that look like stems, to challenge your students' abilities to apply the criteria for identifying stems that you discuss. For example, students usually think the stipe of a fern frond is a stem, rather than part of the leaf itself (the stem is an underground rhizome). Similarly, the flattened stems of *Opuntia* cactus or *Asparagus* fern are frequently mistaken for leaves. Mauseth gives several other good examples.

Mauseth treats cell types before tissues, which is logical because tissues are made up of particular arrangements of specific cell types. You may want to talk about tissues in general before specific cell types (a meaningful discussion of xylem and phloem as specialized tissues, however, must wait until after you cover specific cell types). This follows the order from the known and/or easily observed external structure to the easily observable internal patterns, and finally to the conceptually more difficult cell specializations that are not visible to the unaided eye. At each level, relate function to structure; function will serve as the unifying theme between levels.

You can stimulate critical thinking and observational skills by asking students to observe the features of a particular part, then hypothesize what the function might be. For example, the outer surface of a stem will appear "skin-like" in surface view and be readily distinguishable as a boundary layer in cross section. Like the skin of an animal, the primary function of this tissue is protection, from invasion by other organisms and from desiccation. Similarly, relate the arrangement of vascular bundles in a stem to their function in support - like "rebar" in reinforced concrete. Later students can relate the structure of individual cell types making up these tissues to the function already hypothesized from macroscopically visible structure.

In a single lecture, you will not have time to go into detail on the structural differences between parenchyma, collenchyma, and sclerenchyma. The text explains these well and they are summarized in Table 5.3. A quick review of the table should be sufficient in class, with the provision that you specifically tell students that they are responsible for this information and should read the section carefully in their text.

Xylem and phloem are more complex tissues, structurally, functionally, and conceptually; therefore, plan to spend additional time in lecture covering these topics. Both tissues have relatively undifferentiated parenchyma cells as well as cells specialized for support - fibers. The key features, however, relate to the cells specialized for transport. In xylem these cells, vessels and/or tracheids, are a type of sclerenchyma; in phloem, sieve tube members (or sieve cells) and companion cells are specialized parenchyma. Be sure your students understand the porous nature of the primary cell wall and the structural and functional differences between pits and plasmodesmata.

Pre-lecture Review - Students must understand the nature of the cell wall, including the structure and composition of the various layers and the specialized transport areas, pits and plasmodesmata, before they can understand the difference between the three major cell types and the structure/function of vascular tissues. Time initially spent reviewing this material will be rewarded by students' better understanding of the distinguishing features of these cell and tissue types.

Post-lecture Review - Use figure 5-38 as a flow chart for students to explain primary growth and structure of the stem. Ask one student to explain the relationship between the apical meristem and the three primary meristems. Have a second student describe the relationship between one of the primary meristems and the tissue it gives rise to. A third student can describe the different cells types that typically make up that tissue, how they can be recognized and what their functions are. Repeat for the other two tissue regions.

LABORATORY - EXERCISES

1. Supermarket Morphology

Chapter 5

The produce department of the local grocery stocks a good sampling of roots, stems, and leaves familiar to students. These organs may be modified for storage, there are also typical examples.. For the study of stems, *Asparagus* is a good monocot and broccoli has a sturdy dicot stem. A celery stalk provides good material for dissecting out shoot apices. The individual leaves peel off in a typical phyllotactic pattern (save them for the leaf laboratory). The shoot apex is large and easily dissected beneath the youngest leaf primordia. Some good examples of specialized stems include: potato (tuber), onion (bulb - short stem and modified storage leaves), and kohlrabi. You can use these materials not only to demonstrate external morphological characteristics, but also to be hand sectioned to study internal anatomy.

2. Anatomy, cells and tissues

Hand sectioning, which was introduced in chapter 3 to study cell structure, is a very useful technique for studying the anatomy of plant organs. Typically we use greenhouse grown material, *Coleus* or geranium for dicots and wandering Jew or corn for monocots, but fresh *Asparagus* and smaller branches of broccoli are excellent choices. Both are large enough for students to hold firmly, yet soft enough to cut easily (large broccoli stems get too woody). Even thick sections are adequate to distinguish general tissue regions and wedge-shaped slices with a thin edge extending across the stem is adequate to examine specific cell types. Take advantage of the materials collected for part 1 and experiment. Students enjoy working with, and learning more about, the food they eat.

Begin with a dicot stem. Students can usually identify general tissue regions, vascular tissue, epidermis, and cortex and pith (if present), even without staining. Polarization will make these areas stand out if the section is reasonably thin. After students have made an appropriate sections, or even using prepared sections, have them first make a page-sized sketch outlining only the general regions observed. While they are doing this you can circulate through the class to check on student progress. A glance at a student's sketch will quickly let you know if the student understands what s/he is seeing. You will also be able to pick out a "volunteer" to copy her/his sketch onto an overhead transparency or on the chalkboard to use in discussion. As soon as most people are obviously on the right track you can go to your donated sketch to identify and label tissue regions.

With the general tissue regions identified, the students are now ready to concentrate on cellular characteristics. Even in unstained material collenchyma will stand out as patches of cells beneath the epidermis with thick, "silvery" walls. Have some students draw a drop of phloroglucinol under the coverslip and look for some cell walls to begin turning pink or red. Phloroglucinol reacts with lignin, therefore sclerenchyma will be stained differentially. Students must be patient here, the reaction may take as long as five minutes. Other students should draw a drop of toluidine blue under their coverslip. This will differentially stain many different cell types. By comparison with the phloroglucinol stained material (or polarized light), students will distinguish xylem, fibers and sclereids. Phloem will appear as darker-stained cells in patches associated with the xylem. Students should then add detailed sketches of five or six adjacent cells from each tissue to their overall sketch of the stem section, indicating the general location of each. Again, ask individuals to "volunteer" detailed sketches for class demonstration. Once this stem examination is complete, repeat the same process for the monocot stem. After examining both typical dicot and monocot stems, you can proceed to examine some specialized stems or some "unknowns".

3. Stem Growth

The function of intercalary meristems in stem elongation is a difficult concept for students to understand (we need a good time-lapse video of a terminal bud "breaking" in the spring). Longitudinal sections of the tip of *Elodea,* however, may be used to help visualize this process. Developing intercellular spaces between rows of aerenchyma cells makes it easy to follow changes in cells size within each internode as you proceed from tip toward base.

Each student should have a longitudinal section of *Elodea* to observe under her/his microscope. Begin by finding the elongated shoot apical meristem and identifying the leaf primordia. Next identify nodes and internodes. Ask students to make an outline sketch of the shoot as it appears under medium power. Ask one of the quicker students to reproduce her/his sketch on the board or overhead.

When the class is ready, review the general morphology of the shoot using your student-provided drawing. Be especially careful to point out nodes and internodes (the latter will be conspicuous because of the elongate intercellular spaces). Now ask the students to examine the cells in one file of aerenchyma cells, the cells between elongate intercellular spaces, at the internode nearest the apical meristem. Estimate the length of the top and bottom cells in one file. Record this data in a table similar to the one below. Make similar measurements of the top and bottom cells of rows in at least the next five internodes proceeding toward the base of the shoot.

NODE	TOP CELL	BOTTOM CELL

Chapter 5

Now plot this data on a graph with cell position (node 1: top, bottom; node 2: top, bottom; etc.) along the "x" axis and cell length along the "y" axis. The result will be a saw-tooth pattern where the cells at the top of the internode, the intercalary meristem for that node, will all be about the same length, while the cells at the bottom of each internode file will be progressively longer. Within each file there will be a gradual increase in cell length. It will now be easier for students to visualize that as a shoot elongates, there is simultaneous contribution from the intercalary meristems of a number of successive internodes.

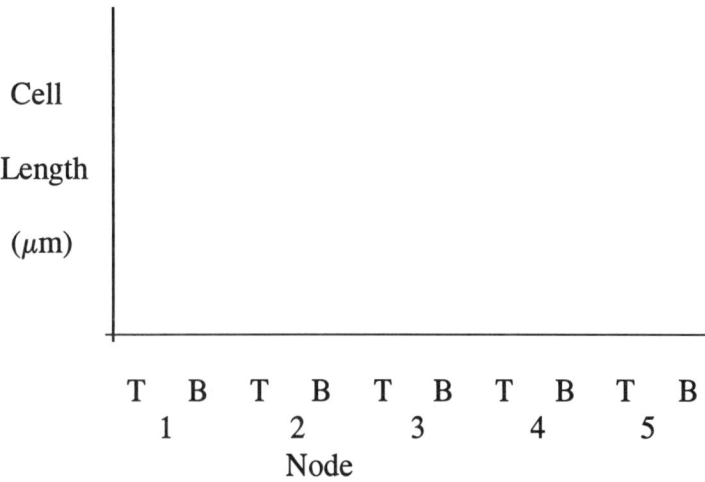

LABORATORY - MATERIALS, EQUIPMENT, AND PREPARATION

1. Supermarket Morphology
- variety of vegetables, with edible stems, from the supermarket including: *Asparagus,* broccoli, celery (for shoot apex, edible portion is leaf), kohlrabi, and potato.
- dissecting microscope or hand lens

2. Anatomy of cells and tissues
- fresh materials, selected from above, or: *Coleus,* geranium, begonia, etc.
- compound microscope
- slides and coverslips
- razor blades
- dropper bottles of:
    - water
    - toluidine blue (0.05 g toluidine blue / 100 ml H_2O)
    - phloroglucinol (saturated aqueous solution in 20% HCL)
- polarizing filters

3. Stem Growth

- compound microscope (calibrated field of view)
- prepared slide of *Elodea* shoot tip

REFERENCES

General:

Berrie, Alex M.M., 1977. *An Introduction to the Botany of the Major Crop Plants.* Heyden and Son Ltd., London. 220 p. Although technical, this is a compendium of useful information about the growth of crop plants.

Bianchini, F. and Corbetta, F. 1975. *The Complete Book of Fruits and Vegetables.* Crown Publishers, Inc. New York. 303 p. A beautifully illustrated introduction to common fruits and vegetables.

Harlow, Wm. M. 1959. *Fruit Key and Twig Key to Trees and Shrubs.* Dover. New York. 56 p. A useful winter key to woody plants of the northeastern United States which relies of vegetative stem characters.

Resources:

Foster, A.S. 1949. *Practical Plant Anatomy, 2nd ed.* D. Van Nostrand Co, Inc. Princeton, New Jersey. 228 p. This is a classic sourcebook for anatomical features of flowering plants.

Hayward, H.E. 1938. *The structure of economic plants.* Macmillan. New York. 674 p. Now considered a classic, this contains a detailed survey of the anatomy of many of the most important crop plants.

Mauseth, J. 1988. *Plant anatomy.* Benjamin/Cummings Publishing Co. Menlo Park, California. 569 p. This is a good introduction to the study of internal plant structure.

Muir, W.H., A.C. Hildebrandt, and A.J. Riker. 1958. The preparation, isolation and growth in culture of single cells from higher plants. *Amer. J. Bot.* 45:589-597.

CHAPTER 6: LEAVES

The leaf is the second component of the shoot produced by the shoot apical meristem. As Mauseth points out, the structure of leaves is quite plastic with adaptations optimizing for photosynthesis balanced against those with a protective function. A study of the morphology and anatomy of leaves is a study of the tradeoffs that evolved to meet these conflicting demands. Leaves, however, are an obvious part of plants with several readily observable variations familiar to most students; they will thus be a relatively straightforward topic of study.

OBJECTIVES

1. Describe the characteristic morphology of simple and compound leaves.
2. Describe the internal anatomy of leaves and relate this structure to the corresponding structure of stems and the conflicting demands for gas exchange required for photosynthesis and protection.
3. Relate leaf initiation to growth of the shoot apex and describe leaf development.
4. Explain characteristic modifications that adapt leaves to special conditions or functions.

CONCEPTS

Students understand that a primary function of leaves is photosynthesis and that the structure of the leaf is related to this function. What they do not understand so intuitively are the tradeoffs required for protection against dessication or specialization for other functions. At this point students will not be able to appreciate fully the significance of these tradeoffs. Make them aware of the problems.

Leaf morphology and anatomy are easily observed and present few conceptual problems. A notable exception, however, is the pinnately compound leaf. A key feature, which is useful to distinguish between leaflets of a compound leaf and a series of simple leaves along a stem, is the presence or absence of axillary buds where the appendage attaches to the axis. Axillary buds are usually conspicuous at the base of a leaf, but are never found at the base of a leaflet. This structure thus serves as a key indicator for distinguishing a leaf from a leaflet.

A final concept that frequently presents difficulty to students is the relationship between the vascular systems of leaves and the stems to which they are attached. The previous chapter introduces students to the structure of vascular bundles in monocot and dicot stems. Given that both the stem and leaf originate in the shoot apical meristem, it is logical that there is some continuity in their tissues. In the stem, vascular bundles develop with a characteristic spatial arrangement of xylem and phloem, xylem to the inside of the bundle, phloem to the outside. A bundle that moves out to supply a leaf, then, should maintain this arrangement as it moves into the leaf. As a consequence, vascular bundles in leaves typically have xylem located toward the upper epidermis and phloem located toward the lower epidermis.

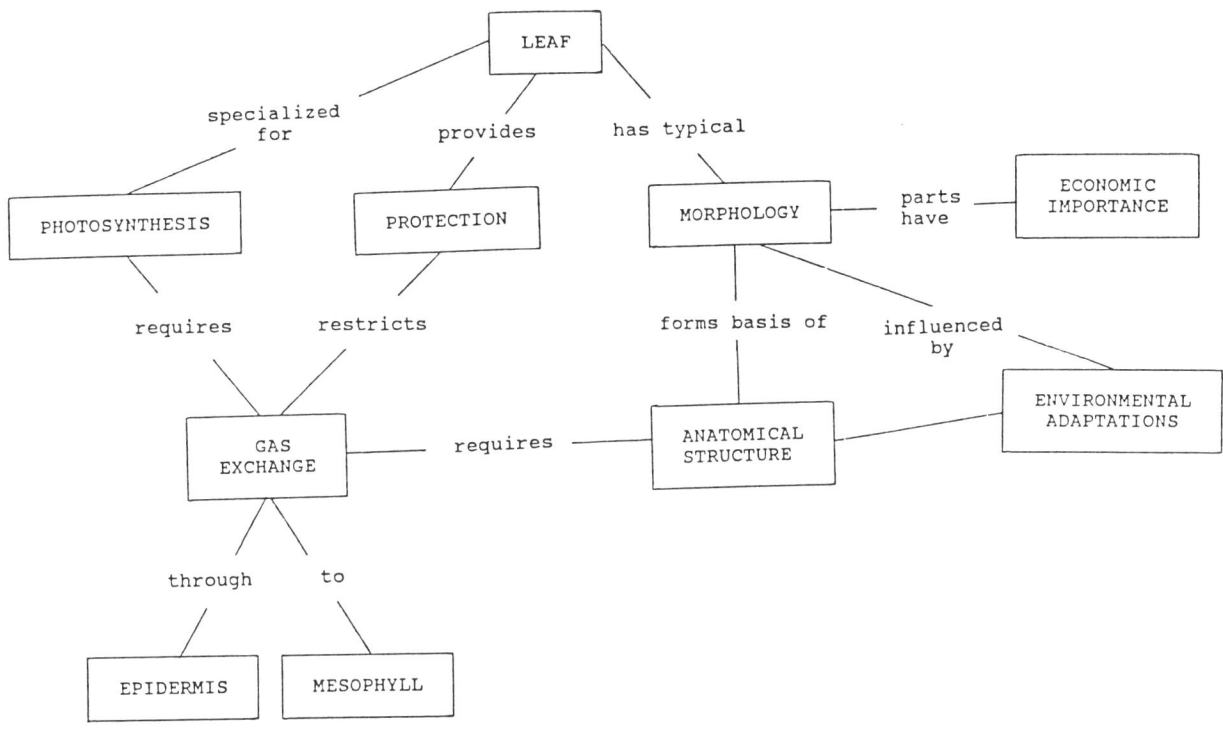

INSTRUCTIONAL AIDS

Lecture - A "field trip" outdoors is equally effective for studying leaf morphology as it is for stems. A survey of the landscape plantings around and near your classroom building will allow you to plan a tour to include examination of most of the common structural features of leaves, including: lamina, petiole, stipule, and axillary bud; monocot vs dicot; simple vs compound; pinnately vs palmately compound; various lobed and toothed margins; and different venation patterns. Although you can easily strip leaves from stems to pass among the class, it is desirable to have a knife or pruning shears so that students can examine intact shoot segments.. It is also desirable to have your clipboard, pad and pen handy to illustrate quickly those features you want to point out and to explain terminology. Be sure to point out axillary buds and find an example of a pinnately compound leaf - again, this is one of the more difficult concepts concerning leaf structure for students to comprehend.

If thick leaves are available, you also might want to pass out hand lenses; positional arrangement of spongy and palisade mesophyll and vascular tissues are easily observable in a cut section through a thick lamina. Again, a less satisfactory substitute is to bring samples into the classroom or to depend on slides to illustrate these concepts.

You can demonstrate the positional arrangement of xylem and phloem in a leaf vascular bundle, and the relationship of leaf bundles to stem bundles, by using yourself as a model. Ask

Chapter 6

your students to imagine the trunk of your body to be a segment of stem tissue. Extend your arms above your head, palm to palm to represent just two vascular bundles on opposite sides of the stem. Quiz a student about the relative position of xylem and phloem in one of these "bundles". Xylem, of course, will be on the inside (palm-side) of the bundle (arm). Now ask the students to imagine that there is a leaf growing out from the stem at about the position of your shoulder. The most efficient method of supplying a vascular bundle to that leaf would be to have the bundle directly in line with the leaf move out to supply that leaf. Illustrate this arrangement by dropping your arm to shoulder height, without rotating your hand. The xylem (palm-side) that was previously oriented toward the inside of the "stem" is now positioned toward the top side of the "leaf". Visualizing this relationship will not only help students understand the close association between stem and leaves, but will help orient them when they examine cross sections of various leaves.

Mauseth does a good job of explaining the selective advantage of not only typical leaf structure, but also of various adaptations. Reinforce the structure/function relationship concerning the benefits and disadvantages of particular leaf structure when discussing the morphology and anatomy of typical leaves. Students will then be better able to understand how peculiar variations are adaptive to specific conditions. This sort of understanding is critical for students to eventually understand the theory of evolution.

Pre-lecture Review - The term "shoot" emphasizes the close association between stems and leaves and the effect one organ has on the other. Begin your discussion of leaves by reviewing what students know about stems and their growth. What is the role of the shoot apical meristem? What characters differentiate between nodes and internodes? What is the anatomical relationship of tissues in the stem and how might this affect the structure of the corresponding leaves?

Post-lecture Review - Now is an opportune time to challenge students to integrate their understanding of both stems and leaves by constructing a concept map for the shoot system. There is too much information for you to expect students to get it all into a single map; rather, concentrate on a specific aspect such as: growth and development; tissue systems; adaptation to stress; etc.

LABORATORY - EXERCISES

1. Supermarket Morphology

Although leafy vegetables are well known to most students, there are some items in the produce section that may challenge their knowledge of leaf morphology. The onion is a classic example of a bulb - a shoot whose leaves are greatly enlarged and modified for storage. Surprisingly, cabbage and brussels sprouts sometimes confuse students. Another classic example is celery, whose elongate, fleshy petioles are frequently mistaken for stems. Don't forget to look at the spices and herbs. Whole bay leaves are easy to recognize as a leaf, but rosemary is more of a challenge while oregano and basil will test their observational and critical thinking skills. Some clues with the latter are specialized epidermal cells such as hairs, guard cells, and glands as well as occasional veins traversing pieces of the herbs. Rehydrating dried herbs in alcohol, then water (with a bit of detergent to act as a surfactant) facilitates making usable water mounts of this material.

2. Field Identification

Tree identification is an exercise usually saved for a laboratory on taxonomy or systematics. Many keys to local trees, however, are based almost exclusively on leaf characteristics, therefore, doing field identifications now will reinforce the terminology used in describing leaf morphology, as well as give a practical example of how to apply an understanding of this terminology . Ideally you should choose a key for as small a local area as possible; even better is to make one of your own specifically for the species that are likely to be encountered on and around your campus. Appendix 1 provides two examples, one suitable for the North-Central states, the other for the South-Central. If you haven't already taken the class out for a field trip to examine leaf morphological diversity do this now and walk the class through the key for the first species or two. Group competitions work well once most of the students seem to understand how the key functions. Divide the class into groups, then walk to an "unknown" and let each group attempt to make a correct identification. Award points, based on a preannounced scale, to students in each group as they make their identification. The group with the most points at the end of the exercise "wins."

3. Leaf Anatomy

Fresh leaves are much more difficult than either stems or roots to section with a razor blade. Although it can be done, it usually leads to much frustration and you should therefore not have your students attempt it. You will have to depend on prepared leaf cross-section slides of various species; fortunately there are a lot to choose from. Begin with a typical dicot. *Ligustrum* is frequently used). As with the stems during the previous week, begin by having the students simply find the material and make a large sketch of the general tissue regions. Have a volunteer copy her/his sketch on the overhead and use it to review the general tissue regions and the adaptive significance of this particular structure. Now have them find, and sketch, more detailed structures, indicating the proper location on the overall outline sketch: stomata with guard cells in the lower epidermis; slightly sunken multicellular glandular trichomes mostly on the upper surface; the arrangement of xylem and phloem in a vein. Ask the students to tell you at least two different clues as to which side of the leaf is "up."

Now repeat this procedure for a typical monocot such as *Zea*. Stomata are located on both surfaces but groups of swollen bulliform cells occur only on the "upper" surface. Bundle sheaths are prominent here with bundle sheath extensions on the larger veins.

Chapter 6

With the above two examples for reference, supply several different leaf types. They may be monocot or dicot, mesomorphic or adapted to extreme environmental conditions. Again you can work this as a competition by covering the labels with a piece of paper identified only by number or letter. For each of the "unknowns" ask for identification as to either monocot or dicot and whether the leaf is adapted to dry (xeromorphic), aquatic (hydromorphic), or typical (mesomorphic) conditions. They must also list two features of the leaf that support their choice. Depending on the materials available, people can work in groups of different sizes with each group responsible for the entire set of "unknowns."

LABORATORY - MATERIALS, EQUIPMENT, AND PREPARATION

1. Supermarket Morphology
   -variety of vegetables, with edible leaves, from the supermarket, including: artichokes, cabbage, celery, collards, lettuce, onions, parsley, spinach, etc., and a variety of herbs and spices.

2. Field Identification
   - key to common local trees, hand lens optional.

3. Leaf Anatomy
   - compound microscope
   - prepared leaf cross-sections of *Ligustrum* and *Zea*
   - prepared leaf cross-sections of additional plants:
        xerophytes - *Nerium oleander, Ficus elastica, Yucca*
        hydrophytes- *Nymphaea, Typha*
        mesophytes - *Acer,*

REFERENCES

General:

Lott, John N.A. 1976. *A Scanning Electron Microscope Study of Green Plants.* C. V. Mosby Company. St. Louis. 170 pp. This is an intriguing collection of SEM micrographs surveying the plant kingdom but concentrating on flowering plants.

Sandved, Kjell B. and Ghillean T. Prance. 1985. *Leaves: the Formation, Characteristics, and Uses of Hundreds of Leaves Found in All Parts of the World.* Crown Publishers, Inc. New York. 244 pp. Superb photographs with accurate botanical descriptions make this an appealing reference.

Resources:

Juniper, B.E. and C.E. Jeffree. 1983. *Plant Surfaces.* Edward Arnold Publishers, Ltd. London. 93 pp. This is a small monograph concentrating primarily on the leaf epidermis.

Preston, Richard J. 1976. *North American Trees, 3rd ed.* Iowa State University Press, Ames, Iowa. 399 pp. An easy-to-use tree key based primarily on leaf characters.

Elias, Thomas S. 1980. *The Complete Trees of North America.* Van Nostrand Reinhold, New York. 947 pp. Similar to Preston, but a bit less technical.

Sundberg, M.D. 1984. A quantitative technique for beginning microscopists. *J. Coll. Sci. Teaching.* May: 417-419. A simple morphometric comparison of sun and shade leaves written for introductory-level students.

CHAPTER 7: ROOTS

The root is the simplest of the three vegetative plant organs, yet it is less familiar to students simply because of lower visibility. Because the root is an underground extension of the plant axis, it will have much in common with the stem: apical growth, primary meristems, and basic tissue regions. Unlike the shoot system, though, there are no lateral appendages equivalent to leaves. Thus, the root apical meristem involves only axis elongation. There is no distinction of node or internode-like units with associated intercalary meristems. Furthermore, the origin of branches will be deep within the root tissue, unlike the superficial position of shoot branches in the axils of leaves.

Like the other plant organs, there is a close association between structure and function. Structural variation correlates with specialized function or environmental conditions. A major theme of the textbook, and this manual, is the selective advantage of variation in structure and function. Emphasize this relationship whenever possible, but be careful not to slip into the phraseology of the plant "doing something" (eg., increasing the amount of parenchyma in the cortex for storage) "because of something" (eg., the need for long term storage). Evolution is not a conscious process by the organism; it is a fortuitous event for an organism if it happens to result in variations that have selective advantage under a particular set of conditions. This organism will be more likely to pass on its genes, including those responsible for the advantageous traits, to the next generation and, so, effect evolution of the species.

OBJECTIVES

1. Describe three principle types of root systems: fibrous, tap, and adventitious.
2. Describe the internal anatomy of a typical dicot and monocot root.
3. Explain root growth in terms of the apical and primary meristems and the origin of lateral roots.
4. Describe characteristic modifications that adapt leaves to special conditions or functions.

CONCEPTS

Roots are in many ways mysterious to students, in large part because they are usually out of sight. A fundamental concept to understanding the function of roots is an understanding of surface area relationships. Surface area is important not only for the absorptive function of roots, but also in providing a solid anchorage system. Many students will be familiar with the villi and microvilli of the small intestine; relate this structure/function relationship to the root hairs behind the growing root tip. In both cases, increased surface area promotes increased transport. Similarly, some students will have experience weeding in a garden or around the home. They will have observations on the relationship between extensiveness of the root system and difficulty to uproot the plant.

One of the distinguishing anatomical features between root and stems is the presence of an endodermis in the former. It is sometimes difficult for students to visualize the position of casparian strips around the radial walls of endodermal cells and the spatial relationship of one endodermal cell to another. The function of the casparian strip is also difficult to interpret unless the student realizes that much of the movement of water and dissolved minerals through the root cortex literally occurs through the cell walls without ever entering the cytoplasm of the cortical cells. Students generally have the misconception of the plant cell wall being solid and impermeable

Root structure is not difficult to understand, especially after already having covered stems, but growth and the penetration of roots through the soil is frequently a problem. Mauseth does a good job explaining elongation of the root through the soil, and especially the role of the ever-growing root cap.

A final concept that some of your more alert students may raise is the inherent problem in making vascular connection between root and stem.
You will have already discussed the characteristic anatomical differences between monocot and dicot stems; monocots and dicots are also distinguishable by their root anatomy. The problem is that the arrangement of tissues in roots, particularly the vascular tissues, is much different from that in stems. The vascular tissues of these organs must be continuous, therefore a pattern transition must occur somewhere. As they might expect, after comparing root and stem sections from the same species, the transition must be very complex. It occurs in the hypocotyl region of the plant.

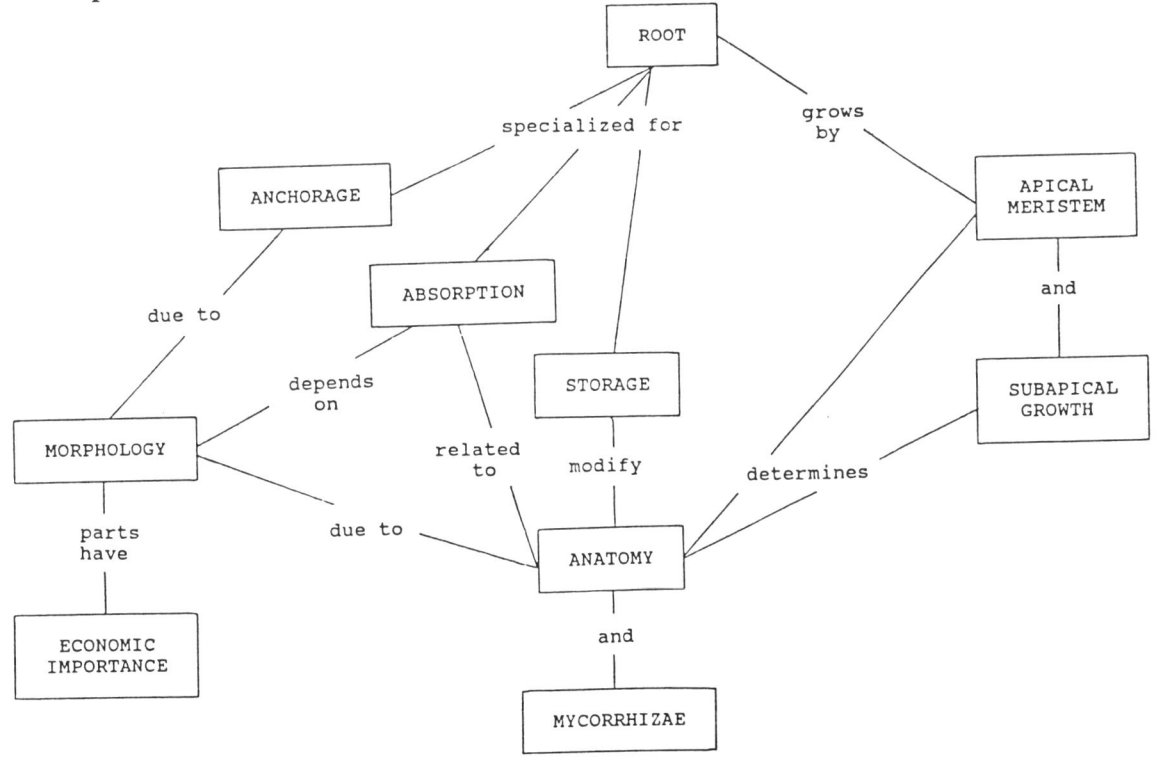

Chapter 7

INSTRUCTIONAL AIDS

Lecture - On the way to lecture, collect an example of a taproot (e.g., dandelion) and a fibrous root (eg. quackgrass). Placed on an overhead these produce a silhouette which leads to an initial discussion of basic root types and structure/function relationships of roots, particularly anchorage, absorption, and storage. Preferably, do not wash the roots clean before projecting their image. A comparison of attached soil particles on the two types will illustrate the other side of the anchorage problem - the role of plants, especially fibrous- rooted sod formers like grass, in anchoring the soil against erosion. It is also sometimes useful to have a petri dish available with about 5-day-old germinating radish seedlings on moist filter paper. One such seedling placed on an overhead will illustrate the tremendous increase in surface area due to development of root hairs. A useful analogy may be drawn here between the structure and function of root hairs and the villi lining the small intestine of the human digestive system. Most students will have had at least one high-school class in biology and this course probably stressed human anatomy and physiology. They will therefore be familiar with the role of increased surface area in absorption of food and the production of numerous small finger-like projections to increase surface area.

Growth of the root through the soil is difficult to visualize. Ask your students to imagine that one of their fingers is a growing root tip - it is the right shape after all. That root tip (finger) must now grow (be pushed) through the soil. What problems can they anticipate? One problem will be damage to the tip. A protective layer, the root cap, helps to minimize damage to the tip of the root proper. The root cap itself must grow through the soil, a process aided by division of meristematic cells at the base of the root cap, adjacent to the root apical meristem and subsequent elongation of files of cells toward the tip of the cap. The cap thus "paves the way" for the root proper which continually pushes its way through the older root cap cells. A second problem is to produce a driving force to permit penetration through the soil. Part of the answer, of course, is the action of the apical meristem itself, continually producing files of new cells that then elongate to push the meristem further ahead. But the root must first anchor itself well or the plant will be pushed up and out of the soil rather than the root begin driven further into the soil. Thus, a second role of root hairs is to anchor the root solidly just behind the growing tip.

There are several aids that you can use to clarify the role of casparian strips and the endodermis. A chalkboard eraser is a good model of a plant cell. Use a strip of masking tape to simulate a casparian strip around a single endodermal cell. Mask several erasers and then hold them up side to side, top to bottom, etc. to illustrate how Casparian Strips force materials to be transported through endodermal cells. Again, use your body to represent the stele of the root, the central core of vascular tissues. Your masked erasers placed around your torso visually illustrates the positional relationship of endodermal cells to stele and emphasizes how the barrier of adjacent casparian strips forces all transport into the vascular tissues to pass through endodermal cells, thus providing a means of control.

Pre-lecture Review - The anatomy of dicot and monocot roots is as distinctive as that of dicot and monocot stems. Begin by reviewing the characteristics of the two types of stems. What features do they have in common (general tissue types and xylem and phloem in the same bundle) and what features are unique (a ring of vascular bundles surrounds the pith in dicots and bundles are scattered in monocots).

Post-lecture Review - Students should now be able to recognize four basic anatomical patterns in plant axes - roots and stems of dicots and roots and stems of monocots. What features characterize roots? - endodermis and separate xylem and phloem bundles. What is the anatomical distinction between monocot and dicot roots? - pith in monocots, central xylem in dicots. At this point, if students have not already raised the question, you might ask what problems there might be concerning transport through the xylem and phloem of a plant. The problem is that the vascular tissues in roots and stems do not "match up" and there must be an orderly transition between the two patterns.

LABORATORY -EXERCISES

1. Supermarket Morphology

The classic vegetable example of a root is the carrot. If possible, get home-grown carrots with the shoot intact. Most students have never seen an intact carrot plant, complete with rosette of leaves sprouting out of the top - it really is a plant. Store bought carrots are usually cleaned and most of the tiny branch roots are washed off. A few may be visible, though, so be sure to point them out. Cross sections of the tap root at this level may show small traces of xylem moving across the cortex toward the small lateral branch roots. Other common roots in the produce department include: sweet potatoes, beets (both red and sugar), horseradish, ginger roots, turnips, rutabagas, and radishes. The last three are unusual in that the edible "root" consists not only of enlarged primary root, but also and enlarged portion of hypocotyl.

There are several alternatives to making several trips to the market. The obvious one is to bring purchased materials into lab. Another alternative, which I prefer, is to make a handout for students to take on their own to the market. Such a list includes the names of vegetables students should observe, what part of the plant is edible, a justification for their decision as to plant parts, and some specific questions about certain parts such as: botanically speaking, what are the "eyes" of a potato? or does celery have simple or compound leaves - how can you tell?

2. Root Anatomy

Have students examine cross sections of dicot and monocot roots in the same way they examined stems. First, have them observe a dicot stem, such as *Ranunculus,* under low power

Chapter 7

and make a large sketch of the general tissue regions seen. Ask one of the faster students to copy her/his sketch onto an overhead transparency for you to use in identifying tissue regions and pointing out areas for closer examination. Take a closer look at epidermis and the subepidermal cortex (use polarizers to confirm the presence of starch grains in the cortical cells, endodermis with casparian strips, and vascular bundles. Have the students sketch a group of cells from each of these regions to identify specialized cells types. Repeat this procedure with a monocot such as *Zea* or *Smilax*. In these the endodermis has distinctive "U"-shaped casparian thickening; a hypodermis with inverted "U"-shaped thickenings may also be evident. Be sure that students note the alternating bundles of xylem and phloem in a ring beneath the endodermis.

3. Root Growth

A good way to localize the root apical meristem and demonstrate the role of elongation in root growth is to do a marking experiment with very young seedings. Soak some pea seeds, have them just barely covered with water when you start, then cover the container to maintain humidity, for three or four days prior to the lab. Primary roots should be a centimeter or two long when you begin. Have students quickly mark equidistant intervals along the length of their root, sketch the marked root, and record the distance between intervals in their notes (be sure to use a waterproof marker - India ink and a fine tipped marking pen works well). While one student is marking the roots, a partner can prepare a culture apparatus - a drinking straw cut to fit within a test tube with room for the pea seed on top and a strip of filter or chromatography paper cut to fit within the straw to serve as a wick. Add about two cm of water to the bottom of the tube to act as reservoir to keep the wick moist and the growing root from drying out. After students have marked the root and the tube prepared, insert the taproot into the top of the straw, place the seedling and straw into the tube, and put a test tube cap (or aluminum foil) over the end. Every day, observe growth of the taproot down the straw and record displacement of the original marks. The distance between marks near the seed itself will be relatively unchanged, while those nearest the tip will have spread considerably.

To correlate this morphological record of root growth with meristematic activity and cell elongation within the root, supplement the above exercise with a quantitative examination of longitudinally sectioned root tips. The same onion root tip slides used to estimate duration of mitotic stages are used for this. Again, center the field of view under high power in the root apical meristem so the junction with the root cap is positioned at the bottom edge of the field. Count the number of dividing cells in the field (all stages of mitosis) and estimate the average cell length. Record the data, then carefully move the slide down so the cells that were at the top edge of the field are now at the bottom edge. You should now be looking at a field of cells one field of view behind the root tip. Again count the number of dividing cells and estimate the average cell length. Repeat this procedure for at least two more fields of view sequentially further from the root tip.

Field	# Dividing Cells	% Dividing	Avg. Cell Length
1			
2			
3			
4			

After data collection, students should plot both the data for percent of cells dividing and average cell length against field number. The resulting graph will illustrate the relative contributions of cell division and cell elongation, at increasingly greater distances from the root tip, to overall root growth. Correlate this information with the gross measurements made on the pea taproots.

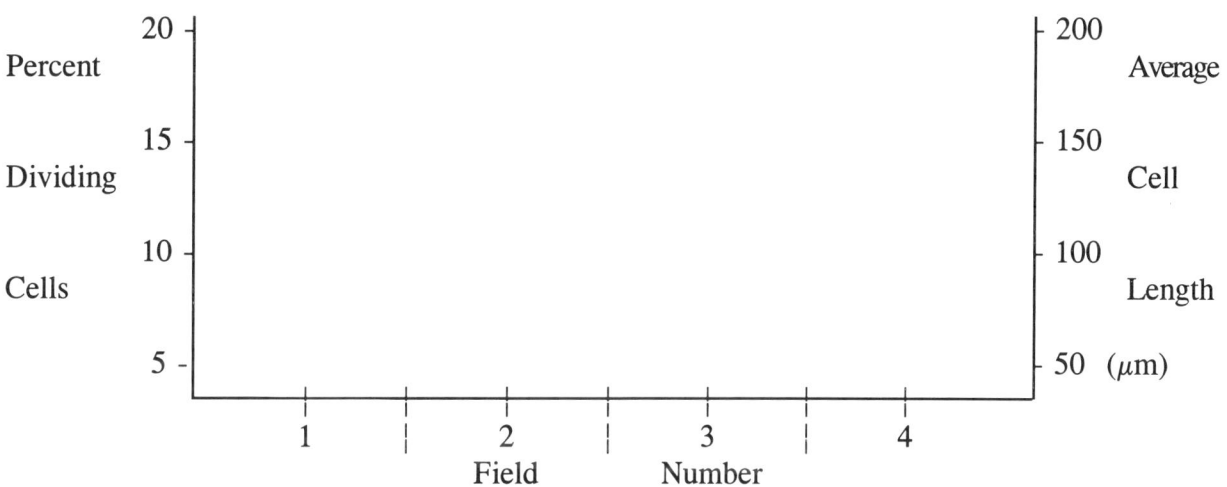

LABORATORY - MATERIALS, EQUIPMENT, AND PREPARATION

1. Supermarket Morphology
- variety of vegetables from the produce department including: carrots, sweet potatoes, beets (both red and sugar), horseradish, ginger roots, turnips, rutabagas, and radishes.

2. Root Anatomy
- compound microscopes
- overhead projector, clear transparencies, and marking pens
- prepared slides of *Ranunculus, Zea,* and/or *Smilax* root cross sections.

Chapter 7

3. Root Growth
- pea seeds and covered finger bowls or dishpans for germination.
- test tubes with closures or aluminum foil to form a cap.
- plastic straws
- scissors
- filter paper or chromatography paper to act as a wick.
- fine-tipped india ink marking pen
- compound microscopes
- onion root tip slides

REFERENCES

General:
Emboden, W.A. 1974. *Bizarre Plants: Magical, Monstrous and Mythical.* Macmillan. New York. 214 p. Several of the examples presented in this interesting volume are of roots.

Resources:
Sundberg, M.D. 1981. Making the most of onion root tip slides. *Amer. Biol. Teach.* 43:386-388. Details of procedure to investigate role of mitosis and cell elongation in the growth of roots.

Torrey, J.G. and D.T. Clarkson. 1975. *The Development and Function of Roots.* Academic Press, London. 618 p. The single best source of information on root biology.

CHAPTER 8: WOODY PLANTS

Secondary growth characterizes woody plants. This type of growth in girth is due to the action of two lateral meristems, the vascular cambium and the cork cambium. As expected, there is both a structural and functional correlation between secondary growth and primary growth, growth in length due to the action of apical meristems. As described in previous chapters, three primary meristems, including procambium, arise directly below the apical meristem in both roots and stems. Within procambial bundles, cells on the periphery are the first to differentiate into functional vascular tissues. For instance, in a stem bundle, the outermost cells will differentiate into phloem (protophloem) while the innermost cells form xylem (protoxylem). Undifferentiated procambium remains between these differentiated cells. In progressively older regions, additional phloem differentiates from the residual procambial bundle in a centripetal direction while xylem differentiates simultaneously in a centrifugal direction in the same bundle. The very last procambium to differentiate does so near the center of the original bundle.

All the while this process of differentiation is occurring, the plant continues to grow in length as the apical meristem produces additional cells. The selective advantage of this is that the plant continues to grow taller and will thus be at a competitive advantage for maximizing exposure to sunlight. The disadvantage is that as the stem grows continually taller it becomes increasingly less stable and prone to breakage. This phenomenon is known as "lodging" in agriculture. The stems of elongate cereal grains tend to snap in response to heavy wind or rainstorms.

The greater the height of the plant, the greater the selective advantage of providing reinforcement in the stem to resist breakage. Thus, physical pressure is one stimulus involved with differentiation of specialized cells, such as fibers, from the xylem and phloem. Continued production of such supporting cells, as the stem grows, is advantageous. Because both cell types arise from a common meristematic precursor, it is not unexpected that the residual procambium in the vascular bundles of wood plants is the origin of the vascular cambium of woody plants. This cambium, originating within the bundles, is called fascicular vascular cambium. The vascular cambium thus provides a continual source of additional supportive cells increasing the girth of the plant as it continues to elongate.

The evolution of a vascular cambium to solve the problem of support, and transport as described by Mauseth in the text chapter, created an additional problem - one of protection. An epidermal layer covers the primary body of both stems and roots. This layer protects the internal tissues against desiccation and invasion by pathogens. As the volume of secondary tissues increases internally, this boundary layer must either grow to compensate or be stretched and broken. Epidermis, however, is a mature tissue composed of already differentiated cells that normally will not resume cell division and growth. There is, thus, a selective advantage for production of a new protective layer, the periderm, capable of continued expansion to compensate for the proliferation of cells from the vascular cambium. Cork cambium is the secondary meristem responsible for production of periderm.

Chapter 8

OBJECTIVES

1. Describe the origin of the vascular cambium and how it produces secondary xylem and secondary phloem.
2. Explain the structural and functional differences between fusiform and ray initials and their derivatives.
3. Describe the structural basis of growth rings in wood and correlate these structural differences with growth conditions.
4. Describe the origin of the cork cambium and how it produces cork and phelloderm.

CONCEPTS

Two concepts tend to provide particular difficulty to students. First, how doe one of the daughter cells produced by division of a vascular cambium cell differentiate into either a xylem cell or phloem cell, but not both? Second, why is there only a single ring of vascular cambium in a woody plant that may have many growth rings. These two related problems trace back to the origin of vascular cambium in the plant. As described above, vascular cambium originates from residual procambium. A single row of procambial cells remains between the primary xylem and primary phloem that differentiated toward each other within the original procambial bundle. Each cell in this row thus has an adjacent xylem cell on its internal tangential face and a phloem cell on its adjacent external tangential face. This relationship is maintained as the plant grows. Thus when a vascular cambium cell divides into two daughter cells, one daughter cell must remain cambial and between the xylem and phloem. The other daughter cell will differentiate into either xylem or phloem, depending on whether it is internal or external to the cambial daughter cell. In either case, a single vascular cambium cell separates a file of xylem cells from a file of phloem.

An axiom of developmental biology is that the fate of a cell is a function of its position. That is, neighboring cells influence each others development. This is certainly true of the vascular cambium. Although each vascular cambium cell produces a single file of daughter cells, extending radially on either side, it does so essentially in synchrony with the cambial cells of adjacent files. This lateral influence, in fact, results in differentiation of interfascicular vascular cambium from parenchyma cells between the original vascular (procambial) bundles. Thus, a ring of vascular cambium gradually advances outward laying secondary xylem behind and pushing secondary phloem ahead. It does not matter if the rate of advance changes, or even if it goes through periods of halt and advance. The single layer of cambium remains between the most recently formed xylem and the most recently formed phloem. The pattern resulting from successive growth periods provides a record of the rate of growth each season. These are the growth rings characteristic of familiar woods.

A minor conceptual problem has to do with the term "bark." Wood and bark are technical terms that have widespread colloquial usage. In the case of wood there is no problem because both usages refer to xylem, particularly the secondary xylem produced by woody plants. In a colloquial sense, however, the term bark usually refers to what is more properly termed "periderm", or outer bark. In technical usage, bark also includes the phloem and any residual primary tissues external to the phloem.

Finally, one misconception many students have about secondary growth is that it is a characteristic of stems only. This probably arises because students see woody stems but seldom notice woody roots. We tend to reinforce this misconception when we use stem examples for most discussion of vascular cambium and the characteristics of wood. After discussing secondary growth in stems, you should at least mention that the same processes occur in roots - in essentially the same way. The major difference is that the patterns observed in root wood are usually less pronounced than those found in stems.

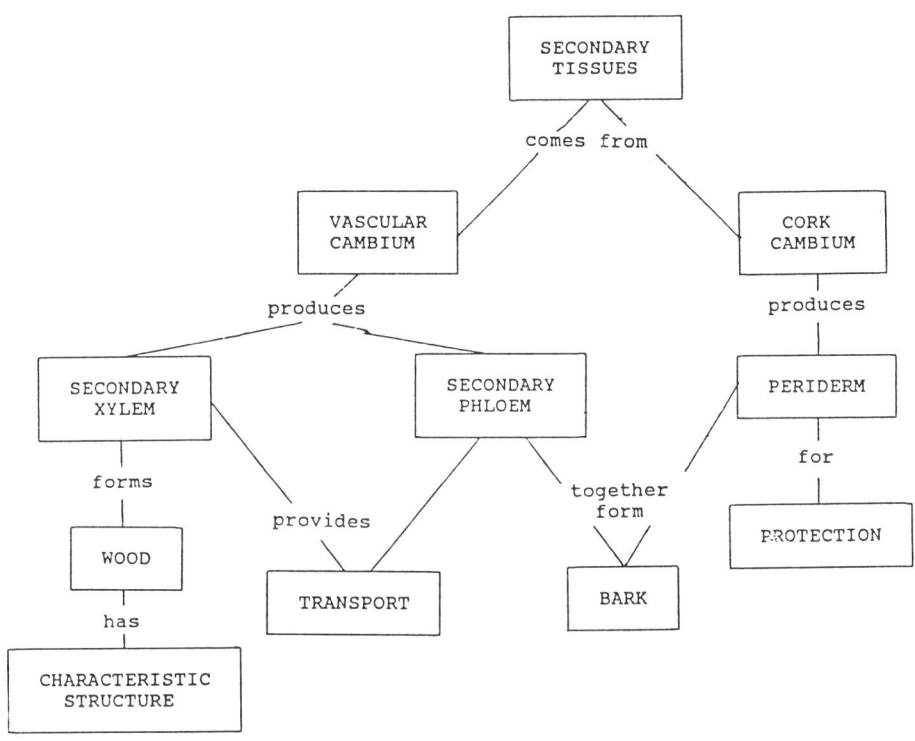

INSTRUCTIONAL AIDS

Lecture - An interesting way to start your lecture on secondary growth is with a demonstration of etiolated seedlings or seedlings, such as sunflower, that were treated with gibberellic acid a week or so in advance. In either case you will have long, spindly plants. Now pose some questions to the class. What disadvantages might this plant have in competition with other plants growing in nature? What kinds of modifications or adaptations would allow this plant to grow competitively with other plants? Students will come up with several possibilities for each

Chapter 8

question. Make a list on the board for each. Several of the suggested answers for the second question will relate in some way to secondary growth, either directly - "form wood," or indirectly - "increase the strength of the stem." In turn, plants with secondary growth are at an advantage in dealing with many of the problems related to the first question.

Before moving into the details of secondary growth, review the structure and primary growth of stems. Sketch the general primary tissue regions of a dicot stem on the board. Relate this structure to some of the problems recognized above. For example, what regions of the stem, or specific cell types in the stem, are particularly effective in providing support for upright growth? Developmentally, what is the source of this tissue or these cells? Introduce the procambium and procambial bundles. From there you can begin a discussion of the origin and function of the vascular cambium as outlined in the introduction and concepts sections above.

Concentrate on growth of the fascicular cambium at first, emphasizing how the vascular cambium divides and the potential fates of the two daughter cells. Begin by enlarging one of the vascular bundles of your board diagram of a dicot stem. Emphasize the single row of vascular cambium between the already differentiated xylem and phloem. Use this enlargement to illustrate division of a vascular cambium cell and the possible differentiation pathways that may result. When the class appears to understand the process and the gradual migration of the vascular cambium outward in the stem as it produces files of secondary xylem from the inner face of cambial cells, ask what the resulting stem will look like after a period of growth. If secondary growth occurs only in the original vascular bundles, the stem will quickly assume a ridged appearance. This is, in fact, what happens in many woody vines (lianas). This anatomy enhances their flexibility in the same way stranded electrical wire, as in an appliance cord, is more flexible than solid wire of the same gauge. In most woody plants, of course, this does not happen because interfascicular cambium begins to form between the bundles when the fascicular cambium becomes activated. Thus, a continuous ring of vascular cambium forms quickly and the stem uniformly increases in diameter. Connect the vascular bundles of your original diagram to illustrate the now complete ring of vascular cambium.

The main functions of woody stems are to support physically the ever enlarging shoot system and to provide the necessary vascular tissue to transport an adequate volume of water and nutrients. Indeed, most of the elements of secondary vascular tissue are axial, running vertically up and down the axis. Yet many of the cells along this route remain alive. How are they supplied with necessary metabolites etc.? This question can lead into a discussion of the structure and function of rays and the differentiation of ray initials from fusiform initial in the cambium. What happens to the ratio of rays to vascular tissue as the tissue increases in girth? If it stays the same, and it does, how can this ratio be maintained? New ray initials must differentiate within the ring of vascular cambium.

Your diagram of a primary dicot stem section now has a ring of vascular cambium connecting the original vascular bundles. Your students should understand how this ring gradually moves outward through the stem as the vascular cambium produces more and more secondary xylem toward the inside. Ask what the eventual problem will be for the plant if this enlargement process continues? The problem is that the cortex and epidermis also must grow or their integrity will be disrupted and they will become nonfunctional. The problem, then, is to replace these layers with a new tissue that can assume the same functions but continue to enlarge in pace with the underlying tissue. This discussion provides an introduction to the cork cambium and periderm. Compare their structure and development directly to the vascular cambium and its derivatives.

A good-sized section of tree trunk, complete with bark, is the classic visual aid when lecturing about secondary growth. Plantation-grown pine is an excellent example because it will have large, distinctive growth rings. In large classes you will have to resort to slides, but have some of the real thing available for interested students to examine after class. After covering the structure of wood, it is interesting to give some examples of how to put this knowledge to use in a practical sense. For instance, in evaluating the quality of wooden furniture, how can you recognize veneer or laminated wood from solid wood (mirror image grain patterns in face is veneer as is similar grain pattern in both face and edge of a cut piece)? Or why are pieces constructed of quarter-sawn boards (radial faces) more expensive than plain sawn (tangential faces)? Furthermore, understanding the structure of wood is useful in identifying wood type. It also has practical application in evaluating the quality of wooden articles.

Pre-lecture Review - In addition to providing a sketch of a typical dicot stem in cross-section to illustrate initiation and later development of the vascular cambium, as suggested above, it is also useful to review how apical and/or intercalary meristems produce files of cells. Recall that daughter cells cut off on one face of a meristematic cell either forces the meristem itself to move continually forward, or extends a growing file of cells ahead of the meristem. Secondary meristems produce the same sort of growth pattern in their derivatives.

Post-lecture Review - Ask your students to sketch a five-panel cartoon illustrating the result of five successive divisions of a vascular cambium cell. In each panel, label the cells to correspond to the same cell in previous and later panels. For example, label the first derivative of the cambium "#1" in the first panel and then continued to label it "1" in all subsequent panels. After completing the cartoon, the student should write a paragraph describing the process illustrated.

LABORATORY - EXERCISES

1. Microscopic Structure

A. Early Development.
Prepared microscope slides of *Helianthus* are good material to examine the early development of vascular cambium. Primary bundles will be obvious and students should begin

Chapter 8

by sketching the general tissue regions of the stem, emphasizing these bundles. Direct students to look closely at the thin layer of cells within a bundle between the xylem and phloem. If it appears that these cells have started to divide (if there are columns of narrow cells lined up - even if only 2 or 3 cells deep), then we should properly call this layer fascicular vascular cambium. If such cell divisions are not apparent, we call this layer residual procambium.

The tissue composed of parenchyma cells between vascular bundles forms a pith ray. If there appears to be a line of parenchyma cells connecting the fascicular cambia of adjacent vascular bundles, we call these cells interfascicular vascular cambium.

B. Later Development.

We commonly use slides of *Tilia* (Basswood) to examine later stages of secondary growth, but the phloem of this species is unusual and tends to confuse students. Basswood is a soft wood good for carving, therefore it is easy to make good microscope slides with this material. Nearly any other woody plant would be a better choice to show typical phloem. Two examples are described below.

Pine, *Pinus,* demonstrates typical secondary growth in a gymnosperm. The cells of the secondary xylem are very uniform because there are no vessels; the only conducting cells are tracheids. Growth rings are clearly visible. Ask students to sketch a group of 12-15 adjacent cells on the boundary of a growth ring, then label spring wood and summer wood appropriately. Emphasize that there is no vascular cambium between the summer wood of one season and the spring wood of the next - the boundary is due simply to the difference in cell size. Vascular cambium occurs between the outermost xylem and the phloem. Periderm will also be present. The large cavities found throughout a pine stem are resin canals. Special parenchyma cells, epithelium, line each canal. Resin canals occur in all tissues of gymnosperms.

Maple, *Acer,* is a typical woody dicot. In a slide of a one-year stem, you can find and identify: pith, primary xylem, secondary xylem, vascular cambium, secondary phloem, primary phloem, cortex, and periderm. Students also can locate and identify vessels, fibers, tracheids, sieve tube elements and companion cells. Xylem and phloem rays will form contiguous files; the progenitor cambial cell will be a ray initial. In cross section, ray initials will be indistinguishable from fusiform initials except by their position.

2. Gross Structure

In trunk sections, you may be able to distinguish between heartwood and sapwood, depending on the age of the stem when it was cut. Growth rings will be visible and have obvious differences in thickness. The thickness of a ring correlates to the growing conditions during the season that ring was produced. In this way tree rings are useful in providing a historical record of the climate of a region. Assuming a single growth season per year, as in temperate regions, rings also allow us to age the stem when it was cut.

Wooden blocks, e.g., cut from 2X4's, are useful for demonstrating the appearance of wood in different section planes. In transverse, or cross-section, concentric growth rings are visible. Growth rings will also be visible in radial and tangential sections, but as longitudinal lines of cells. In radial section (quarter-sawn), the lines will run parallel with blocks of rays scattered checkerboard-like across the cut face. In tangential section, growth rings will converge toward themselves like perfect parabolas and rays will appear as small dark streaks running parallel to the growth rings. (Veneers are made by peeling continuous sheets, tangentially, from a stem - thus the mirror image patterns of growth rings.)

3. Wood identification.

Students can identify many common building and furniture woods with little or no magnification once they are familiar with general wood structure. A hand lens is useful for examining some of the finer features - if a hand lens is not available, the ocular of a microscope, turned backwards, works well as a hand lens!

A. Wood without vessels - "nonporous" (gymnosperm or softwood) .... B
    B. Axial resin ducts present
        C. Epithelial cells of resin ducts with thin walls,
            rays with tracheids .... Pine
        CC. Epithelial cells of resin ducts thick walled,
            rays without tracheids .... Fir
    BB. Axial resin ducts absent
        D. Wood reddish in color .... Redwood
        DD. Wood white or white and red banded .... Cedar
AA. Wood with vessels - "porous" (angiosperm or hardwood) .... E
    E. Wood "ring-porous" - Pores, vessels, in distinct bands .... F
        F. Summer wood with distinct radial lines or patches
            of pores and light colored ray parenchyma .... Oak
        FF. Summer wood without distinct radial lines .... G
            G. Pores in summer wood much smaller than in
                springwood .... Ash
            GG. Spring and summer wood pores similar in
                size but much less frequent in spring wood .... Hickory
    EE. Wood diffuse porous - vessels scattered throughout wood .... H
        H. Some pores visible without magnification .... Walnut
        HH. Pores minute I
            I. Pores solitary or in small clusters .... J
                J. Rays narrower than pores, inconspicuous .... Birch

Chapter 8

      JJ. Rays as large or larger than pores ....Maple
    II. Pores very numerous, crowded ....Cherry

  Depending on the material you have available several you can devise different games or competitions between student groups using trunk sections. These might include guessing ages of log sections or identifying wood types from scraps of lumber.

LABORATORY - MATERIALS, EQUIPMENT AND PREPARATION
1. Microscopic Structure
   - compound microscope
   - prepared slides of: *Helianthus, Pinus,* and *Acer*

2. Gross Structure
   - hand lens, dissecting microscope, or microscope ocular lens
   - tree trunk sections
   - blocks of 2x4 cut to show transverse, radial, and tangential planes

3. Wood Identification
   - wood blocks of various timbers - scraps from lumberyard are sufficient

REFERENCES
General:
Cutler, D.F. 1978. *Applied Plant Anatomy*. Longman, London. 103 p. A concise introduction to all aspects of microscopic plant structure.

Harlow, W.M. 1970. *Inside Wood* The American Forestry Association, Washington, D.C. 120 p. This well illustrated volume concentrates on the structure of wood in relation to its functional properties and uses.

Walker, A. 1989. *The Encyclopedia of Wood: A Tree-by-Tree Guide to the World's Most Versatile Resource*. Facts on File, Oxford. 192 p. The first third is wonderfully illustrated with photos and art work on the botany and application of woods; the remainder of the book is a compendium of information on the characteristics of specific woods.

Reference:
Jane, F.W., K. Wilson, and d.J.B. White. 1970. *The Structure of Wood*. Adam and Charles Black. London. 478p. The authoritative text.

Record, S.J. 1934. *Identification of the Timbers of Temperate North America*. John Wiley. New York. The reference keys for identification of native woody species.

CHAPTER 9: FLOWERS

Flowers are short shoot systems, characteristic of angiosperms, modified for sexual reproduction. Many structural features are shared between flowers and vegetative stems and leaves. Floral meristems, for instance, frequently arise from transformed apical meristems. They, in turn, produce floral appendages: sepals, petals, stamens, carpels, and sometimes additional accessory structures, in much the same manner as a shoot apical meristem produces leaves. The vascular system of the flower is continuous with that of the subjacent shoot system, with traces supplying the lateral floral organs in the same way that stem bundles supply leaves.

Unlike vegetative shoot growth, which is usually indeterminate (growth will continue so long as conditions are favorable), floral development is determinate. Flowers produce a specific number of floral parts, in precise order, but this production uses up the floral apical meristem in the process. Plants that produce a show of flowers over an extended period must continue to produce new floral meristems from axillary buds of a growing shoot system.

Closely tied to flowers and flowering is the life history of specific animals, especially insects. So closely are they tied that in some instances flowers will mimic the shape, size and scent of a female insect so faithfully that male insects attempt to copulate with the flower (pseudocopulation). Even the timing is matched so the flowers open at the same time the male insects hatch (several days before the females)! Pollination biology, the study of flowers and their pollinators, is an extremely interesting area of study that includes many of the classic cases of co-evolution between plants and animals.

OBJECTIVES

1. Explain the role of alternation of generations in the life history of plants.
2. Describe the typical structure of a flower and the role of each organ in the reproductive process.
3. Describe the process of gamete formation in flowers, the events leading to syngamy, and embryo development.
4. Relate typical modifications of flowers to pollinators or pollination mechanisms.
5. Relate fruit type to dispersal mechanisms.

CONCEPTS

The most difficult concept in this chapter, and one of the most challenging in the course, is alternation of generations. Most sophisticated college students think they already know all there is to know about sex (until you start teaching human sexual biology - but that's in another course). Males produce sperm and females produce eggs; put them together and that's it. We even reinforce this oversimplification when we teach meiosis with the products being four sperm or an egg and three polar bodies (check to see how your general biology text describes meiosis). That a haploid cell can divide and grow, forming a multicellular body composed of haploid cells is completely foreign to students. Worse is that this haploid gametophyte is only one phase of

Chapter 9

an organism that also has a multicellular sporophyte body composed of diploid cells. At least the latter is "normal" (like us)! In addition to the conceptual difficulty of understanding plant life cycles, there is a glossary of unfamiliar terminology used to describe various aspects and details of the process. Concentrate on the major concepts first, details and terminology can come later.

Many students fail to distinguish between pollination and fertilization, in the same way that students frequently equate copulation and fertilization when they think about animal reproduction. Pollination is an intricate process that delivers pollen from anther to stigma. Even in the seemingly most simple cases, such as wind pollination, elaborate mechanisms have evolved to optimize the success of this process. Today, the field of pollination biology is buzzing with activity as researchers discover new subtleties in the process. Similarly the process of fertilization, the union of sperm and egg that seems so straightforward, is a phenomenon of renewed interest as it is finally possible to maintain viable plant sperm and egg cells in culture to study the process in detail. The details are not important at this level, but we must make students realize that so much remains to be learned, even in the most basic fields of botany.

One unique aspect of flower fertilization is the process of double fertilization. The concept of a second fertilization process, not involving an egg cell, but essential for successful growth of the offspring, is difficult for students to comprehend. Perhaps this is because animals require so many sperm to insure a single fertilization event. In flowering plants, only two sperm are delivered to the ovule and both have a specific role to play - one to fertilize the egg, thus initiating the zygote, the other to unite with both polar nuclei to form the primary endosperm cell.

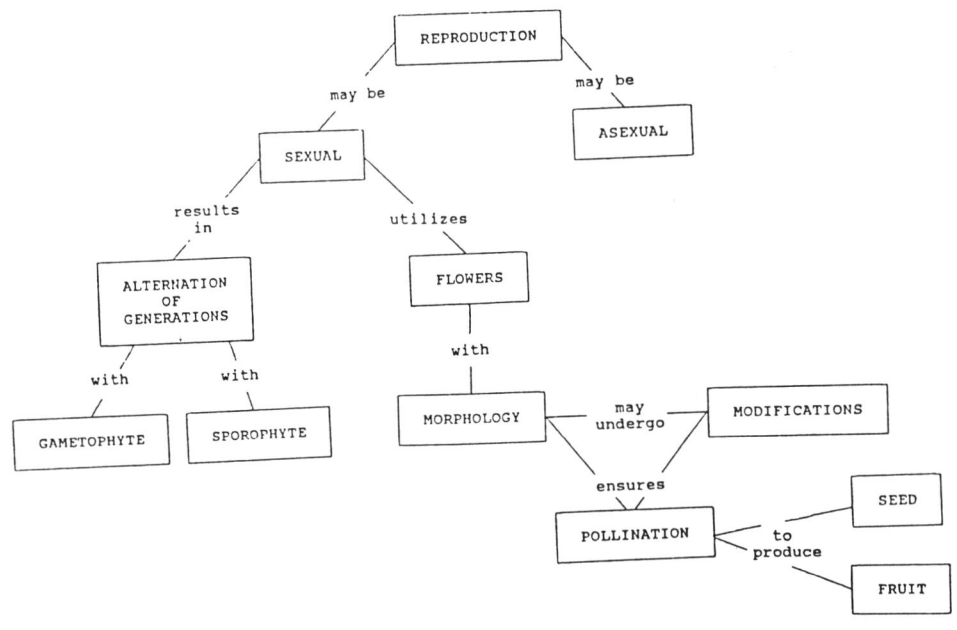

INSTRUCTIONAL AIDS

Lecture - An effective way to approach the concept of alternation of generations is to introduce the basic life cycle as a quadripartite scheme. Start with a large circle on the board divided by diagonal lines into four equal quadrants. Label the top Syngamy, then explain the term - fusion of gametes. Moving clockwise, label the next quadrant "2n, Diploid," to represent the product of syngamy. In the simplest case, the original 2n cell will immediately undergo meiosis. Thus, label the lower quadrant "Meiosis." The products of meiosis must be 1n, Haploid; so label the last quadrant. In the simplest case. these haploid cells will act as gametes and immediately fuse, thus completing the cycle and returning to the original position.

In the simplest case scenario, outlined above, there is no mitosis and the hypothetical unicellular organism continually cycles back and forth between the haploid and diploid condition. In real organisms, however, growth in cell number, either forming a multicellular body or population of asexually reproducing unicells, always occurs. You must insert mitosis into the simplest case outlined above. In the view of most students, the logical place to insert mitosis is in the diploid quadrant, thus producing a multicellular sporophyte body. This, after all, is a condition similar to that found in animals. But there is no reason mitosis could not occur in the haploid quadrant, producing a multicellular gametophyte with a zygote being the only diploid cell in the life cycle. This condition is found in many algae and fungi. Similarly, mitosis can occur in both the haploid and diploid quadrants resulting in a typical plant life cycle.

Basic floral structure is straightforward. Use a floral sketch on the board to tie structure to microsporogenesis, megasporogenesis, pollination and fertilization. When illustrating the structure of stamens, enlarge a diagram of a single stamen on one side of the board, along with a diagrammatic cross-section through the anther so you can diagram microsporogenesis there later. Similarly, when discussing the carpel, make an enlarged diagram off to the other side of the board, including a single enlarged ovule. Later you can diagram megasporogenesis within the ovule and diagrammatically transfer a hypothetical pollen grain from the stamen to the stigma of the carpel, etc.

A good time to introduce fruits is in association with growth of the pollen tube through the style toward the ovary. Without going into the details of the substances involved, you can mention that chemicals produced within the pollen tube mediate pollen tube growth. Small amounts of these chemicals diffuse out into the surrounding tissue and may stimulate growth there as well. Similarly, following fertilization the embryo and endosperm begin a period of rapid growth and also produce small amounts of chemicals that promote growth in the surrounding tissue. These surrounding tissues are the ovary wall, and sometimes additional tissue as well. The proliferation of this tissue results in the formation of the fruit wall.

Chapter 9

Angiosperms exhibit much variation in floral structure, usually associated with pollination mechanisms. Examples include: changes in floral symmetry; fusion, loss, or multiplication of parts; production of special parts such as nectaries and staminodia; aggregation of flowers into inflorescences; patterns of color; production of scent; etc. Rather than simply listing a catalog of adaptations, with corresponding terminology, an effective approach is to provide specific examples, either as specimens to hand out or slides to project. For each example discuss the selective advantage of this modification over typical floral structure. An integral part of this discussion is the correlative adaptation of the pollinator resulting in co-evolution.

Pre-lecture Review - Before beginning a discussion of alternation of generations, it will be useful to review the processes and purposes of mitosis and meiosis. The role of the latter in alternation of generations will be obvious to students as essential ultimately to form gametes that can fuse during syngamy to form a new diploid individual. Only a diploid cell can undergo meiosis. It will not be immediately apparent, however, that both diploid and haploid cells are equally capable of undergoing mitosis. When students understand that haploid cells have this capability, it is much easier to understand how the haploid product of meiosis can begin to divide and grow to form a multicellular gametophyte.

Post-lecture Review - Ask your students to diagram the life cycle of a flowering plant by drawing a generalized quadripartite scheme on a full sheet of paper then placing the appropriate parts of a flower or vegetative plant into the various quadrants. They can then use this diagram to help organize a two paragraph description of the typical angiosperm life cycle.

LABORATORY - EXERCISES

1. Floral Morphology

If there is a florist in town, you have the potential to offer a fantastic laboratory on floral diversity. Florists will frequently donate an interesting selection of cut flowers that have passed their prime for retail value but are more than adequate for classroom dissection and examination. There are also some teaching "standards" that are easy to maintain under lights or in a small greenhouse, such as: *Pelargonium* (geranium), *Tradescantia* (spiderwort), and *Petunia*. Select some "typical" flowers for everyone to examine and dissect. Students should make a sketch of the parts, as they dissect them, making notes as to size, color, position in the flower, etc. Also provide a variety of flowers showing different specializations. Students can work in groups with individuals dividing responsibility for examining different flowers and setting up demonstrations for the other members of their group. Unfortunately, many of the varieties in commercial production have unusual or specialized morphology which, although showy, are confusing to students who are trying to understand basic floral structure. If possible, avoid double flowers, flowers clustered into compact heads (such as composites and proteas) and flowers with additional whorls of parts (such as milkweeds and passion flowers.

2. Floral Anatomy

Virtually every laboratory manual of biology and botany includes examination of prepared cross-sections of *Lilium* anthers and ovaries to follow microsporogenesis and megasporogenesis respectively. The following are some hints to help avoid some typical student problems:

A. Orientation. Have an example of a lily, preferably in bud, available (or at least a diagram or model of the flower) so you can show students how a section made at the level of the anthers to show pollen development will have the anthers arranged in a ring around the style of the fused carpels. Similarly, a section through the ovaries will not include the anthers. A general rule of microscopy for students is that they will understand what they are looking at more easily if they first orient themselves to where they are looking.

B. Microsporogenesis. Lily pollen is one of the frequently used materials to illustrate meiosis. Unlike onion root tips, where the cells are oriented in files and the plane of cell division is always perpendicular to the axis, the developing microsporocytes in the anther are free to rotate. The plane of orientation is not uniform and, thus, two cells in the same stage of division might appear very different. Illustrate this with the fingers of your hands spread but touching the corresponding fingers of the other hand (resembling a spindle apparatus or the chromosomes during anaphase. Now rotate your hands from a side view, to oblique, to polar, etc. (recall the discussion of meiosis in Chapter 4).

C. Stage of Meiosis. Students can distinguish meiosis I from meiosis II by the presence or absence of a crosswall between dividing nuclei. If the crosswall is present, the cells are in meiosis II.

D. Megasporogenesis. Lily is an unfortunate example to illustrate megasporogenesis and the development of an embryo sac because it is very atypical of angiosperms (three of the daughter nuclei of meiosis fuse to form a triploid "restitution" nucleus before continuing to divide in synchrony with the remaining haploid nucleus to form an eight-nucleate embryo sac). It is very unusual for the nuclei of four- and eight- nucleate stages to be lined up in a plane parallel to the plane of section when the slide was prepared. As a result, frequently only three nuclei are visible in embryo sacs of slide labelled four-nucleate, and only 5, 6, or 7 nuclei are visible in slides labelled eight-nucleate. You must assure your students that the "missing" nuclei were present in the embryo sac, but were simply not included in the section as it was being cut.

3. Pollen

A. Pollen Morphology. Some of the most intricate and beautiful structures of all living things are the pollen grains of seed plants. The variety of structure is so great that botanists can identify many species based solely on the structure of their pollen grains. Have students make

Chapter 9

a wet mount of pollen from each of the flowers they examine, stained with a drop of 1% safranin, and examine this slide under high power with a compound microscope.

B. Pollen Tube Growth. Many botany laboratory manuals suggest placing pollen of *Impatiens* or *Tradescantia* in a drop of 10% sucrose to stimulate pollen tube growth. Students also can try other plant materials, but frequently they will have different germination requirements. This opens up the possibility for a number of experiments utilizing different concentrations of sugars, different sugars, different pH, different temperatures, inhibitors, etc. Each student group can be responsible for designing an experiment utilizing either different sets of conditions or different plant materials under the same conditions.

4. Pollination Biology. - Depending on the time of year, an interesting exercise is to go out in the field to an area with masses of wild flowers in bloom - asters and goldenrods are common in the fall, while several flowering shrubs work well in the spring. Students should concentrate on recording the visiting pollinators of a particular plant or group of plants and include observations on behavior of visitors to the same flower, searching behavior of individual visitors, rewards collected, method and amount of pollen collected, etc. As an alternative to a class activity, assign this as an individual student project.

5. Supermarket Fruit Morphology. - All of our common fruits, and many of our common "vegetables" as well are fruits in the botanical sense of the term. The nature of the fruit wall is the primary characteristic distinguishing between different fruit types and provides an interesting and tasty exercise in careful observation. Obtain a selection of fruits from the market including one or two "exotics" such as kiwi, papaya, or pomegranate, for added interest. Don't forget some of the dry-walled fruits such as peanuts or true nuts and some of the "vegetables" such as tomatoes, green beans, peppers, or pod peas. Place your variety of fruit types at stations around the room and include a label at each station. On the front of the label, include the common and scientific name of the plant; on the back of the label, indicate the botanical fruit type.

Mauseth includes a table of fruit types (Table 9.5) that students may use directly or you may wish to rearrange this table into a dichotomous key for your students. Instruct students to examine carefully the external features of the fruit before doing any dissection to look for clues as to the original position of the ovary in the flower and the possibility of other tissues being involved in formation of the fruit wall. For instance, on apples the sepals and stamens will still be visible on the end opposite the stem, which indicates that the apple has an inferior ovary. In oranges, however, the remains of sepals will be visible at the base of the fruit where it attaches to the stem. Stamens may also still be visible in each flower (segment) of a pineapple, suggesting the composite nature of this fruit.

In small groups of two or three, students should move from station to station and attempt to identify the correct fruit type using their table or key. They can check their own answers on the back of the label. Of course, in many cases they will have to do some simple dissection to make closer examination. For larger fruits, such as apples or pineapples, you may want to provide a knife, but for smaller fruits such as grapes or berries, simply suggest that they can use their teeth (ask them to dispose of these succulent morsels properly when they have finished their examination)!

LABORATORY - MATERIALS, EQUIPMENT, AND PREPARATION

1. Floral Morphology
- variety of flowers from florist, nursery, greenhouse, or collected in the wild.
- hand lens or dissecting microscope (the ocular lens of a microscope, inverted, may substitute for a hand lens.)

2. Floral Anatomy
- lily flower in bud, or diagram or model of a lily flower, to place sectioned material into perspective.
- prepared microscope slides of lily anthers and lily ovaries in various stages of development.
-compound microscope

3. Pollen

A. Morphology. - pollen, anthers, of a variety of flowers.
- compound microscope, high dry magnification is minimum - one or more demonstration microscopes with oil immersion capability is desirable to examine smaller pollens or pollen with intricate wall sculpture.
- microscope slides and coverslips
- dropper bottle of 1% safranin and water

B. Pollen Tube Growth - pollen, anthers, of a variety of flowers, including *Impatiens,* where pollen is known to germinate readily in 10% sucrose.
optional:
- 10 % solutions of other sugars
- volumetric glassware to make known dilutions of sugar solutions

4. Pollination Biology
- no special equipment is required although a camera with close-up capability is desirable to document pollinators.

5. Supermarket Fruit Morphology
- Simple fruits such as: Achene (sunflower "seeds"); Berries (banana, blueberry, cranberry, grape, pepper, pomegranate, tomato); Caryopsis (popcorn, wild rice); Drupe

Chapter 9

(apple, mango, pear, pistachio, prune, quince); Hesperidium (lemon, lime, orange, tangerine);Legume (green bean, pea pod, peanut); Pepo (pumpkin, squash, cantilope, watermelon); Nut (pecan, walnut); Pome (apricot,cherry, coconut [the "fruit" in the store in only the endocarp, the outer fibrous and tough layers are missing], olive, peach, plum).
- Compound fruits such as: Aggregate (blackberry, raspberry, strawberry); Multiple fruit (fig, pineapple).
- knife
- hand lens

Smaller fruits can be purchased in sufficient quantity that each student is able to sample a complete fruit. Divide larger fruits so each student can sample a piece. Be sure to include some of the more "exotic" fruits that many students will have never tasted.

REFERENCES

General:

Meeuse, B. 1961. *The story of pollination.* The Ronald Press Co., New York. 243 p. A delightful introduction to pollination biology.

Meeuse, B. and Sean Morris. 1984. *The Sex Life of Flowers.* Facts on File, New York. 152 p. This is the book companion to "Sexual Encounters of the Floral Kind" with outstanding photography by Oxford Scientific Films, available on videocassette as part of the "Nature" series, and on videodisk from "VideoDiscovery".

Sattler, R. 1973. *Organogenesis of Flowers: A Photographic Text-atlas.* University of Toronto Press, Toronto. 207 p. Images and descriptive text of early floral development in a variety of angiosperms.

Resources:

Eames, A. 1961. *Morphology of Angiosperms.* McGraw-Hill, New York. The classic volume on floral structure.

Pijl, L. van der. 1982. *Principles of Dispersal in Higher Plants, 3rd ed.* Springer-Verlag, Berlin. 214 p. A compendium of fruit modifications for dispersal.

Pijl, L. van der. 1979. *Principles of Pollination Ecology, 3rd ed.* Pergamon Press, Oxford. 244 p. A compendium of structural modifications for pollination.

Proctor, M. and P. Yeo. 1972. *Pollination of Flowers.* Taplinger Publishing Co., New York. 418 p. The closest thing to a textbook of pollination biology.

CHAPTER 10: PHOTOSYNTHESIS

This chapter begins the unit on physiology and development, a unit that will be conceptually more difficult for students simply because much of what is discussed is of a nature not easily visualized. It will be critically important for you to concentrate on your explanations of key concepts and let your students fill in the details from the text later.

Once again scale is an important consideration. On the scale of cells, living things are highly ordered, indeed, this ability to maintain orderliness is a characteristic of living things called homeostasis. To maintain this order requires the expenditure of large amounts of energy. When we look at living things, and the history of life on earth, we see an increasing degree of complexity in both structure and function. From physics, however, we know that the general state of the universe is moving toward disorder. This phenomenon becomes apparent in organisms when they die. What is it that allows organisms to grow and develop in spite of the natural tendency for entropy to occur? Ultimately, of course, the answer is the ability of photosynthetic organisms to harvest energy from an external source, the sun, and use that energy to maintain order within the body. In turn this energy passes on to other organisms.

Most students will have a general understanding of the role of photosynthesis in allowing the plant to harness energy (make its own food) and pass that energy on to other organisms (be eaten by herbivores, etc.). It is less likely that they will realize that plants have another essential function in the evolution and maintenance of life, as we know it, on earth. That function is to produce molecular oxygen, O_2, by splitting water. The gradual buildup of oxygen, and ultimately ozone, O_3, was probably the single most important modifying effect of life on the environment.

OBJECTIVES

1. Describe the electromagnetic spectrum and the relationship of energy level to wavelength; explain the role of pigments in absorbing light energy.
2. Relate photosynthetic processes to the structure of chloroplasts and where in the leaves they occur.
3. Describe the light dependent reactions of photosynthesis: photosynthetic units and reaction centers involved, electron source and various electron carriers.
4. Describe the cyclic nature of the C_3 stroma reactions; CO_2 acceptor, role of ATP and NADPH, and ultimate use of acquired carbon.

CONCEPTS

The nature of light is a very difficult concept to get across to students. On the one hand there are physical properties, photons, that are important for a pigment's ability to absorb and use light energy. Simultaneously there are wave properties that are important in determining the amount of energy carried in a photon.

Chapter 10

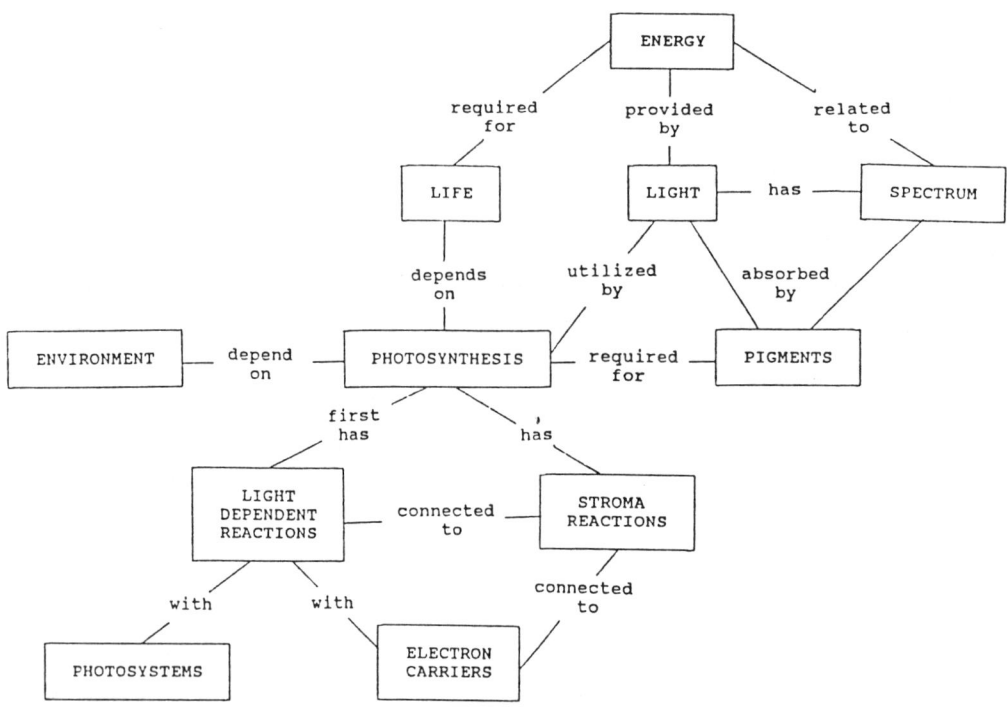

Hand-in-hand with the nature of light is the nature of the pigments that absorb light. Most students intuitively understand that pigments are molecules that somehow absorb light. The problem is an apparent reverse in what seems intuitively logical - if a pigment is red, then it must absorb red light. Of course, the opposite is true, red is the color NOT absorbed. Instead it is reflected, or transmitted, thus it is the only color seen. Once students understand the nature of light, it is easier to understand the nature of the pigments that absorb light.

Since light of different energies, colors, is available and because different pigments absorb light of different energies, it is an easy step to understand the general concept of a photosynthetic unit. It is less apparent how individual pigment molecules, each embedded in the thylakoid membrane, can pass the energy they absorb to the single central chlorophyll a molecule that acts as the reaction center for that unit.

Non-cyclic photophosphorylation, the "Z-scheme" of many textbooks, is not a difficult concept for students to follow - as long as you do not burdened them initially with too many details, e.g., names and sequence of electron acceptors. Providing an ultimate source of electrons, however, is a different matter. You must remind students that electrons hold atoms together into molecules. When water, H_2O, is split into its components, $1/2\ O_2$, $2\ H^+$ ions, and $2\ e^-$ result. Only the oxygen is not used further in photosynthesis. It bonds covalently with another 1/2 O_2 to form a molecule of O_2. This is a fortunate waste product for us and all other aerobic organisms.

The stroma reactions, frequently termed the "light-independent" or "dark" reactions, are more difficult to understand. Cover three related concepts - and avoid details. The first is that overall the reactions are cyclic so the reaction continually regenerates starting product, RuBP and the process repeats. Second is that these reactions link to the light-dependent reactions through their requirement for NADPH and ATP to convert PGA to the higher energy form, PGAL (phosphoglyceraldehyde). Third is the process produces virtually all of the important molecules required for the cells metabolism. This is done by drawing off intermediate compounds from the cycle.

INSTRUCTIONAL AIDS

Lecture - Entropy is a measure of the disorder of a system. Like all concepts involving a negative relationship, entropy is difficult for students to understand. For our purposes, however, this is not a great problem. Our concern is that students understand that in a closed system, there will be a tendency toward disorder. In open systems, only an input of energy from outside the system maintains order. A useful analogy is this: dormitory rooms naturally have a strong tendency toward entropy - they seem to get cluttered and messy even when no one is around. It is only through a student's inordinate expenditure of time and energy that he or she can maintain order in the room. More seriously, it is only the enormous amount of energy provided by the sun, and trapped by photosynthetic organisms, that maintains the necessarily orderly structure of cells and organisms.

A good way to begin a description of the process of photosynthesis itself is to put the overall equation for photosynthesis on a far corner of the board (see text figure 10.6). Leave this equation on the board throughout your discussion so that students can keep track of the overall picture. I also take this opportunity to discuss briefly how students can measure the rate of photosynthesis. This provides background material for laboratory exercises, but it also sets the students up for some data that suggests at least two separate processes are involved.

During the early part of this century, the English plant physiologist, Blackman, discovered that up to a point, as you increase the intensity of light, the rate of photosynthesis also increases. After a point of light saturation, however, further increase in light intensity has little effect. Graph this phenomenon on the board and ask the class to list some hypotheses that might explain this curve. Blackman went on to discover that different temperatures produced similar curves, but at lower temperatures the curves leveled off sooner, while at higher temperatures, up to a point, the curves leveled off later. To Blackman this suggested that at least two different processes are involved in photosynthesis. One is light intensity dependent (he called these the light reactions). The others are more typical chemical reactions affected by temperature, etc., but unaffected by light (these were the dark reactions). With this introduction

Chapter 10

you are ready to begin a discussion of the light-dependent reactions by briefly reviewing the physics of light.

The nature of light is a difficult concept to teach in a biology class, but it is essential for an understanding of the process of photosynthesis. A part of the problem is the dual nature of light - particle and wave. A useful analogy is a rotary-type lawn sprinkler. With the pressure on low, successive waves of water drops arc in a lazy trajectory from the sprinkler out to the lawn. Each drop behaves as a particle, yet these particles are thrown out in waves; the distance between successive drops in a single radius, their wavelength, relates to the energy pushing them out. High water pressure still releases individual drops, but because the rotor turns more quickly, successive drops along a radius are spaced closer together than before - their wavelength is shorter. Similarly, light energy leaves the sun in particles, photons, but some photons have more energy than others. Longer wavelengths have less energy than shorter wavelengths.

At this point it is useful to place an overhead of the electromagnetic spectrum on the board to review the inverse relationship between wavelength and energy. In addition to elaborating on the spectrum of visible light, you can use this opportunity to introduce some of the other parts of the spectrum which will be important later in the course, eg., infrared (heat), ultraviolet (tanning, burning, mutating, and killing), x-rays, etc. Reviewing the visible spectrum of "white" light provides you with an opportunity to discuss pigments. Look for a student wearing an especially bright shirt or sweater and ask this "volunteer" to stand up for the class. Now begin a series of questions - what color is it? Why is it that color? What is a pigment? What happens when light strikes a pigment molecule, etc? With an understanding of the nature of light and the role of pigments in absorbing light, you are now ready to begin a discussion of the light-dependent reactions of photosynthesis.

I find that it works well to generally move backwards through the "Z-scheme." Begin by describing the organization of a photosynthetic unit for photosystem I in a thylakoid membrane. The pigment molecules are precisely arranged to channel their energy, through resonance or fluorescence, to the P700 reaction center molecule. (A dramatic demonstration that intact membranes are necessary is to hold a flask of plant extract, the kind used for chromatography, which you can hold in front of a slide projector as you turn off the room lights. Green light is transmitted, as the students expect, but the sides of the flask will appear red due to fluorescence of pigment molecules from the ground-up plastids.) In the intact system, electrons from P700 will be excited out of their orbitals and picked up at the higher energy level by electron acceptor molecules. The electrons may then be used in forming NADPH. This process converts light energy into chemical energy, but the P700 molecule cannot function again because it is missing an electron. It will not function again until the missing electron is replaced. We say it has an "electron hole." The role of photosystem II is to absorb light of different wavelengths, thus improving overall efficiency, excite electrons and pass them down an electron transport chain to replace those lost from photosystem I.

Photosystem II fills the hole in photosystem I, and generates ATP in the process, but it leaves a similar problem as before - P680 is now missing electrons. It cannot function again until they are replaced. With water readily available and easily split, there is no shortage of electrons to fill the "holes" in P680.

If any time remains in your first photosynthesis lecture, ask a few students to repeat back to you, and for the class, the order of key events in the light-dependent reactions. Have one volunteer give the first step, then get a second person to pick up and carry on the next step, etc.

Before beginning your lecture on the stroma reactions, go back to the original formula for the overall reaction of photosynthesis. Identify which components are accounted for by the light dependent reactions. The remaining parts involve the stromal reactions. Illustrate the general concepts of the stromal reactions by using a diagram of the carbon backbones of the involved molecules. For instance, represent RuBP as a 5-carbon chain and CO_2 as a single carbon. When RuBP picks up CO_2 the result is a symmetrical 6-carbon molecule. The instant it forms, however, the enzyme Rubisco splits the molecule into two identical 3-carbon PGA molecules. Before these 3-C molecules can be used to synthesize more stable molecules, they must be raised to a higher energy level. This is where the stromal reactions couple to the light dependent reactions using the ATP and NADPH produced in the latter.

From this point metabolism may follow several alternative synthetic pathways; it is unimportant to provide the details of any one pathway. In general terms two 3-C molecules will join to form a 6-C intermediate. One possibility is that the two reduced PGAL molecules could join to each other, but they also may react with other 3-C intermediate molecules present in the stroma. In any case, after passing through a series of intermediates, a 5-C RuBP molecule is regenerated and you are ready to begin the cycle again. In a general sense students can think of the 6th carbon as being temporarily stored as the cycle repeats itself. On the second cycle, a second CO_2 is brought into the system and stored, a third CO_2 on the third cycle, etc. Six cycles will bring enough carbon into the system to form a single molecule of glucose, thus explaining the $6CO_2$'s in the overall formula.

Pre-lecture Review - Your lectures on cell physiology, particularly photosynthesis and respiration, require an understanding of the general principles of biological chemistry introduced in Chapter 2. This was probably some time ago, however, so it will be important to review some salient concepts such as: exergonic vs endergonic, the relationship of ATP to ADP and their role in coupling reactions, and the relationship of NAD (NADP) to NADH (NADPH).

Post-lecture Review - Photosynthesis is an excellent topic in which to emphasize the relationship between structure and function. Relate the reactions of photosynthesis to the structure of the chloroplast where they occur, to the structure of the cells containing chloroplasts, and to the structure of leaves. You can also describe the special modifications associated with C4 and

Chapter 10

CAM photosynthesis. If your students understand the general concepts of the C3 cycle, they should be able to follow Mauseth's description of C4 and CAM on their own.

LABORATORY - EXERCISES

1. Chromatography

A standard exercise is to spot plant extract on chromatography paper then separate the major plant pigments by developing the chromatogram in appropriate solvents and identifying the pigments by color and relative position. A simple variation can make this more than a "cookbook" exercise.

A. Identification of Pigments

In a particular solvent system, pigments are known to travel a certain distance up chromatography paper relative to the movement of the solvent front. The ratio of the distance the pigment moved relative to solvent movement is termed Rf. Students can identify unknown pigments by calculating the Rf value of the unknown pigment, and comparing this to a table of Rf values for a particular solvent system. Only two modifications need be done to adapt usual directions for chromatography. First, mark, in pencil, the position of the original pigment application to serve as the origin for measurement. Second, mark the position of the solvent front as soon as the paper is removed from the chromatography chamber. The solvent will evaporate quickly and once it has, it is not be possible to measure the distance the solvent travelled.

After marking the solvent front, mark the average distance from the initial source of each of the visible pigment bands. Then measure to the closest mm the distance travelled by each of the pigments and the solvent. Calculate Rf for each pigment by dividing the distance it travelled by the distance of solvent movement. Students can identify pigments by comparing their values to a table of Rf values. For instance, in 9:1 petroleum ether/acetone, the Rf of the common photosynthetic pigments are: B-carotene, 0.98; lutein, 0.71; violaxanthin, 0.62; neoxanthin, 0.25; chlorophyll a, 0.19; chlorophyll b, 0.13.

2. Determination of Absorption Spectra

A. Qualitative Determination

The simplest method of determining the absorption spectrum for plant pigments is to observe the effect of these pigments on the spectrum of visible light passing through them. Have students observe the spectrum of white light through a spectroscope and make a sketch illustrating the pattern observed. Place a series of colored filters (at least a red and a blue filter) between the light source and spectroscope and observe the effect on the original light pattern.

Now place a vial of dilute plant extract between the light source and the spectroscope and observe the effect.

B. Quantitative Determination

Use a spectrophotometer to quantify the absorbance of light by plant extract. Absorbance readings should be taken at regular intervals from 400 to 700 nm. Fifty nm intervals will show the general trends and moves fairly quickly, but 10 nm intervals gives better resolution. In most cases this will require that you have at least two spectrophotometers, one with a photocell for the blue end of the spectrum, the other with a photocell and filter for reading in the red range. Using a blank tube containing the solvent used in extracting the pigments, zero the spectrophotometer at the first wavelength to be measured. Record the reading, then move to the second wavelength, rezero and take the second reading. Repeat this procedure for each wavelength examined. Have students graph their data and compared to the qualitative determination.

C. Absorption by Photosynthetic Pigments

Students can determine the absorption characteristics of individual photosynthetic pigments by combining some of the above procedures. First, separate individual pigments from plant extract using chromatography. Fairly large amounts of each pigment will be required, so rather than using a "spot" application of extract on a narrow strip of paper, students will have to streak a line across a wider sheet of paper. A 6 x 6 in. square sheet is about the size that will roll into a tube and fit inside a wide-mouth quart jar for developing. A pasteur pipette, drawn out to a finer point, makes a good applicator. Pencil a line, about 1 in. from the bottom of the paper to serve as a guide for streaking pigment. Load at least ten overlapping layers of pigment, the thinner and straighter the better, on each sheet. Each layer must be dry before adding the next - a hair blower speeds this process considerably.

Once pigment is loaded, roll the sheets and stapled them so the pigment streak is at the bottom of the resulting tube and the two sides nearly touch. Add about 1/4 in. of solvent to the bottom of a quart jar, insert the tube of loaded paper, and cover with half of a petri dish. When the solvent front nears the top of the tube, remove the chromatogram, quickly mark the solvent front, in pencil, and identify the various bands by their Rf values.

Three pigment bands will be obvious, B-carotene and the two chlorophylls. The three xanthophylls will be fainter, particularly lutein. Divide the class into an appropriate number of groups so each group is responsible for determining the absorption spectrum for a single pigment. Once identified, carefully cut out the individual pigment bands are and distributed them to the appropriate group. For instance, one group will receive all of the chlorophyll "a" strips, a second group gets all the B-carotenes, etc. Place the pigment strips in a large test tube with enough acetone to elute, with shaking, the pigment from the paper. Transfer the resulting pigment extract to a spectrophotometer tube and determine the absorption spectrum as in part B above. Have individual groups plot their results on a single graph on the board to compare the characteristics of individual pigments and the overall absorption and action spectra of photosynthesis.

3. Effect of Environment on Rate of Photosynthesis

A standard experiment in botany and biology laboratory manuals is to measure the rate of oxygen production by aquatic plants, or submerged leaf tissue, under different light intensities. A simple variation is to discuss initially some of the factors that may affect photosynthetic rate (recall Blackman's experiments) then ask individual groups to formulate and test a hypothesis concerning one factor of their choice. Some of the more obvious factors include: light intensity (have plants at different distances from a light source); temperature (plants are already in water, increase the volume to serve as a water bath, or place in a larger water bath); light quality (use colored filters in front of the light source, or colored cellophane wraps around the plant containers - don't forget to account for different intensities in designing the control); different CO_2 concentrations (add bicarbonate to the solution to saturate CO_2 levels - use different concentrations).

LABORATORY - MATERIALS, EQUIPMENT, AND PREPARATION

1. Chromatography
-plant extract. Spinach is the traditional choice, but even a collection of grass and weeds from the lawn outside the building will work. Place in a blender with enough solvent to at least half cover the plant material (acetone works best but there is danger of fire if the blender's motor is not adequately shielded - alcohol will work) and blend away. Filter the extract through several layers of cheesecloth. Although this works best if used fresh, extract will keep for months if stored covered in a refrigerator.
-chromatography paper
-pasteur pipette applicator. Heat the tip in a flame until red, then quickly draw out the tip with a forceps, cool, and snap off the tip. This capillary tip will allow thin layers of pigment extract to be applied.
-covered cylinder, tube, or jar of sufficient diameter to hold loaded chromatography paper.
-chromatography solvent. 9:1 petroleum ether/acetone.

2. Determination of Absorption Spectra

A. Qualitative Determination
-spectroscope and light source
-colored filters (cuvettes, or square sided dropper bottles, of dilute stain, eg. safranin, methylene blue, etc. will work)

-cuvette (or dropper bottle) of plant extract
-cuvette of solvent used to prepare extract - the control

B. Quantitative Determination
-spectrophotometer with cuvettes (if not equipped with a wide range phototube, at least 2 spectrophotometers will be required, one with the blue tube and one with the red tube and filter. Be sure the instruments are labelled as to which has which tube.)
-plant extract
-solvent
-test tube rack
-Kimwipes™ or towels to wipe spec. tubes before inserting into instrument

C. Absorption by Photosynthetic Pigments
-6x6 in. squares of chromatography paper
-quart jars with petri dish covers
-materials as in 1 above.
-large test tubes to extract pigment strips
-acetone
-scissors
-materials as in 2.B. above

3. Effect of Environment on Rate of Photosynthesis
-*Elodea* for direct measurement of oxygen evolution (some manuals use evacuated leaf disks and observe time required for disks to refloat)
-beaker
-sodium bicarbonate (balance and graduated cylinder to prepare solutions of known concentration)
-ring stand and clamps to hold:
-pipette with rubber tubing and clamp over delivery end
-light source
-meter stick
-colored filters or cellophane
-water baths and/or hot plates
-ice

REFERENCES

General:
Asimov, I. 1968. *Photosynthesis*. George Allen & Unwin. London. 193 p. An informal, often anecdotal, enquiry into the process of photosynthesis by a biochemist most noted for his science fiction.

Chapter 10

Bassham, J.A. The Path of Carbon in Photosynthesis. *Sci. Amer.* 89-100. This is a readable account of the Calvin-Benson Cycle and the production of fats, amino acids, etc., as well as carbohydrates.

Bjorkman, O. and J. Berry. 1973. High-Efficiency Photosynthesis. *Sci. Amer.* 229:80-93. This is a readable elaboration of the C4 pathways described by Mauseth in the textbook.

Cathey, H.M., L.E. Campbell, and R.W. Thimijan. 1978. Comparative Development of 11 Plants Grown under Various Fluorescent Lamps and Different Durations of Irradiation with and without Additional Incandescent Lighting. *J. Amer. Soc. Hort. Sci.* 103:781-791. An informative comparison of the relative effectiveness of a variety of light sources - specialized (and more expensive) growth lights are not necessarily better!

Govindjee, and R. Govindjee. 1974. The primary events of photosynthesis. *Sci. Amer.* 213: 68-82. An elaboration of the basic processes during the light-dependent reactions.

Reference:
Black, Durhan and Zweig. 1958. *Manual of Paper Chromatography and Paper Electrophoresis.* Academic Press, New York. This contains a compilation of Rf values for many organic compounds, including photosynthetic pigments, in a wide a variety of solvent systems.

Blackman, F. 1905. Optima and limiting factors. *Ann Bot.* 19:281. Based on the experiments described in this paper, Blackman hypothesized that there must be separate light and dark reactions of photosynthesis.

Sundberg, M.D. 1989. *Biol 1208: Biology for Science Majors.* Burgess International Group, Edina, Minnesota. Contains detailed procedures for these and other laboratory exercises on photosynthesis.

CHAPTER 11: RESPIRATION

Having already covered photosynthesis, the reactions of respiration will be much easier for students to follow - but it still won't be easy. The amount of detail is still overwhelming for a student reading the material for the first time. Your main job will be to make your students "see the forest through the trees." Overview the entire process, then take one step, glycolysis, Kreb's Cycle, and Electron Transport, at a time; concentrate on concepts and processes, let the reading provide details. Perhaps the biggest cause of anxiety among students is being faced with a seemingly insurmountable number of chemical names and formulas. Make it clear from the outset exactly what your expectations are!

OBJECTIVES

1. Describe the overall process of glycolysis and the location and conditions where these reactions occur.
2. Explain how the presence or absence of oxygen affects the outcome of cellular respiration.
3. Describe the overall process of the Kreb's Cycle, including carbon input and output, energy output, and location in the cell.
4. Describe the electron transport chain, including the electron donors, ultimate electron acceptor, and production of ATP.
5. Summarize net ATP production during aerobic vs anaerobic respiration.

CONCEPTS

It seems inefficient for a cell initially to have to supply energy to a glucose molecule in order eventually to get energy back out, yet this is the very first step of glycolysis. As you go through the processes of respiration, it is useful to keep a running energy balance sheet on the side of the board to keep track of energy inputs and outputs.

A second thing to keep track of throughout the process is the number of carbons at each step of the way. Students should account for all six. The single biggest problem is to remember that during glycolysis, the single 6-carbon sugar is split into two 3-carbon molecules. Both pass through all the subsequent reactions to form two pyruvates. Each of these, in turn, is processed further, either aerobically or anaerobically. Thus every reaction you talk about, or diagram on the board, occurs twice for each initial glucose molecule.

Overall, the equation for respiration is the reverse of photosynthesis. This is a useful concept not only in comparing the overall processes now, but even more importantly in describing the energetic and chemical cycles in the ecosystem later. Reinforcing this concept, as well as the conservative nature of evolution (adapting structures and processes that are already available to new conditions) is the similarity of molecules and reactions between glycolysis and the stromal reactions of photosynthesis.

Chapter 11

The Kreb's Cycle can be daunting at first, so concentrate on one aspect at a time. First is its cyclic nature - follow the number of carbons. Reducing the number is easy, loose a CO_2. But how about getting back to 6 carbons from 4? Tie this back to glycolysis. Now look at energy - add to your tally from glycolysis. Finally, be sure to tell your students which names they will be responsible for.

The most difficult concept in respiration is electron transport, but this should be easier having already covered a similar process in the light reactions of photosynthesis. The membranes and, in particular, the space between the membranes, are important here, so a quick review of the structure of mitochrondria is useful. Emphasize the stepwise loss of energy as electrons move from one carrier to the next. Finally they reach the ultimate electron acceptor molecule, oxygen. It is difficult for students to understand how the very last step in a long series of reactions can actually affect whether pyruvate feeds into the Kreb's cycle or ferments.

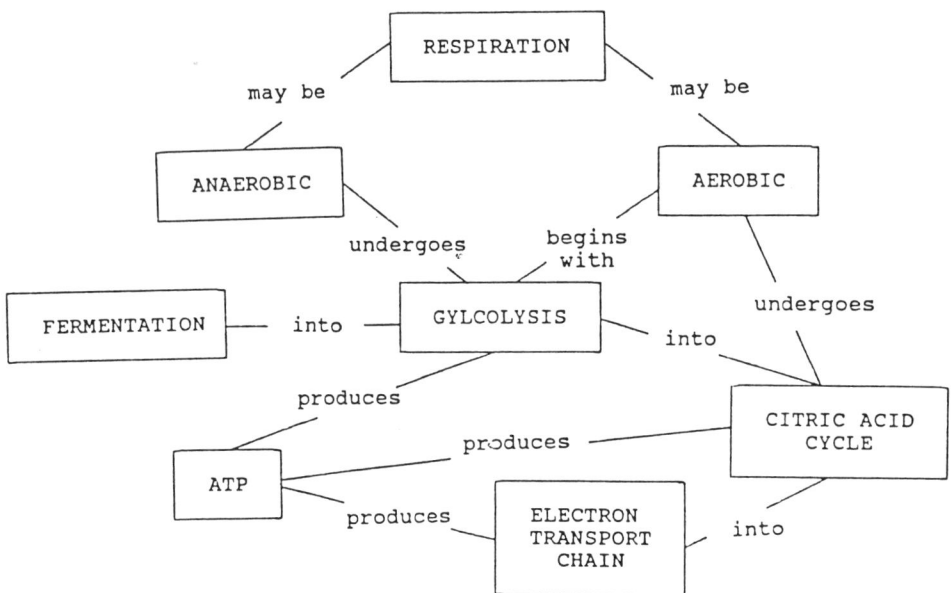

INSTRUCTIONAL AIDS

Lecture - Respiration is an energy-releasing process, so it seems inconsistent to students that the first steps actually require an input of energy. The problem is that glucose is a very stable molecule, which makes it good for storing energy. To release the stored energy, the molecule must first be made more reactive. This requires an initial input of energy. An analogy which is useful to explain the energy input of ATP required to initiate glycolysis is the necessity to "prime" a pump. Although they are becoming increasingly rare, some students will be familiar with the old-fashioned water pump whose leather must be wet before it will pull a column of water. If the pump has not been used recently, you must first pour some water in through the top, then start pumping, before any water comes out.

The net effect of glycolysis is to split a 6-carbon glucose into two, 3-carbon pyruvates, and capture some of the released energy as ATP. Students tend to get lost in the details and forget there are two of everything after sugar cleavage. It is useful simply to outline the important steps, using carbon skeletons, so students can keep track of the carbons.

The same is true for the Kreb's Cycle. To emphasize the cyclic nature of this process, begin by looking at carbon in the cycle itself. Another name for this process, of course, is the Citric Acid Cycle as the 6-carbon molecule citric acid is the first compound formed. Sketch a T-shaped backbone of six carbons on the board. It is several steps before a carbon is lost, but when it is there are five carbons left and the carbon lost was a molecule of CO_2. Represent this by a series of small arrows from your 6C compound to a new 5-carbon compound. From the last arrow, diagram a second diagonal arrow to show CO_2 leaving. The next step looses another CO_2 and leaves a 4-carbon molecule. There is, then, a series of 4C molecules (diagram as a series of arrows only) until oxaloacetic acid (OAA, also 4C) is formed. The next step is back to citric acid, but there's a problem. You must add 2 carbons. You do not have an immediate source of 2-carbon molecules, but you do have 3-carbon molecules, the pyruvate from glycolysis. Removing one carbon from pyruvate, as CO_2, leaves two carbons that can be added to OAA to reform citric acid. This completes the cycle. As long as pyruvate is available, the Kreb's Cycle can go round and generate three CO_2's for each revolution. Of course two pyruvates form for each glucose we started with,, so the Kreb's Cycle will turn twice and release 6 CO_2's.

With an understanding of what happens to carbon, your students are now ready to go back to the Kreb's Cycle and consider the selective advantage of this process. Additional usable energy is produced compared to glycolysis alone. Add to your tally sheet from glycolysis the additional energy carrying compounds produced here.

Your students should now be familiar with electron transport chains from your discussion of photosynthesis, but review again the stepwise nature, in terms of energy level, of successive electron acceptors. Again, the membranes, and the lumen between membranes, will be important for the ultimate production of ATP. The final electron acceptor is oxygen with water being produced. These are the last two factors of the overall equation for respiration. You will have to remind your students that way back at the end of glycolysis, two alternative pathways were possible, aerobic or anaerobic. It is the presence or absence of oxygen as an electron acceptor at the end of the electron transport chain that actually determines which pathway will be chosen. How can that be? A useful analogy is the cars coming off an assembly line. As long as there is a driver at the end to drive off the last car, the line will function smoothly. But as soon as the next driver fails to appear, the last car will sit at the end of the line. This, in turn, will back up all of the uncompleted cars in that line until the whole line is shut down.

Chapter 11

Pre-lecture Review - The orderly transfer of electrons, via electron transport chains, is an essential component of both photosynthesis and respiration. It is a very difficult concept for students to grasp. Single out this area of the light-dependent photosynthetic reactions to review by putting the appropriate overhead on the screen and calling on students to explain what it means.

Post-lecture Review - Use an outline diagram of the overall process of aerobic respiration to summarize the energy yield, in terms of ATP. Reveal one segment at a time, e.g., glycolysis, Kreb's Cycle proper, acetyl-CoA pathway, etc., and ask a different student to summarize what happens in that portion of the cycle. In particular note energy gain.

If time permits, this is also a good time to ask the questions, "What is the significance of photosynthesis and respiration to all living things?" and "How do you think they came to be linked?" Put the overall equations of these processes on the board, one above the other, and raise these questions with your students. As you list their responses on the board, they will begin to realize the cyclic pattern of carbon and oxygen in the ecosystem - a point you will return to in a later chapter.

LABORATORY - EXERCISES

1. Anaerobic Fermentation in Yeast

Production of CO_2 by yeast during fermentation is a traditional, and easily quantifiable, respiration exercise. Make a yeast suspension of known concentration. Place equal volumes of the yeast in fermentation tubes or pipettes. Read changes in gas volume, CO_2 production, periodically.

An easy variation is to ask student groups to hypothesize about what factors might affect the rate of the reaction, and how they might test their hypothesis. Typical suggestions include: source of carbohydrate (different sugars at same concentration); concentration of carbohydrate (from 0-20%); concentration of yeast (0-5%); temperature (5-40°C); pH (2-12).

2. Aerobic Respiration

Aerobic respiration is more difficult to measure because gas volume change is the net result of CO_2 production <u>and</u> O_2 consumption.

A. CO_2 Production - Indirect

A common technique is to take advantage of the fact that CO_2 in water forms carbonic acid. As a result pH decreases as more CO_2 forms. A simple qualitative demonstration of this is to place germinating seeds in a test tube of phenol red and observe color change after a period of time under specified conditions. Vary the conditions, as for fermentation, to test hypotheses concerning the effect of environmental factors on rate. A quantitative approach is to measure change in pH directly over time.

B. CO_2 Production and O_2 Consumption - Direct

The actual volume change in O_2 may be measured directly. Change in CO_2 is calculated by subtracting O_2 change from net volume change in a respirometer. The respirometer is a closed system, containing the test tissue. A calibrated gauge, to measure volume change, is attached to the system. In practice, a stoppered flask connected to a pipette works well. Fluid is drawn into the pipette to "close" the system and to facilitate reading gas volume change on the flask side of the system. As gas volume decreases, fluid is drawn toward the flask. An increase in volume pushes the fluid toward the open end of the pipette. Provide an additional outlet to allow the system to equilibrate prior to beginning measurements. This outlet can also be used to adjust the level, if necessary, during measurement. A glass tube with a clamped rubber hose works well.

The test tissue, e.g., germinating seeds, is placed in the flask, the apparatus assembled, fluid drawn into the pipette, and the conditions allowed to equilibrate for about 10 minutes. To begin measurement, note the position of the fluid in the pipette. It should be near the middle to measure movement in either direction.) After equilibration, close the vent tube with the clamp. Record the volume change (net of CO_2 produced minus O_2 consumed) at periodic intervals. (Note: if the tissue produces exactly as much CO_2 as it consumes O_2, there will be no net movement! If it produces more CO_2 you will have a positive value. If it consumes more O_2 you will have a negative value.)

At the end of measurement, open the apparatus, suspend a cheesecloth sack containing NaOH or KOH pellets in the flask, and reassemble. Tie the cheesecloth with a string, then suspended it in the flask by pinning the string to the underside of the stopper. The strong base reacts with CO_2 removing it from the air. As a result, once the experiment is restarted, any gas volume change is due solely to O_2 consumed. Repeat the measurements as above to determine directly the amount of O_2 consumed by the tissue (this will be a negative value). Now calculate the actual volume of CO_2 produced by subtracting the net change from the O_2 consumed.

An added benefit of this procedure is that you can determine the respiratory quotient of the tissue. Respiratory quotient, RQ, is the value of CO_2 produced divided by O_2 consumed. RQ = 1 indicates that carbohydrates are the food source being utilized; RQ < 1 indicates that either proteins or lipids are being respired; RQ > 1 means that either intermediate metabolites are being oxidized or that respiration is anaerobic.

LABORATORY - MATERIALS, EQUIPMENT, AND PREPARATION

Chapter 11

1. Anaerobic Fermentation in Yeast
-yeast
-sugars (at least table sugar - sucrose)
-fermentation tubes (or test tube and pipette with tip closed)
-water baths
-balance and graduated cylinders (to mix known concentrations)
-clock or timer
-phosphate buffer solutions
    -stock solution A: dissolve 12.37 g H_3BO_3 in about 500 ml dist. H_2O.
        add 10.51 g citric acid and dilute to 1 liter.

    stock solution B: dissolve 38.01 g $Na_3PO_4 \cdot 12H_2O$ in dist. H_2O and
        dilute to 1 liter.

pH	Stock A (ml)	Stock B (ml)
2	975	25
4	775	225
6	590	410
7	495	505
8	425	575
10	270	730
12	85	915

2. Aerobic Respiration

A. CO_2 Production - Indirect
-germinating seeds (soak 2-3 days before lab)
-test tubes and racks
-phenol red (add enough phenol red dye to color water noticeably. If the color is yellow instead of red, add drops of NaOH solution until the color changes to red)

B. CO_2 Production and O_2 Consumption - Direct
-germinating seeds (corn seeds utilize carbohydrate, sunflowers use lipid, beans utilize protein)
-erlenmeyer flask with two-hole stopper
-1ml pipette
-one straight glass and one bent glass tube per stopper
-short rubber tubing on end of each glass tube, bent tube connects to pipette, straight tube is closed by clamp
-clock or timer
-ring stand and clamps to hold apparatus

-cheesecloth
-string
-KOH or NaOH pellets
-pin to hang pellets from bottom of stopper

REFERENCES

General:

Webb, A.D. 1984. The Science of Making Wine. *Amer. Sci.* 72:360-367.

White, Robert M. 1990. The Great Climate Debate. *Sci. Amer.* 263:36-45. The effect of CO_2, and other greenhouse gasses, on the environment.

Resources:

Krogman, D.W. 1973. *The Biochemistry of Green Plants*. Prentice-Hall, Englewood Cliffs, N.J. 239 p. A fairly detailed, but readable, description of respiration and other biochemical processes.

CHAPTER 12: TRANSPORT

Ask students how living things transport materials like food and water through the body and most will answer that a heart pumps them by a heart through a circulatory system. But then ask how water and nutrients can reach the cells at the top of a growing redwood tree more than 367 ft. tall (an Australian *Eucalyptus*, cut in 1868, was reported to be 464 ft. tall!)? The remarkable efficiency of plants in transporting water and nutrients may be explained by a relatively few basic concepts. Yet, these concepts tend to give students an inordinate amount of trouble. In part, this is due to students' attempts to learn the material solely by memorization, rather than attempting to understand the underlying concepts.

OBJECTIVES

1. Explain diffusion and osmosis in terms of free energy gradient.
2. Describe the pathway of water from the soil, through the xylem, and into the atmosphere.
3. Explain the transport processes involved in water movement through each component of the pathway described in 2.
4. Explain the movement of water and solute through the phloem from source to sink.

CONCEPTS

Ask your students for a definition of diffusion and you will get: the movement of something from a region of high concentration to a region of low concentration. This textbook definition is usually adequate for the kind of examples given in freshman courses. It also is useful in emphasizing the concept of a gradient, but technically it is not correct. As Mauseth points out, concentration (osmotic potential) is only one factor that influences the rate of diffusion. In fact, diffusion is movement of a substance from a region of higher free energy to a region of lower free energy; thus, any factor that influences the energy of the system will have an effect on diffusion.

Water is a particularly important molecule in living cells, so important that its free energy is given a special name: water potential. Like other molecules, the water potential (free energy) of a solution may be due to several components, osmotic potential being one. Make students aware that other factors may be involved. For the sake of simplicity, consider water potential to be equivalent to osmotic potential.

As botanists began to study diffusion, it soon became apparent that they needed a convenient standard against which to compare different solutions. The standard chosen was pure water. By definition, the osmotic potential of pure H_2O is 0. Because adding a solute can only decrease the free energy of water, the osmotic potential of the resulting solution is a negative number. The more solute added, the more negative is the resulting osmotic potential. Concepts involving negative relationships are very difficult for freshmen to understand - repetition is essential.

In addition to dealing with negative numbers, diffusion also causes some difficulties because it is relative, not concrete. A solution with an osmotic potential of -1 bar may have either a higher or lower free energy than another solution (or it may be the same). Knowing the osmotic potential of a single solution will not tell you if it will diffuse. You must know the relative free energies of two solutions. This leads to the concepts of hyperosmotic, hypoosmotic, and isoosmotic (note: these are different from the terms hyper-, hypo-, and isotonic). Hyper- (as in hyperactive) osmotic means the solution has a more negative osmotic potential than another. Conversely, hypo- (less) osmotic means the solution has a less negative osmotic potential than another. With the two solutions adjacent to each other, water will move from the region of higher free energy (the hypoosmotic region that has a less negative osmotic potential) to a region of lower free energy (the hyperosmotic region that is more negative). With these concepts in hand, both diffusion and osmosis become more easily understood processes.

The pathway of water is relatively straightforward, but a couple of concepts do cause problems. The first is that students tend to think of cell walls as barriers to transport. In fact, in many regions the apoplast is more important in water movement than the symplast. The role of the Casparian thickenings is, thus, to force movement through the cytoplasm of the endodermal cells that then have a regulatory function. The second is the manner in which stomata function. This process involves active transport of potassium ions into or out of the guard cells; water simply follows to reestablish equilibrium.

If the students have a grasp of the factors effecting free energy, and the resulting effect on water potential, the concepts of long distance transport by mass flow through the phloem and cohesion/tension through the xylem will be relatively straightforward.

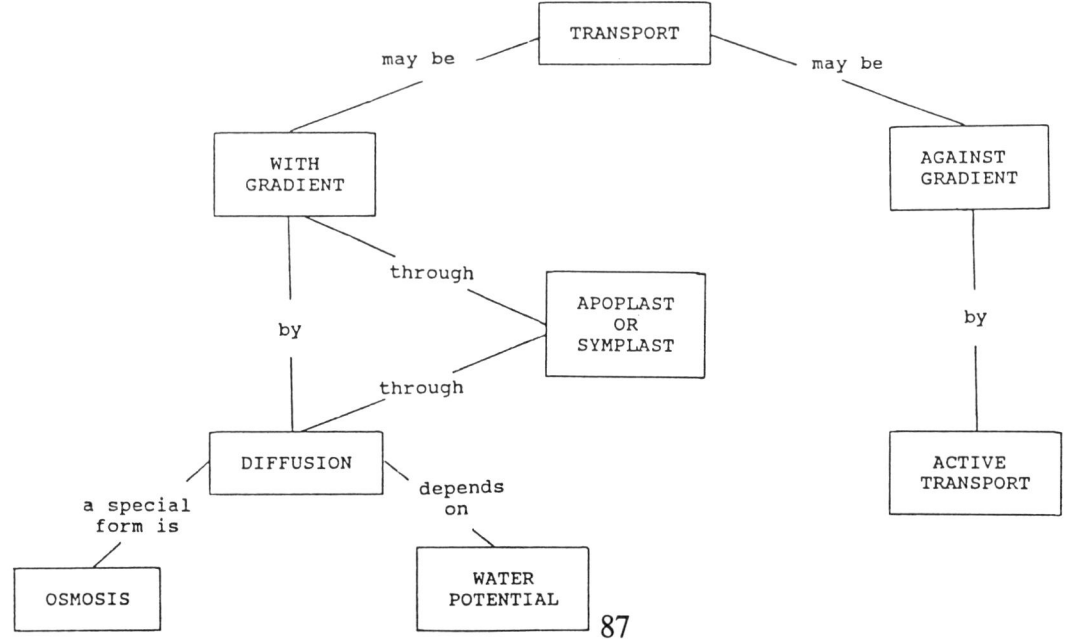

Chapter 12

INSTRUCTIONAL AIDS

Lecture - Bring a jar containing a towel saturated with an aromatic (I use amyl acetate) compound to lecture, set it on the front desk, open it, and ask students to raise their hand when they can detect the smell. A wave of hands will soon be evident moving out from the students closest to the jar (pockets along aisles and back corners of the room may also be evident, depending upon air currents in the room). Now ask for someone to explain what happened. This will lead to the standard textbook definition of diffusion based on relative concentrations of substances and the random movement of molecules out of regions of higher density.

Now ask students what they expect will happen if you heat the container prior to opening it? Someone will suggest that diffusion will be faster because the heated molecules will move faster. Now what if you heated the room air but cooled the container? Diffusion will slow. But you haven't changed the concentrations of the molecules, so why would the rate of diffusion have changed? Other factors, like heat, also must affect diffusion. This is your lead into a discussion of free energy; diffusion is the movement of molecules from a region of higher free energy to a region of lower free energy.

The concept of negative values for the free energy of water in solutions is very difficult for students to understand. It is helpful if you first describe how the addition of solute lowers water potential. Use more familiar terms, percentage of water, initially. Pure water is 100% water; adding solute to water lowers the percentage water. The more solute added, the lower the percentage water in the resulting solution. In the same way, pure water has maximum water potential. Adding solute lowers the water potential of the solution proportionally. The major difference is that for convenience in actually making measurements, we define water potential of pure water as 0, not 100.

Make your examples as concrete as possible; use and talk through many examples to explain osmosis and the terms hyperosmotic, hypoosmotic, and isoosmotic. There is much less likelihood of these terms getting mixed up in your students' minds if they understand the concept behind the names.

The pathway of water movement through the plant is rather straightforward. Start with the root and review the anatomy with a diagrammatic cross-section. Given an osmotic potential of soil water of [some value] and the corresponding value of the root hair cell as some lower value, which direction will water flow? Assuming that the root hair and the underlying cortical cell originally had the same osmotic potential, what would you expect to happen now that water has moved into the root hair from the soil? How does this affect the osmotic potential of the root hair? How does this affect the osmotic potential of the next cortical cell in line toward the stele? This line of questioning will establish the concept of a concentration gradient within the tissue, a concept that is useful to describe movement both in the root and in the leaf. Don't forget to bring in apoplastic transport and the role of the endodermis.

You have now discussed water movement into the xylem of the root. Before "moving it" through the xylem, take a look at what is happening at the other end of the system. Review leaf anatomy in the same way as you did the stem. Given that water vapor saturates the internal air in the leaf , while atmospheric air is seldom saturated, what will be the effect of having open stomata? What happens to mesodermal cells as the surrounding air looses moisture? Students will see that this establishes another concentration gradient, this time from the xylem to the exterior of the plant.

You are now ready to connect the two gradients already described. Here is the time for a dramatic demonstration if a vacuum source is available. Bring a shoot cutting of a two month old castor bean or sunflower plant to lecture. By this time it will be wilted (sunflowers already in flower are particularly dramatic). Insert the cut stem through a one hole stopper so it fits snugly. Insert the stem/stopper into a side arm flask half filled with water so the cut stem is well below the water surface. Now pull a vacuum on the flask (don't forget a water trap between the flask and your vacuum pump). Ask one of your students in the front row to describe to the class what is happening in the flask. Bubbles will be coming out of the cut surface. Now go on with your lecture.

Review briefly the structure of xylem, in particular the vessels. The vessel elements, stacked end-on-end, essentially form a tube through which water, and anything dissolved in water, can flow. Use the analogy of a straw. What are the characteristics of the straw and the beverage that allows you to drink through a straw. The diameter must be small (try drinking through a hose), there must be a continuous column of fluid (recall the trick of putting a pin hole in the straw), and there must be a force, tension, on the top end (provided by the person sucking). [By now there should be no more bubbles coming from the bottom of your cut stem, so turn off the vacuum pump and disconnect the hose.] How does this straw analogy apply to transport of water through the xylem of a plant, the plant on your desk for example? As you begin to recount the similarities, some of the students in front may begin to notice your plant starting to move. It is slow at first, one leaf beginning to uncurl at a time, but soon it is very obvious that your previously wilted specimen is "coming back to life"! Thus you have a dramatic demonstration of the cohesion/tension theory in action.

With xylem transport already covered, the basics of phloem transport are fairly easy to comprehend. If sugar is pumped in at one end, water must follow to maintain osmotic equilibrium. If cells pump sugar out at the other end, again water must follow, for the same reason. If solution moves in at one end of a column of cells, and moves out of the other end, there must be a flow from source to sink. The problem is that the rate of movement is

Chapter 12

considerably faster than we calculate it should be. But that's what makes science so exciting: there's still so much more to learn!

Pre-lecture Review - You will want to review several previous topics during the course of your lecture on transport. These might include the anatomy of roots and leaves and the structure of xylem and phloem. Before beginning to lecture, you will want to review the peculiar properties of water - particularly cohesion and adhesion. A thorough understanding of these principles will be essential to your students' understanding of the concepts of water potential and the movement of water within the plant.

Post-lecture Review - Present your class with this scenario: One homeowner never fertilizes his lawn, which looks pretty scrawny compared to his neighbor, Mr. Jones, who applies fertilizer religiously according to the package directions. A third neighbor, trying to outdo Mr. Jones, puts down twice as much fertilizer as recommended (but his lawn doesn't look any better). The fourth neighbor put down four times the recommended rate and his lawn turned brown and died. Given what you know about transport in plants, what do you think happened to the fourth neighbor's lawn?

LABORATORY - EXERCISES

1. Osmosis in Pseudocells

Construction of pseudocells using dialysis tubing bags is a traditional exercise for examining osmosis. The advantage of this exercise is that the pseudocells provide a concrete visualization of what actually happens to real cells under similar situations. Students must handle the bags in the process of weighing them, and thus they can feel the difference in turgidity directly, as well as making an indirect determination of water uptake or loss.

Cut strips of dialysis tubing, about 20 cm long, and soak them in water until they are flexible and can be opened into a tube. Tie one end by folding it back on itself then wrapping and tying tightly with string. Fill the resulting tube about half full with the appropriate solution, then fold and tie the other end of the bag. Weigh the resulting pseudocell to determine an initial mass, then placed in a test solution. By varying the initial concentration of solution in the pseudocell, as well as the solution in the beaker, students can test a variety of different hypotheses..

There are only a few potential difficulties with this system. The most obvious is that the ends must be tied tightly so the bags do not leak. It is usually sufficient to fold the end back on itself before tying. A second problem is that the string used to tie off the bags will adsorb water. Thus even a bag that is isotonic might gain weight. Similarly, if the initial weighing is of a dry bag while all subsequent weighings are of bags that have been soaking in solution, then blotted dry, there will be an apparent weight gain with the first reading.

Transport

2. Plasmolysis and Determination of Osmotic Potential

Plasmolysis is the shrinking of the cell membrane away from the cell wall. The point of incipient plasmolysis occurs when the cell is isoosmotic with the solution bathing it. For practical purposes, we define incipient plasmolysis as the point when 50% of the cells of a tissue are plasmolyzed. Because at this point the osmotic potential of the cells must be the same as the solution, we can use this relationship to determine the osmotic potential of cells.

Treatment	Molarity (Sucrose)	Osmotic Potential (bars)	Cells Plasmolysed	# Cells Counted	% Plasmolysed
1	0.1	-2.6			
2	0.2	-5.4			
3	0.3	-8.2			
4	0.4	-11.3			
5	0.5	-14.5			
6	0.6	-18.0			
7	0.7	-21.8			
8	0.8	-25.8			
9	0.9	-30.2			
10	1.0	-35.0			

Prepare a series of solutions ranging from 0.1 M to 1.0 M in 0.1-M increments. Place some of the cells in question into each of the solutions (red onion leaf peels work particularly well, but *Elodea, Tradescantia* stamen hairs, and other leaf peels also work). After 30 min. remove the tissue and examine for plasmolysis under the microscope. Record the % plasmolysis for each solution in a table as given below. At the lowest concentrations none of the cells will be plasmolyzed, whereas, high concentrations plasmolyze virtually all of the cells. The transition may occur abruptly, so you will not see the actual point of incipient plasmolysis. You will, however, be able to determine a narrow range where incipient plasmolysis must occur.

In practice you would want to prepare a new range of solutions, at smaller intervals, between the two concentrations that bracket the point of incipient plasmolysis. Alternatively, you could plot your data and graphically determine the approximate osmotic potential. The osmotic potential of different tissues varies considerably in the plant. Even in a single onion, the osmotic potential of epidermal cells may vary from one scale to another (this in itself is an interesting experiment).

LABORATORY - MATERIALS, EQUIPMENT, AND PREPARATION

1. Osmosis in Pseudocells
- dialysis tubing
- scissors and string
- sugar and balance
- distilled H_2O and graduated cylinder
- beaker
- clock or timer

2. Plasmolysis and Determination of Osmotic Potential
- 10 petri dishes, label tape or china marker
- compound microscope, slides and coverslips
- red onion or other plant material
- sugar, dist. H_2O, balance and graduated cylinder
- clock or timer

REFERENCES
General:

Moorby, J. 1981. *Transport Systems in Plants.* Longman. London. 169 p. A brief text written for undergraduates.

von Blum, R. 1981. Experimental studies of permeability in red blood cells. pp 63-119.in: Glase, J.C. (ed.) *Tested Studies for Laboratory Teaching, Vol 2.* Kendall/Hunt, Dubuque, Ia. In addition to an interesting exercise examining hemolysis of red blood cells, von Blum presents an excellent self-tutorial on the principles of osmosis.

Resources:

Nobel, P.S. 1983. *Biophysical Plant Physiology and Ecology.* W.H.Freeman and Co., San Francisco. 608 p. A technical treatise concentrating on water relations.

Stadelmann, E.J. 1966. Evaluation of turgidity, plasmolysis, and deplasmolysis of plant cells. in: Precot, D.M. (ed.) *Methods in Cell Physiology, Vol. II*. A more detailed procedure for making precise determinations of osmotic potential in plant cells.

Sundberg, M.D. 1988. *General Botany, BOTY 1202*. Alpha Editions, Burgess International Group, Inc. Edina, Minn. 106 p. A quantitatively oriented laboratory manual for freshman botany.

CHAPTER 13: SOILS AND MINERALS

By far, plants derive the greatest proportion of their bulk from elements abundant in the atmosphere. The process of carbon fixation was covered in the chapter on photosynthesis. Except for nitrogen, plants obtain the other essential elements from the soil. Nitrogen is a special case; fixation of adequate amounts of atmospheric nitrogen and recycling of nitrogenous compounds both depend on the activity of prokaryotic organisms. Similarly, the availability of other nutrients may depend on other symbiotic organisms or the physical and chemical characteristics of the soil itself.

OBJECTIVES

1. Explain what is meant by the term "essential" element and be able to differentiate between macro- and micro-essential elements.
2. Describe the symptoms of nutrient deficiency and relate the symptom to the role of the nutrient.
3. Describe how the physical and chemical characteristics of soil affect nutrient availability.
4. Describe nitrogen fixation and the nitrogen cycle.

CONCEPTS

The concept that an essential nutrient is necessary for a plant to complete its life cycle seems straightforward and easily testable. However, as Mauseth points out, it is often difficult, in practice, to determine that there is no alternative substitute or that the nutrient is actually acting within the plant, not outside it.

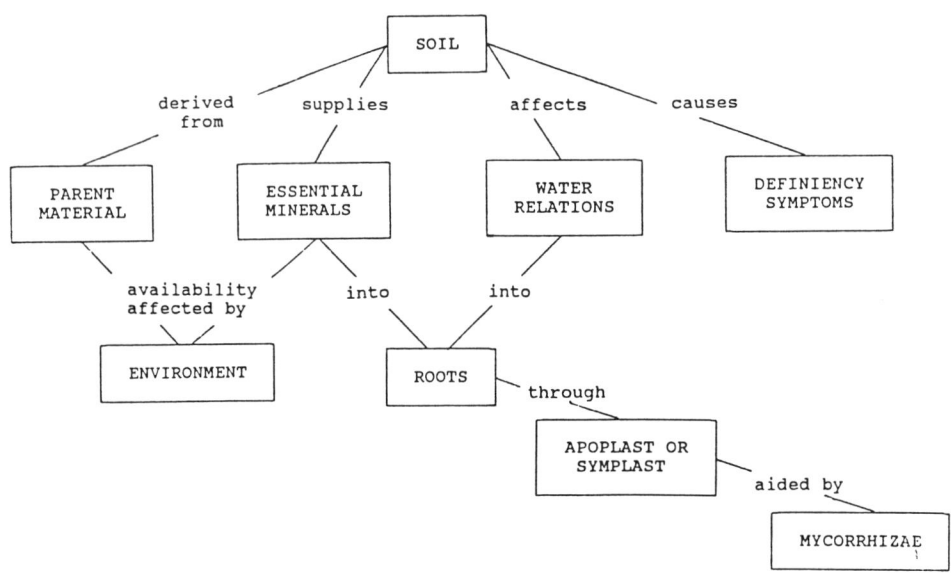

Confounding the latter situation is the symbiotic associations between plants and various microbes. Symbiosis is a term familiar to many students, but most will not understand the concept well, nor will they appreciate the widespread occurrence of symbiotic associations. Both nitrogen fixation in root nodules and mycorrhizal associations are excellent examples.

Freshman students, used to thinking in concrete terms, will not understand the concept of availability. They will assume that it is enough to have a nutrient, or water, merely present in the soil. If a plant shows a deficiency symptom, it must be because that element is missing. Corrective action, then, is merely to supply the missing element. The idea that the properties of the soil may determine whether or not an element, otherwise present in sufficient amounts, is available to the plant is an important concept as it has significance in some of the environmental problems, such as acid deposition we face today.

INSTRUCTIONAL AIDS

Lecture - Ask students to recall the chemical composition of some of the important molecules they have been studying this semester. Have an overhead of a carbohydrate, lipid, amino acid, and nucleotide ready to put up as examples. Which atoms are the most commonly found? C, H, O, and to a lesser degree N will be obvious from your examples. The essentiality of these elements will be obvious from their structural presence in common compounds, but what about some of the other essential elements? List the macroelements on the board and ask for an essential function of each. A few will be fairly easy, e.g., P in ATP and S in some amino acids. Some will be more difficult, e.g., Mg in chlorophyll and Ca in pectin, and Fe in electron transport chains, while some will be obscure - K, in maintaining osmotic equilibrium.

What is the natural source of the elements listed above? The atmosphere, including water, and the processes of photosynthesis and water uptake will be ready responses to explain the first three. The ultimate source of nitrogen is also the atmosphere, but the process of uptake is not so evident.

The above exercise leads you to a discussion of nitrogen fixation and the nitrogen cycle. Here is a good opportunity to begin to attack a commonly held misconception about bacteria - that they mostly cause diseases. Nitrogen fixation, reduction, and assimilation are all dependent on particular groups of bacteria without which amino acid metabolism by plants, and animals, would not be possible. Nitrogen fixation also allows you to introduce the concept of symbiosis. You can reinforce this with a discussion of mycorrhizae. The latter is especially useful in emphasizing the widespread occurrence of symbiotic associations - inform your students of the thousands of symbiotic microbes in and on their bodies as they sit in class!

Pre-lecture Review - Review the general equation for photosynthesis. What is the source of the raw materials required for this process? When students examine the individual steps of the process, do any other elements appear to be essential, e.g., Fe, N or P? What must be their

Chapter 13

source? The elements listed obviously are used in photosynthesis, but are they essential? You are now ready to discuss the concept of essential nutrients and the source and availability of these nutrients.

Post-lecture Review - We apply fertilizers to compensate for nutrient deficiencies in the soil. Does it make a difference whether we use organic or inorganic fertilizers? Most students will respond that it does make a difference, because organic fertilizers are somehow easier for the plant to take up. Of course, the chemical form in which an element is absorbed will be the same regardless of source, so this is not a difference. There are advantages, and disadvantages, to both chemical and organic forms of fertilizer. We can tailor chemical fertilizers to specific needs and apply them at precise rates for rapid incorporation. Organic fertilizers are slow-release forms and improve the texture of the soil as well as providing nutrients.

LABORATORY - EXERCISES

1. Mineral Nutrition

Traditionally we use hydroponic culture, usually with tomato plants, to demonstrate the effects of various nutrient deficiencies on plant growth. This requires a considerable amount of time, a month or more from the time seeds are planted. It also requires enough space to handle a series of quart-size jars. Fast growing *Brassica* seed ("Wisconsin Fast Plants") is now available and provides a considerable savings of time and space.

Prepare nutrient solutions in the same manner as usual except 100 ml volumes, rather than 1 liter, are sufficient. Students must clean and rinse their glassware thoroughly before preparing deficiency stocks. Cover extra solution, and keep it in the dark to prevent algal growth, so that students can replenish solution in the growth containers as needed over the course of the experiment.

Rather than using the soil mix and fertilizer pellet provided in Fast Plant Kits, fill the growth containers with washed sand or perlite. Place each container in an individual petri-dish half that will act as a nutrient solution reservoir. Place one seed in each square and add the appropriate nutrient solution. Check the nutrient level at two- or three-day intervals. Within one week, there will be obvious deficiency symptoms in some containers and within three weeks the control plants will be in flower.

LABORATORY - MATERIALS, EQUIPMENT, AND PREPARATION

Per Class
-"Wisconsin Fast Plants" kit (Carolina Biological #15-8702) One kit provides
enough material for groups of students.

-1M solutions of: KNO_3, KH_2PO_4, $Ca(NO_3)_2$, $MgSO_4$, KCl, $CaCl_2$, Na_2SO_4, NaH_2PO_4, $NaNO_3$, $Mg(NO_3)_2$, and $MgCL_2$.

-iron solution: dissolve 0.745 g Na_2EDTA in about 80 ml dist. H_2O
 add 0.557 g $FeSO_4·7H_2O$ and bring to 100 ml with dist. H_2O

-micronutrient solution: begin with about 400 ml of dist. H_2O, add the appropriate amount of each of the nutrients listed, then dilute to 1 liter (such a large volume of stock solution must be made in order to measure accurately the minute amounts of the last two nutrients!).

Micronutrient	g
$MnSO_4·4H_2O$	2.2300
$ZnSO_4·4H_2O$	0.8600
H_3BO_3	0.6200
KI	0.0830
$Na_2MoO_4·2H_2O$	0.0250
$CuSO_4·5H_2O$	0.0025
$CoCl_2·6H_2O$	0.0025

-pipette (labeled) and pipetter for each stock solution
-distilled H_2O, preferably glass distilled

Per Group
 -eight petri-dish halves
 -eight containers (to hold at least 100 ml) for deficiency stock solutions)
 -fluorescent light fixture
 -100-ml graduated cylinder

Preparation of deficiency solutions. Add the indicated volume (ml) of the appropriate stock solution to a labelled container, then bring the volume to 100 ml with distilled H_2O. Each solution should also receive 1 ml of micronutrient stock solution and all but -Fe receive one ml of iron stock solution.

REFERENCES
General:
Faulkner, E.H. 1943. *Plowman's Folly*. University of Oklahoma Press, Norman. A classic little book which introduced the idea of surface tillage.

Resources:
Hambidge, G. (ed.) 1941. *Hunger Signs in Crops: A Symposium.* American Society of Agronomy and National Fertilizer Association. Washington, D.C. 327 p. Text and photographs illustrating the symptoms of nutrient deficiencies on a variety of crop plants.

CHAPTER 14: MORPHOGENESIS

Morphogenesis, the development of form, is one of the most interesting and least understood areas of botany. All organisms must respond to their environment to be successful. Similarly, the individual cells of the body must respond to other cells to integrate growth and function. In animals, because of the nervous and endocrine systems, there are straightforward means of studying such interactions. Plants, however, do not have obvious organs specialized for sensing the environment and communicating within the body. Morphogenesis is therefore difficult to study. We have learned a great deal about morphogenesis during the past 50 years, but there remain more questions than answers. The results of "spray and pray" experiments, although often providing convincing support for hypotheses, are not conclusive. Even the most basic hypotheses continue to come under review as we gather new information. For instance, the traditional explanation of the role of auxin in phototropism has been challenged during the past decade. There is a tendency, especially when discussing plant hormones and growth regulators, to teach the material as dogma - it is not.

OBJECTIVES
1. Describe five environmental stimuli which may affect development and morphogenesis.
2. Describe four classes of response to environmental stimuli.
3. Explain the source, transport and function of each class of plant hormone.
4. Describe the interaction of two or more hormones in producing a response.
5. Explain the role of photoperiodism in flowering

CONCEPTS
That plants respond at all to their environment will be a revelation to many students. Although response to stimulus may be one of the characteristics of living things listed on the first day of class, most students associate this character with animals. Furthermore, to most students, response to the environment will imply conscious control by the brain. For instance, birds must know that it is time to fly south for the winter when the temperature starts to drop, the days get shorter, and they put two and two together. Not only can plants respond to their environment, at least as effectively as animals, but they also respond to each other and to other organisms. This is truly amazing, much more so than in animals, because plants lack the specialized receptor organs and communication systems that are so prominent in animals.

Chemical control is a difficult concept for students, even in animal systems where there are definite source and target organs. In plants, many different tissues respond to the same hormone, simultaneously and in different ways. Thus, the same auxin may stimulate the shoot top to bend toward light while it inhibits lateral buds from growing. It also promotes differentiation of secondary xylem and inhibits branch root elongation but it promotes branch root initiation. You must first clarify the concept that a relatively few molecules, produced in one part of the plant, can have a profound effect on other parts of the plant. Then you must show that different kinds of molecules can have different effects. Finally you must demonstrate that the relative concentrations of two or more of these compounds produce different effects.

Light is an extremely powerful morphogenetic stimulus, but one for which we have little intuition. Light is essential for plant growth because of photosynthesis. Furthermore, the color and intensity of light affect photosynthesis and, therefore, growth of the plant. This is the limit of most students' understanding of the role of light in plant growth. Of course, color (there are blue, red, and far-red effects) and intensity (phototropism has both low-intensity and high-intensity effects) have morphogenetic effects as well. The most important aspect of light in controlling development, however, is the relative duration of the light and dark cycles - photoperiodism. Photoperiodism is important in animals, including humans, as well as plants, but we simply don't notice it. It is, thus, a difficult concept for students to comprehend.

Tied to photoperiodic cycling are endogenous rhythms. Biorhythms are pervasive in eukaryotic cells and appear to be membrane-associated phenomena that are not affected by parameters such as temperature. Thus they are not a simple chemical process. Although we know endogenous rhythms regulate a wide variety of cellular processes, from cell division, respiration, and photosynthesis to flowering, the mechanism of control is as elusive as the mechanism itself. Students are best able to visualize the process by examining an easily measured phenomenon such as the diurnal leaf movements of legumes..

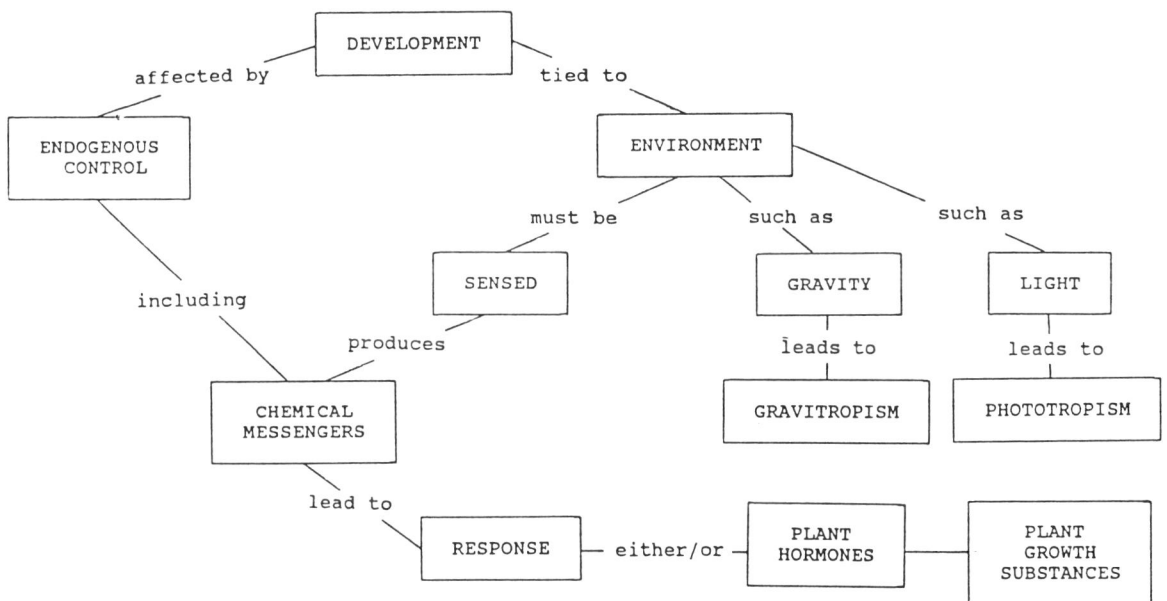

INSTRUCTIONAL AIDS
Lecture
Demonstrate that plants do indeed respond to stimuli. This may be done in several ways. One is to bring a two-week old *Brassica* (Wisconsin Fast Plants) with you to lecture and place it, on its side, on the front desk at the beginning of class. By the end of the hour it will exhibit gravitropic bending with the tip orienting vertically. Ask students for some hypotheses to explain this tropic response. A second demonstration is to place a *Mimosa* (sensitive plant) in

Chapter 14

the lecture room ahead of time (if you carry it in with you, the jostling will stimulate closure and you will have to wait until the end of class to demonstrate response to touch). Depending on the force of stimulus, you can initiate a closing response from a few leaflets, to entire leaves, to the whole plant. A third demonstration is to bring an entrained *Albizzia* plant, about the same size as the *Mimosa* if possible. The two plants are both woody legumes and look quite similar but *Albizzia* does not respond to touch. Its leaves, however, exhibit diurnal sleep movements. For at least three days prior to lecture, place the plant under a light regime which cycles from light to dark at a time corresponding to midway through your class period (a 12:12 light/dark cycle will do). Although *Albizzia* will not respond to you touching the leaves, shortly into the class period the leaflets will slowly begin to close. By the time class is half over they will be completely closed. The movements are too slow to watch, but some student will eventually notice the completed response. This provides a good lead-in to endogenous rhythms.

The best take-home message about the role of individual plant growth regulators is their commercial application. Mauseth mentions 2,4-D, which is a selective auxin herbicide. At relatively low concentrations, it affects dicots but not monocots. It is, thus, the usual "weed" component in "weed-and-feed" lawn fertilizers. A broad-spectrum auxin herbicide, which kills both monocots and dicots, is 2,4,5-T. "Agent Orange" was a mixture of both. Another commonly used synthetic auxin is IBA (indolebutyric acid) which is found in root promoters, such as "Roottone." IBA stimulates root formation in difficult-to-root cuttings. A commercial gibberellin product, "ProGibb," is frequently sprayed on fruit crops to increase fruit set. Use of cytokinins is restricted to commercial propagation of clonal plants by tissue culture. Fruit ripening is promoted with Ethylene; CO_2 is used to delay ripening. Control of these two gasses is important during shipment and storage of many vegetable and fruit crops. Abscissic acid promotes leaf drop in some crops prior to harvest to facilitate machine picking.

Apical dominance is a good phenomenon to show the interaction of hormones in producing a response. Present it as a series of experiments beginning with grandmother's common sense - to make her plants bushier she "pinches back" the top. List some hypotheses. How could students test them? Once students understand the role of auxin, add a new observation. If you add cytokinin directly to the bud, the bud will begin to grow out whether or not the shoot apical meristem has been removed. Given these experiments and the fact that shoot tips normally produce auxin while cytokinin is produced in root tips, how might these hormones control branching in intact plants?

The historical approach works extremely well when presenting photoperiodism. Begin with Garner and Allard's classic experiments with Biloxi soybeans and Maryland Mammoth tobacco and how they artificially controlled light conditions with "doghouses" in the field during the summer and electric lights in the greenhouse during the winter. When your students grasp the concept of a short day being one with less than a certain critical daylength, and a long day being one with more than a critical daylength, you are ready to introduce Hamner and Bonner's light break experiment. In fact, it is the night length which is critical! Under natural conditions this is not a problem because under a 24-hour day, a relatively short day will always correspond to a relatively long night and *vice versa*. Additional experiments demonstrated that it is red light that is most important in eliciting the response. Subsequently phytochrome was found to be the receptor pigment involved. Unfortunately, we cannot go much beyond this as the time for dark conversion of P_{fr} to P_r is usually much less than the critical night length.

We know that endogenous rhythms somehow fit into the process of photoperiodic induction of flowering, and other physiological processes as well. Use an example of a visible biorhythm, such as sleep movements, to discuss the phenomenon of circadian rhythms. Describe experiments to show how they were determined to be endogenously, rather than exogenously, controlled.

Pre-lecture Review

Recall the properties of light that are important in photosynthesis - quality (color), intensity, and duration. Red, and especially blue, light are important for photosynthesis and, in general, the greater the light intensity, up to saturation level, the greater the rate of photosynthesis. You might point out to students, however, that there can be too much of a good thing. Too high a light intensity will destroy chlorophyll. Shade tolerant species, like African violets, are especially sensitive to high light intensity. In general, photosynthetic output also correlates with duration of light exposure, but again there can be too much of a good thing. Some plants grow poorly when given continuous light.

Post-lecture Review

If your students have not been doing concept mapping regularly, here is an excellent chapter to put it to good use. Choose either a single stimulus or a single response and ask students to map the information related to that topic. With a dozen or more choices here, you can divide the class into teams and assign each team a different topic.

LABORATORY - EXERCISES

1. Tropic Responses

A. Quantification

Phototropism and gravitropism are easy to demonstrate with seedlings. Instructions are presented in virtually all laboratory manuals. As mentioned above, in less than one hour *Brassica* (Wisconsin Fast Plants) seedlings respond to gravity. Go beyond this simple observation of negative gravitropism. Is the response due to cell enlargement or cell division? Are all cells involved or just the epidermis? Or cortex? Or vascular tissue? Students can test any of these assumptions by longitudinally hand sectioning the seedlings in the plane of curvature and counting and measuring cells on the inside half vs the outside half of the curve.

B. Mechanism of Auxin Action

Chapter 14

Auxin activity has long been associated with tropisms. A current hypothesis is that auxin induces proton extrusion from the responsive cells. Protons increase the plasticity of cell walls which then stretch because of the osmotic pressure of the cytoplasm.

Obtain a plate of Bromcresol Purple Agar and a 3-day-old germinating corn seed. Place the seed on one side of the plate so that the emerging root extends toward the middle of the plate. Press the root firmly against the agar. Tape the petri dish together and place it on edge with the root vertical for 10 - 15 minutes. Note the appearance of the agar after this time. Rotate the dish 90° so the seedling is now horizontal. Make periodic observations during the remainder of the lab period and again the next day. The agar will turn yellow where proton efflux occurs (adjacent to the bending cells).

2. Nastic Responses

Measure sleep movements in *Albizzia* plants that have been entrained to light/dark cycles that change during the laboratory period. Each student group should obtain three leaflet segments from both the "light" and "dark" plant at the beginning of the laboratory period and place them in water in a petri dish. At 15 min. intervals throughout the laboratory period, measure the angle between leaflet pairs on each segment by comparing the segment to a template of known angles. Average the angle for the three leaflets from each treatment. During the laboratory period the leaflets from the "dark" plant will gradually open while those from the "light" plant will gradually close. This is in spite of the fact that both plants, and their excised leaflets, are all in the light.

Students will realize that the angles are changing as they measure them and record the results in a table. It will be more dramatic if they then graph the data, average angle vs time, for the two treatments. Of course, the plants themselves make the greatest impression. The one that was "open" at the beginning of lab will have it's leaflets "closed" at the end and the one that was initially "closed" will be "open" at the end of class. The reference cited includes several variations of this.

Morphogenetic Responses

A. Gibberellin

Most laboratory manuals include an experiment on the effect of different concentrations of gibberellic acid on elongation of dwarf peas. Assays use either total stem length, or lengths of individual internodes, to assay growth. Measure the plants at the beginning of the experiment; add one drop of gibberellic acid solution to the apical meristem; then remeasure the plants after one week of further growth. The one potential problem is finding the apical meristem. Peas have compound leaves with a pair of stipules at their base. Place the drop of gibberellin between the stipules of the uppermost expanding leaf.

Photomorphogenesis

Flowering is an impressive and classic photomorphogenetic response. Unfortunately, it requires a considerable time period for the response to be visible. One way around this problem is to plan far ahead and set the conditions for a response to be visible when you get to this point. For instance, at the beginning of the semester plant some *Pharbitus nil* seeds in at least two sets of pots (you must scarify the seeds first!). Grow them under 16-hr light, 8-hr dark conditions. When you get to photosynthesis, bring two of the plants to class and pose the question - 'If one plant is grown with only 8 hr of light and the other is given 16 hr of light daily, will the first plant take twice as long to flower?" Then set up the experiment. *Pharbitus* is a short-day plant - you will at least have flower buds present two weeks after treatment begins.

The classic quick photomorphogenetic response is germination of "Grand Rapids" lettuce seed (light sensitive). An unfortunate byproduct of continued breeding is that the red/far-red response is no longer reliable. Test a new batch of seeds before using them in class.

Divide students into groups to prepare sets of petri dishes for the experiments. Place a piece of filter paper, or cut-out ring of paper towel in the bottom of a petri dish and add 10 lettuce seeds. In the dark, add 5 ml of water to the dish and wrap in foil or place in a light-tight container. (The seeds become sensitive to light following imbibition - if a dark room is not available, at least cover the dishes immediately after adding water). Allow the seeds to imbibe water for 6 - 24 hr before giving the appropriate light treatment. At a minimum, you will want a white-light treatment (5-min. exposure is adequate) and a complete darkness control. Other treatments include various combinations of red and far-red exposures - each for 5 min. For instance, you could have red light only, far-red only, red followed by far-red, or far-red followed by red. Students will have to prepare at least one dish with seeds for each treatment planned.

LABORATORY - MATERIALS, EQUIPMENT, AND PREPARATION
1. **Tropic Responses**

A. Quantification
-*Brassica* seedlings, pot, and soil mix (Wisconsin Fast Plants) 1-3 wk old.
-compound microscope
-slides and cover slips
-razor blades
-dropper bottles of stain (0.1% Toluidine Blue or 0.1% Safranin) or polarizing filters
-dropper bottles of water

B. Mechanism of Auxin Action
-3 to 5 day old germinating corn seeds
-petri dish of Bromcresol Purple Agar

Chapter 14

>   -0.6 g agar
>   -5 ml of 0.7 mM Bromcresol Purple in H_2O
>   -bring to 100 ml with dist. H_2O
>   -adjust pH to 5.0 and autoclave

2. **Nastic Responses**

    -2 pots of *Albizzia,* at least one year old to provide enough leaf material
    -2 growth chambers or light banks with timers in separate dark rooms
    (Timers should be set so that one switches from light to dark, and the other from dark to light, one half-hour into the laboratory period. One *Albizzia* plant should be placed under each light regime at least 3 days prior to the laboratory to entrain the plants to these light cycles)
    -petri dishes half filled with distilled H_2O
    -scissors and forceps to remove leaflet segments from plants and to place them into petri dishes.
    -plastic covered angle template (colored cut-outs of various angles from 10° to 180° against which to compare leaflet segment angles.

3. **Morphogenetic Responses**

A. **Gibberellin**

    -pots of 10-14 day-old dwarf pea seedlings (e.g., "Little Marvel") and similar-aged normal peas (e.g. "Alaska")
    -metric ruler
    -dropper bottles of various concentrations of gibberellic acid
    -dropper bottle of water

B. **Photomorphogenesis**

    -*Pharbitis nil* seeds (to scarify, either file a notch through the seed coat or treat for 1 hr in conc. H_2SO_4, followed by a thorough water rinse.
    -at least two pots of soil
    -two growth chambers, one with 8 hr light, the other with 16 hr light (begin treatment at least two weeks prior to lab - plants can be taken from the chambers when it is time to entrain *Albizzia*).
    -light sensitive lettuce seed
    -petri dishes
    -filter paper or paper towel disks
    -aluminum foil or light-tight containers to store imbibing and germinating seeds
    -light source (filters, optional). Red light is obtained by using a
    fluorescent light and red filter; far-red is obtained by using an incandescent light and both a red and blue filter. Several layers of colored cellophane will work; monochromatic filters may be purchased from science supply houses.

REFERENCES

General:

Galston, A.W. and P.J. Davies. 1970. *Control Mechanisms in Plant Development.* Prentice-Hall, Englewood Cliffs, N.J. 184 p. A good introduction to plant growth and, in particular, the role of plant hormones.

Garner, W.W. and H.A. Allard. 1920. Effect of the Relative Length of Day and Night and Other Factors of the Environment on Growth and Reproduction in Plant. *J. Agr. Res.* 18:553-606.

Resources:

Evans, L.T. 1969. *The Induction of Flowering.* Macmillan, Melbourne. 488 p. A sourcebook of factors affecting flowering for a number of plant species, each species treated in a chapter of its own.

Sundberg, M.D. 1980. Plant Biorhythms: A Laboratory Study for Introductory Botany. *Amer. Biol. Teach.* 92:2-5. Several variations of "sleep movement" experiments using *Albizzia*.

Sundberg, M.D. 1988. *General Botany: Boty 1202.* Burgess International Group, Inc. Minneapolis, MN. 106 p. Includes more detailed procedures for both gibberellin and cytokinin bioassays and proton efflux associated with tropisms.

CHAPTER 15: GENES

Genes, including gene regulation, is another topic that is conceptually difficult because it falls outside the range of student experience. The structural building blocks of genes, nucleotides, were introduced already in chapter two. However, that was weeks ago so a review will be in order. Many students will be familiar with at least some of the vocabulary, but this will only be a definitional familiarity. Many of the words will be confusing, e.g., transcription vs translation, exon vs intron, and codon vs anticodon. The key will be to move slowly and repeat frequently. What this usually means is that you will only have time to cover the basic concepts in class, but that's alright. Teach them to see the forest - they can learn the names of the trees later, from reading on their own.

OBJECTIVES

1. Describe the structure of DNA and the process of replication.
2. Explain the differences between DNA and RNA and describe the process of transcription.
3. Describe the process of translation, the roles of the different types of RNA, and the relationships between codons, anticodons, and the genetic code carried by the DNA.
4. Discuss several different mechanisms to regulate genetic development.
5. Describe the role of restriction and ligation enzymes and vectors, such as plasmids and viruses, in genetic engineering.

CONCEPTS

More biochemistry! And structure is important too! The first challenge will be to help your students visualize the structure of the DNA. Most can tell you that DNA is a double helix, but they will have difficulty explaining what a helix is and how is it double? A good helix analogy is the thread on a bolt where the single grove spirals around the surface of a cylinder. But how is it double? One possibility is that the two helices twist around each other like the ends of two joined electrical wires or two strands of twine twisted to lay a rope. However, this sort of structure would be very difficult to "unwind" when it came time to replicate strands. Instead, the two helices run parallel to each other around an imaginary cylinder.

You can explain the precision of base pairing by returning to the general structure of the molecule. Given the different structures of purines and pyrimidines and the potential for forming either two or three hydrogen bonds, it becomes easy to see how the nucleotides hold the sugar-phosphate backbones of the two helices at a constant distance from one another. It also becomes clear why the proportions of A and T, and C and G are constant.

"Unzipping" is not a problem with the helices running "parallel" to each other. The hydrogen bonds are broken and the sites are free to attach new bases. (The sharp students will want to know where these bases came from). What is a very difficult conceptual problem is that the two strands are actually antiparallel. That is, as you move along the length of DNA, one strand orients from 3' to 5' while the other strand is orients 5' to 3'. Because bases add to a strand in one direction, the mechanism must be different for the two strands.

Replication of the leading strand is straightforward - concentrate on this. Because the enzymes only work in one direction, the trailing strand replicates by connecting short stands of DNA that were formed "backwards."

With an understanding of DNA replication, transcription to form mRNA will be relatively straightforward. First clarify the differences between DNA and RNA and discuss the advantages of each in terms of cell function.

The concept of protein synthesis presents no new difficulties. Codons and anticodons will match in the same way that mRNA bases matched their DNA templates. There will be a t-RNA for each triplet combination with the appropriate amino acid attached. Four of the codons serve as "punctuation," either the signal to start, or the signal to stop.

Perhaps the most significant concept dealing with genetic engineering is that every technique we employ was "learned" from the bacteria, viruses and plasmids that do it naturally.

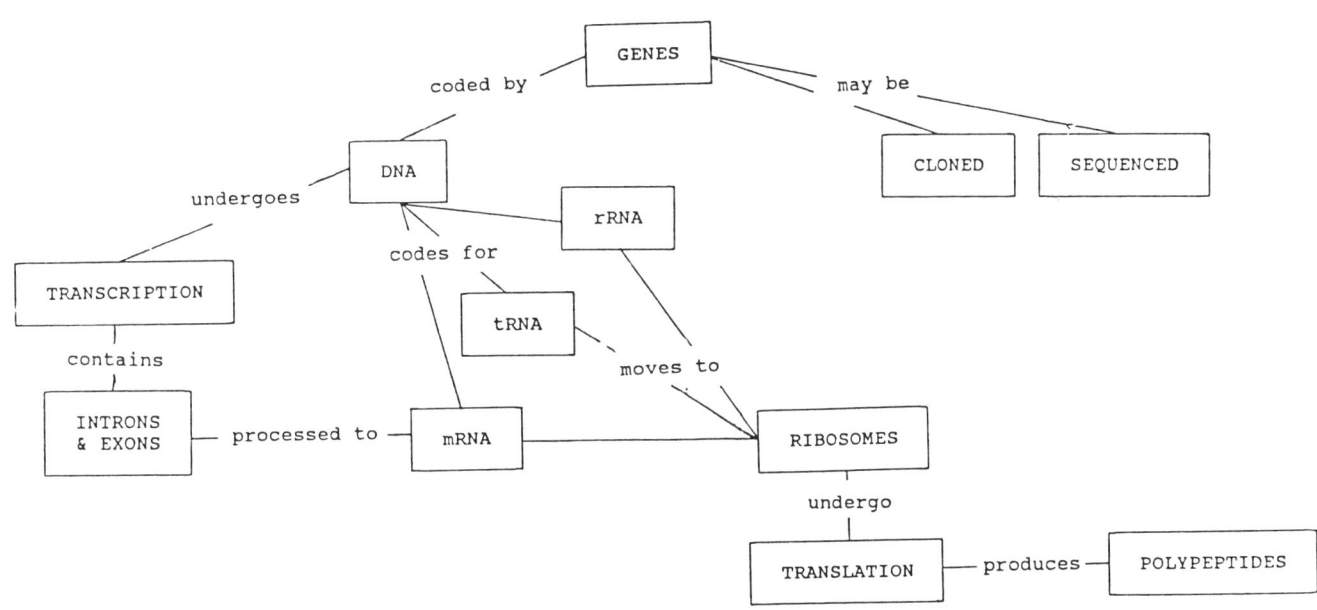

INSTRUCTIONAL AIDS

Chapter 15

Lecture - The key to making molecular biology understandable to your students is good visuals, either a good set of transparencies, or models, or both. The double helix is best demonstrated in three dimensions. For instance, tape two pieces of colored yarn to the bottom edge of a tube from a roll of paper towels. Then lay the yarns, parallel to each other, around the tube for two or three turns. Now students will be able to interpret the twisting illustrated in diagrams of DNA, as well as a function of the histones that occur in the position of the tube.

The structure of the sugar-phosphate backbone of DNA, with its attached sequence of nucleotide bases, is straight forward, but how are the helices held to one another so precisely? Have structural diagrams of the four bases on an overhead transparency, but cut out individually so you can move them. Label each diagram with its name, whether it is a purine or pyrimidine. Be sure to indicate hydrogen bonding sites. Start with the four arranged in a block with purines and pyrimidines together. The nucleotide bases of the two helices interact to hold the double helix together. What are the possible combinations of two bases that could bridge the gap? If purine linked with purine, then pyrimidines also must link with each other; some cross bridges will consist of only two rings, others of four. Demonstrate this by moving the cut out diagrams appropriately. By combining a purine with a pyrimidine, however, each cross bridge will consist of three rings. The resulting double stranded structure will be much more uniform. Now look at the available hydrogen bonding sites; there are either two or three. Thus, the possible combinations are limited to A-T (T-A) or C-G (G-C).

Given the precision with which complementary bases pair, it is now an easy step to move to the basic concept of semi-conservative DNA replication. When the hydrogen bonds between complementary bases are broken, the helices will unwind slightly, exposing the bonding sites. If free nucleotides are available, they may now bind to their complement forming a new strand, identical to the original. This is the "forest" that students must understand in order to understand the mechanism of the cell cycle and reproduction. The "trees" is the difference in replication between the leading strand and the trailing strand.

The move from DNA replication to RNA transcription is easy. First review the differences between DNA and RNA; why are both necessary? Mauseth discusses the need for protecting genes nicely. You also will want to bring in the different kinds of RNA at this point to lead into protein synthesis. How many different RNAs are there? Certainly each gene will produce a different mRNA and you need large and small subunit rRNA. What about tRNA? Ask the class what minimum number of unique letters must be in a code to provide all the instructions to produce a protein? Blank stares! Well, what are proteins composed of? Amino acids, 20 of them, so there must be at least 20 different codes, one for each amino acid. But the only variable along a strand of either DNA or RNA is the bases and there are only four of them. What if every pair of bases was a code for an amino acid? How many different combinations of paired bases are there? Still not enough, but it's close. The bases are analogous to the letters of the alphabet, but there are only four letters. At least 20 "words", formed with these four letters, is required to account for 20 different amino acids. The minimum number of letters that can form at least 20 different words is 3. There are extra combinations, so there must be two or more ways to spell here too. The code must therefore consist of a series of triplet bases. Similarly, the codon of mRNA consists of complementary base triplets; the DNA genetic code and the codon are NOT the same thing, they are complementary. Again, here is the forest; processing mRNA to remove introns and reannealing exons are the trees.

If there are 64 possible code words written in the DNA, there must be an equal number of code words carried in the mRNA. These are the codons. Subtracting out the punctuation codons, 60 must carry information specifying an amino acid. To match amino acids to the code, each tRNA has an anticodon region complementary to the appropriate codon. To form the specified protein, with the correct sequence of amino acids, you simply start at the beginning of the message, translate it one letter at a time, and stop at the end.

Having gone through the processes outlined above, ask your students if they can think of any other ways that the cell might be able to regulate its metabolic activity and development? Remind students that enzymes are proteins. Might there be a correlation between whether or not an enzyme is produced and the reaction it catalyzes? How to control protein synthesis? Ask for a list of possible "control points," given what they know about the sequence of events involved in protein synthesis. You might elaborate on one or two of these, or simply direct the students to the text.

What about genetic engineering? It is too much in the news not to be covered, and there are already serious misconceptions in the public's mind. Probably the single most important concept to get across is that every technique we employ to do genetic engineering was learned from nature. We are mimicking what bacteria, viruses, viroids, and plasmids do naturally. It's just that rather than letting random bits of genome be moved, we are increasing the probability that certain bits, which happen to code for useful genes, are moved. A brief discussion of the use of restriction enzymes and ligases to insert specific sequences, and the role of microbes as vectors and "factories" is appropriate. Probably the second most important concept to bring out is that the government has very tight regulations over the process!

Pre-lecture Review The previous chapter discussed the many ways to control growth and development, but how do external stimuli affect such changes or how does the plant regulate what substances it produces, how much and when? Ultimately, it comes back to genes in the nucleus. After making this connection, review the general structure of nucleic acids.

Post-lecture Review Every week there are articles in the media dealing with genetic engineering - and not just newspapers and news magazines. For instance, the month I am writing this has a feature article on genetic engineering in both *Consumer Reports* and *Business Week*! Ask each student to look for such an article, copy it, and write a one-page "technical" explanation of what it says.

Chapter 15

LABORATORY - EXERCISES

1. DNA Isolation
- coarsely chop about 10 g of frozen calf thymus into small pieces and add to a blender with 50 ml citrate-saline buffer. Homogenize for 3 min.
- decant the solution to centrifuge tubes and spin, at about half speed, for 10 min. (Be sure to balance the tubes in the centrifuge!)
- discard the supernatant. Rinse the sediment (nuclei) into the blender with 100 ml buffer B and homogenize at low speed for 1 min.
- pour the suspension into a beaker and add 4 ml SDS.
- place beaker in a 55°C waterbath while stirring constantly.
- add 4 g NaCl and heat for 10 more min. At this point the solution may be stored in a refrigerator overnight.
- add 50 ml chloroform:amyl alcohol (24:1) to the solution and shake for 10 min. in a separatory funnel.
- pour the solution into centrifuge tubes and spin at half speed for 10 min.
- remove the top viscous layer from the tubes with a dropper and transfer to a clean beaker.
- while stirring the solution with a glass rod, slowly add cold 95% ethanol. The initially jelly-like precipitate will become more fibrous and wind around the glass rod. These are collected fibers of DNA.

2. Electrophoresis

There are now a large number of kits available for molecular biology exercises involving electrophoretic separation of molecules. All come with detailed instructions and all the materials necessary to run the experiment.

3. Paper Cloning

This exercise demonstrates how to construct a recombinant DNA molecule, in this case a plasmid that encodes the functional *lac* operon. The *lac* operon consists of a promoter and three genes, all of which must be present for a cell to utilize lactose. Initially, you have two plasmids, pMC1403 which lacks the promoter and the start site of the first gene, and pUC18 that has the promoter and start site for gene one, but is missing the rest of the genes. The diagrams below indicate the portion of the *lac* operon present in each plasmid by a heavy line. The actual sequence of bases in the region of the arrow is listed on the next page. You are also given a choice of restriction enzymes. Ask students to choose the appropriate restriction enzymes to cut both plasmids and ligate the segments into a recombinant plasmid containing the entire *lac* operon.

```
                                               EcoRI  BamHI   SalI
        TTGACATTCGTTGGCCAGTGTGATATAATAACTATTAGGAGGAGAATATGGAATTCCGGGATCCGGTCGACGAGATTCCTATGGCCCGATATCG
pUC18   **rep**                      lac Promoter ►    \_\    \_\    \_\    ————Apr————
        AAGTGTAAGCAACCGGTCACACTATATTATTGATAATCCTCCTCTTATACCTTAAGGCCCTAGGCCAGCTGCTCTAAGGATACCGGGCTATAGC


                       EcoRI   BamHI                           EcoRI                SalI
        GATTCCTATGGCCCGATATCGTAGAATTCCCGGGGATCCGGGCATTCAGTTCATTAATCCCATGAATTCGACTTACCTCTTGAGGTCGACTATGGCCAG
pMC1403 ————Apr————           \_\    \_\    ————————lac Region—\_\————————       \_\   **rep**
        CTAAGGATACCGGGCTATAGCATCTTAAGGGCCCCTAGGCCCGTAAGTCAAGTAATTAGGGTACTTAAGCTGAATGGAGAACTCCAGCTGATACCGGTC
```

Of the more than 120 restriction enzymes commercially available, the following cut in the region of interest.

Restriction Enzyme	Sequence in DNA (site)	Ends of the DNA formed	
BamHI	...GGATCC... ...CCTAGG...	...G ...CCTAG	GATCC... G...
SalI	...GTCGAC... ...CAGCTG...	...G ...CAGCT	TCGAC... G...
EcoRI	...GAATTC... ...CTTAAG...	...G ...CTTAA	AATTC... G...
PstI	...CTGCAG... ...GACGTC...	...CTGCA ...G	G... ACGTC...

In this experiment, a pair of scissors represents your restriction enzymes. Students must find the appropriate sequence in the DNA and cut it to generate the proper ends. Remember, if there is more than one site on the DNA recognized by an enzyme, the enzyme will cut at all these sites. To ligate or join the ends of DNA molecules, T4 DNA ligase is used. In this experiment, a piece of scotch tape represents the DNA ligase.

Procedure:
1. Use a highlighter to color one of the plasmid sequences on the bottom of the previous page, then cut out both strips. Ligate (tape) each segment into a closed circle, the form in which plasmid DNA normally exists in nature.

2. Examine the pMC1403 sequence for restriction sites that can be used to excise the *lac* operon DNA from the plasmid. Do not use enzymes that cut within the *lac* DNA, otherwise the *lac*

Chapter 15

region will not be "cloned" efficiently. (Hint: you must use two different enzymes, one for each end of the *lac* region).

3. Using your restriction enzymes (scissors), cut the DNA from the plasmid pMC1403. Be sure to make staggered cuts to preserve the "sticky ends."

4. Cut the pUC18 vector with the same two enzymes used to excise the *lac* region from pMC1403. Using the same enzymes will ensure that the ends of the two segments of DNA are compatible.

5. Ligate (tape) the insert DNA (*lac* region of pMC1403) to the vector (pUC18). You should now have a recombinant DNA plasmid with the complete *lac* operon.

You could now introduce this recombinant DNA into a bacterial cell. If it is taken up by the bacterial DNA, it would transform the bacterium and the *lac* genes would be expressed.

LABORATORY - MATERIALS, EQUIPMENT, AND PREPARATION

1. DNA Isolation

A. Materials
-calf thymus (sold by butchers as "sweetbreads"). Order in advance and have frozen immediately - keep frozen until used.
-Citric Acid, Trisodium Salt
-Sodium Chloride
-Sodium Dodecylsulfate
-EDTA, Disodium Salt
-Chloroform
-Amyl Alcohol
-Ethanol (95%)
-conc. HCL

B. Equipment
-clinical centrifuge and centrifuge tubes
-blender
-hot water bath
-500-ml separatory funnel
-100-ml graduated cylinder
-250-ml beaker
-glass stirring rods

-droppers
-pH meter

C. Preparation
-buffer A. 2.9 g trisodium citrate plus 9 g sodium chloride / liter H_2O
-buffer B. 44.1 g trisodium citrate in 900 ml H_2O. Adjust to pH 7.0 with conc. HCL and bring to 1 liter.
-SDS (20% Sodium Dodecylsulfate). 10 g SDS in 40 ml H_2O, heat and stir to dissolve.
-chloroform: Amyl Alcohol (24:1). Add 40 ml amyl alcohol to 960 ml chloroform (store tightly sealed).

2. Electrophoresis
Sources:
CABISCO kits and supplies
    Carolina Biological Supply Co.
    2700 York Road
    Burlington, NC. 27215

EDVOTEK kits, supplies and workshops
    EDVOTEC, Inc.
    P.O. Box 1232
    West Bethesda, MD 20827-1232

GELTEACH kits and supplies
    Wards Natural Science
    5100 West Henrietta Road
    P.O. Box 92912
    Rochester, NY 14692-9021

3. Paper Cloning
-scissors
-transparent tape
-copy of DNA sequences for the two plasmids

REFERENCES

General:
Watson, J.D. 1968. *The Double Helix*. Atheneum, New York. 143 p. This is a fascinating account, by one of the principle players, of the discovery of the structure of DNA.

Chapter 15

Watson, J.D. and F.H.C. Crick. 1953. Molecular structure of nucleic acids: a structure for deoxyribose nucleic acid. *Nature* 171:737. This one-page paper initiated molecular biology as a discipline. It can be read and understood by nonmajors.

Resources:

Ausubel, F.M., R. Brent, R.E.Kingston, D.D. Moore, J.G. Seidman, J.A. Smith, and K.Struhl. 1989. *Short Protocols in Molecular Biology: A Compendium of Methods from Current Protocols in Molecular Biology.* John Wiley Sons, New York. 387 p. A sourcebook of molecular biological techniques.

Watson, J.D. 1986. *Molecular Biology of the Gene, 4th ed.* Benjamin/Cummings, Menlo Park, CA. 494 p. The classic text in the field.

The entire October 1985 issue of *Scientific American* is devoted to "The molecules of life.

CHAPTER 16: GENETICS

Genetics will be either one of the most interesting or one of the more difficult topics for your students. Genetic traits and inheritance patterns fascinate students, but there are several conceptual stumbling blocks. In addition, a significant number of students will have math anxiety to one degree or another. Unfortunately, it is difficult to avoid numbers when dealing with genetics. The most important thing to remember, especially at the beginning, is to move slowly and repeat frequently.

OBJECTIVES

1. Describe the difference between genes and alleles, dominant and recessive, genotypes and phenotypes.
2. Understand the relationship between meiosis, gamete formation, and the genotypes of offspring.
3. Work simple Mendelian inheritance problems involving complete dominance in monohybrid and dihybrid crosses.
4. Understand the role of probability in inheritance.
5. Understand common variations on Mendelian themes: incomplete dominance, test cross, linkage, crossing over, multiple alleles, and multiple genes.

CONCEPTS

The concepts of dominance and recessiveness will cause problems if not defined initially. The implication of these terms for many students is that the dominant allele is somehow stronger and therefore must be better, whereas recessive implies weaker and poorer. The concepts of gene and allele are themselves troublesome. Gene is particularly so because we define it differently in different situations. In the previous chapter we used gene in a narrow sense as a segment of DNA which codes for a particular polypeptide. In our current context a gene is a segment of the chromosome which is responsible for a particular phenetic trait. It may actually contain several molecular genes where each codes for a different enzyme in a pathway where the ultimate product is responsible for the phenotypic trait. A gene, in this broader usage, is the information on a particular segment of one chromosome. The information on homologous chromosomes is a pair of genes. Each form of a gene is an allele. The symbols AA in a problem represent a pair of genes on homologous chromosomes, not a pair of alleles.

Ask your students what process is ultimately responsible for forming the gametes used in making crosses. The majority, after some considerable thought, and probably some coaxing as well, will answer meiosis. This is intuitively obvious when dealing with monohybrid examples. But when you switch to a dihybrid cross, it will become clear to you that most students do not really understand how gene pair segregation relates to meiosis.. Once this relationship is clear, the concepts of segregation, independent assortment, linkage, crossing over, deletions, insertions and mapping of chromosomes will be much easier to understand.

Chapter 16

A very difficult concept for most students is that of probability. Numbers are difficult anyway, even if they are concrete. The idea of probability is even more so. Punnett squares imply very concrete results, as do phenotypic and genotypic ratios. When confronted with a problem where the numbers do not exactly fit the model, many students immediately bog down.

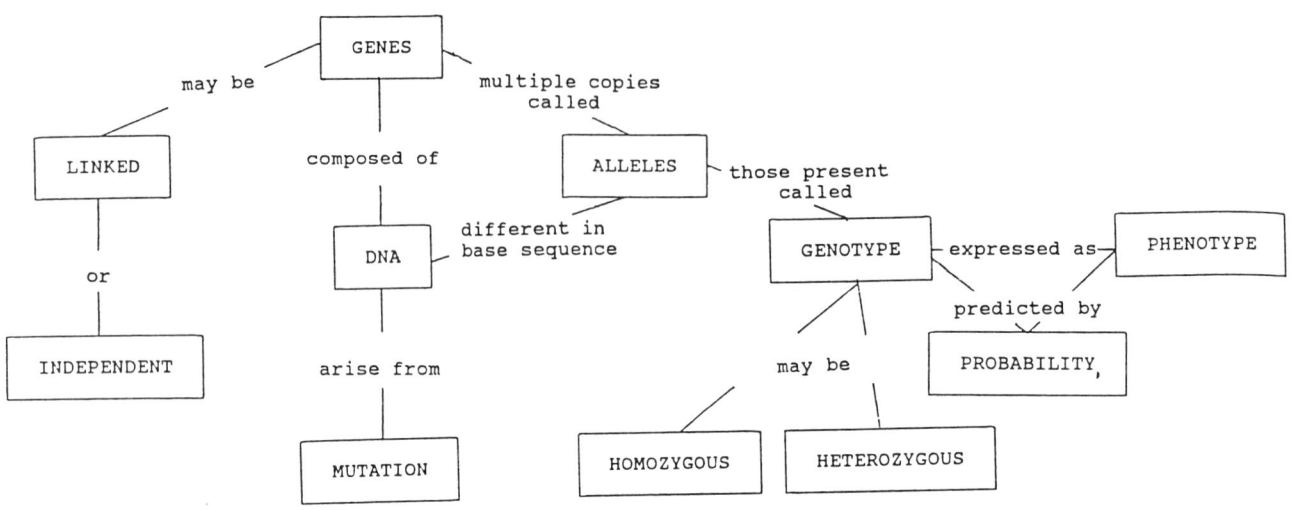

INSTRUCTIONAL AIDS

Lecture - I find it useful to begin a discussion of genetics with an example right from Mendel - a cross between an inbred dominant and an inbred recessive pea. How did Mendel have inbreds to begin with? Peas self-fertilize, so many different inbred lines were available to him in the mid 1800s. The fact that one trait "disappeared" in the first daughter generation but "reappeared" in the next suggested dominance of one over the other. The consistent 3:1 ratio suggested that a pair of factors (his term) must be involved Mendel was trained in mathematics and the fact that he could control inheritance by controlling pollination suggested that these factors must be in the pollen and pistil. With these two crosses on the board, you can introduce the appropriate descriptive terms, P1, F1, F2, gene, gene pair, alleles, genotype, phenotype, homozygous, heterozygous, dominant, and recessive.

The students are at a big advantage over Mendel, of course, because they already know the mechanisms responsible for moving genes from parent to offspring - chromosomes and meiosis. If you have a diagram of the fate of chromosomes during meiosis on the board, from a pre-lecture review, return to it now and add the appropriate genes. If not, repeat Mendel's crosses by diagramming the chromosomes carrying the respective genes.

Does this mean that every cross, produces four offsprings, one of which expresses the recessive genotype? If only one offspring results, probability makes a little sense, but what if three are produced? Introduce the concept of probability, and its implications, at this point with a class demonstration. Ask every student to find a partner and have partners raise their hands together. "Loners" quickly can identify partners. These partners are going to simulate sex - by each flipping one coin, one time. Represent heads by "H" and tails by "h." Each "parent" contributes one gamete with a 50:50 chance of contributing "H" or "h." Ask the class first, how many possible offspring there will be, and second, what proportion of each type so they expect. Now have them flip to produce their offspring. Ask one member of each pair who got "HH" to raise her/his hand - tally the results. Do the same for the other two genotypes. A calculator will allow you quickly to determine the percentages observed. This exercise provides you with data to discuss probability in general and also the effect of sample size on data reliability. It also provides an opportunity to introduce the idea that we can statistically test hypotheses by comparing observed values to expected values. This works especially well in classes of 200 - 400 students. It requires about five minutes to do this exercise in a large class.

A phenotypic ratio of 3:1 is easy to see in monohybrid crosses, but how can you differentiate between the homozygous dominant and heterozygous individuals? Set up a Punnett Square with all the information you know about the cross T? x tt. Most students will quickly see the alternative phenotypic ratios depending on whether "?" is T or t. An understanding of probability reinforces the value of the test cross by emphasizing that a you need only a small number of crosses to confirm a test. A single recessive offspring requires a recessive allele from both parents. The chances are 50:50 for each offspring if the unknown parent is heterozygous.

You are now ready for a dihybrid problem with independent assortment. Reinforce the role of meiosis in forming the spores (ultimately the gametes) by again diagramming the appropriate genes on separate pairs of homologous chromosomes and following them through meiosis. The time spent on this is well worthwhile, as forming the appropriate gametes is the single biggest problem students have in working genetics problems. Your students can simulate this problem by having each person in a pair flip two coins; pennies may represent one pair of genes while nickels or dimes represent the other pair. Class results will very closely resemble the expected 9:3:3:1 ratio.

Once you are convinced that your students understand simple Mendelian inheritance, you are ready to move on to some not-so-Mendelian examples, such as incomplete dominance and linkage. The time spent on the role of meiosis in determining possible gametes enhances student understanding of linkage. Crossing over, as a normal occurrence during meiosis, is a natural lead-in to mapping of genes on the chromosome.

Determining the frequency distribution of student heights in your class is a good way to introduce multiple genes. Put a table on the board with height values ranging from < 5'0",

Chapter 16

5'0" to 6'4" in 2" intervals, and > 6'4", then tally the number of students falling into each size class. In large classes, you will see a fairly normal bell curve, in small classes it may be a bimodal curve (males, females). How to account for this continuous sort of variation? - At least one student will suggest more than one gene with additive affects.

Pre-lecture Review - In the previous chapter (Ch 15), you discussed the concept of gene in terms of DNA coding for a particular polypeptide. Review this concept, then make the distinction between that narrower definition and the broader concept of gene used in this chapter. You will also want to review the chromosomal rearrangements that occur during meiosis. Diagram one pair of homologous chromosomes on the board, then take them through the stages of division. Leave this diagram on the board so you can come back to it later and insert letter symbols for genes to illustrate the chromosomal basis of inheritance.

Post-lecture Review - The single biggest problem students have with genetics problems is determining the appropriate gametes that will be produced by the parents. Give a trihybrid take-home problem asking the students to take a parental set of three pairs of homologous chromosomes, each with a different heterozygous pair of genes, through the process of meiosis to produce one possible set of spores. A quick spot-check of their results, looking for independent assortment, will tell you if they understand the process.

LABORATORY - EXERCISES

1. Probability

The coin-flipping simulations, described above, may also be done as a laboratory exercise. With the additional time, however, you also can go into how we can test if an observed ratio is close enough to the expected ratio to accept it as supporting a hypothesis. For instance, we expect a 3:1 phenotypic ratio in the F2 of a monohybrid cross with complete dominance. After doing an experiment we produce a total of 40 offspring with the ratio of 27:13 - not exactly 3:1, but close. Is it close enough?

To answer this question we determine the "goodness of fit" of the observed values to the expected values by employing a **Chi-square** (X^2) **Test**. The general formula for chi-square is:

$$X^2 = \frac{(O-E)^2}{E}$$

where O is the observed frequency and E is the expected frequency. In the above example the observed values are 27 and 13 while the corresponding expected values are 30 (3/4 of 40) and 10 (1/4 of 40). Thus the value of chi-square is:

$$X^2 = \frac{(27-30)^2}{30} + \frac{(13-10)^2}{10} = 1.2$$

Note that the chi-square calculation uses the actual frequencies, not the percentages or proportions.

Our hypothesis for the Chi-square test is that the observed values are not significantly different from the expected values. This is called the null hypothesis. The larger the difference between the observed and expected values, the larger the resultant Chi-square, and the lower the probability that the null hypothesis is correct. The value of Chi-square above which we no longer accept the null hypothesis is called the critical value. The appropriate critical value depends on the degrees of freedom (number of categories - 1) and the level of significance (α) at which we want to test (usually 0.05 in biological experiments). These values are displayed in the table of critical values below.

Degrees of Freedom	$\alpha = 0.05$	$\alpha = 0.10$
1	3.84	6.63
2	5.99	9.21
3	7.81	11.34
4	9.49	13.28

In the above example there are two categories, therefore degrees of freedom = 2 - 1 = 1. At $\alpha = 0.05$, the critical value is 3.84. The calculated value of Chi-square, 1.2, is considerably smaller than the critical value; therefore, there is no significant difference between what we observed and what we expected. If, however, our calculated Chi-square was 3.84 or larger, we would have to reject our hypothesis that what we observed was a 3:1 ratio.

2. Hybrid Corn

In this exercise students attempt to determine the mode of inheritance of kernel shape and color on the ears of corn provided. First, study the demonstrations of monohybrid and dihybrid crosses for the characters colored or colorless aleurone and starchy or sweet endosperm.

What color is dominant, purple or yellow?

What shape is dominant, smooth (starchy) or sweet (wrinkled)?

The unknown ears will have one or both of these genes determining the phenotype. The first step is to decide if the unknown is a monohybrid or dihybrid. Next, determine the

Chapter 16

phenotypic ratios, i.e., how many of each type are present and what proportion of the total is that. Count the number of each phenotype and calculate the overall phenotypic ratios. It should be close to either a 3:1 ratio (monohybrid) or 9:3:3:1 ratio (dihybrid). Are the observed ratios close enough to the expected to support your hypothesis. Again employ the Chi-square test. Remember, the observed are the actual number of counts for each phenotype, and the expected are the expected proportion of the total counts.

3. Performing Crosses

There are several genetic mutants of Wisconsin Fast Plants *Brassica* available that you can use to do actual breeding experiments in the classroom. The time required is comparable to that required for performing crosses with fruit flies, but the system is easier to work with (no anesthetization), the results are easier to see and score and, of course, they're plants!

4. Genetics Problems

An assortment of genetics problems is available in any biology textbook and most laboratory manuals. Start with simple problems involving gamete formation, then some straightforward crosses before moving on to more elaborate word problems.

LABORATORY - MATERIALS, EQUIPMENT, AND PREPARATION

1. Probability
 -each student should bring coins, at least 1 penny and 1 "silver" coin
 -hand calculator

2. Hybrid Corn
 -demonstration ears illustrating yellow/purple, smooth/wrinkled
 -genetic corn ears illustrating either monohybrid or dihybrid crosses of the above traits. A variety of genetic corn ears are available from most biological supply houses (note: Other crosses may be used. One of my favorites is purple/yellow kernels in a 9:7 ratio where two gene pairs are involved; at least one of each dominant gene is required for purple color, the other [3:3:1] combinations with at most one dominant gene of the pair are recessive.)

3. Performing Crosses
 -Wisconsin Fast Plants, kits and mutant seeds available from Carolina Biological Supply.
 -at least one fluorescent light fixture is required

4. Genetics Problems
 -a variety of problems are available in most biology textbooks and laboratory manuals.

REFERENCES

General:

Keller, Evelyn Fox. 1983. *A feeling for the Organism: The Life and Work of Barbara McClintock.* W.H. Freeman, San Francisco. 235 p. McClintock won the Nobel Prize for discovering that genes in corn "jump around" on the chromosomes. It took awhile for genetics to accept this - and recognize the widespread occurrence of this phenomenon.

Gonick, Larry and Mark Whellis. 1983. *The Cartoon Guide to Genetics.* Barnes and Noble Books. New York. 215 p. "Learning has never been so easy or so much fun."

Mendel, Gregor. 1967. *Experiments in Plant Hybridization.* Harvard Univ. Press, Cambridge, 41p. This is a translation of Mendel's actual paper originally published in: *Verh. naturf. Ver. in Brunn.* Abhandlungen, iv. 1865.

Resources:

Goodenough, Ursula. 1984. *Genetics, 3rd ed.* Saunders College Publishing. Philadelphia. A good general text.

Sundberg, M. D. 1989. *Biology for Science Majors, 2nd ed.* Burgess International Group, Inc. Edina, Minn. 168 p. Contains an elaboration of the chi-square analysis of coin tossing and genetic corn experiments as well as a selection of problems.

CHAPTER 17: POPULATION GENETICS AND EVOLUTION

In a sense, this is the keystone chapter of the entire course. Since the very first chapter, the text emphasizes the selective advantage of various adaptations. The concept of a structure or process being advantageous is intuitively clear, given enough examples. In fact, Darwin devoted much of the "Origin of Species" to presenting such examples. The problem Darwin and Wallace faced, and they realized this, is that they did not have an inheritance mechanism to account for passing these selected traits from generation to generation.

At first, genetics was more of a hindrance than a help. Rather than providing for gradual change, Mendelian genetics accounted for inheritance of discrete characters, in constant and predictable ratios. Only after biologists understood the role of mutations did the reconciliation between these two powerful theories take place. Today the strongest evidence in support of evolution comes from the field of genetics.

OBJECTIVES

1. Describe the roles of overproduction of offspring and variation among individuals in natural selection.
2. Explain what is meant by the term gene pool and describe how equilibrium is maintained in a population.
3. Describe the conditions that must be met for a population to remain at equilibrium.
4. Describe the various pathways that may be followed during the process of speciation.

CONCEPTS

If your students were starting with a "blank slate," the major concepts of evolution would be rather clear and straightforward. Unfortunately, you will not have this luxury. There are more powerful misconceptions held by the general public about evolution and evolutionary theory than about any other area of biology. Your task is a very difficult one, and one for which educational psychologists have not come up with a reliable solution; how to change misconceptions that are already firmly embedded in someone's understanding.

Many of these misconceptions center around the idea of natural selection. We unconsciously reinforce many of these with careless or "sloppy" explanations and examples. Selection acts on <u>EXISTING</u> variation! Purpose, intent, improvement, etc. are NOT involved. Furthermore, although selection does act directly on the phenotype of an individual, the individual does not evolve; <u>POPULATIONS EVOLVE, INDIVIDUALS DO NOT!</u> Evolution by means of natural selection drives changes in the gene pool of a population. Thus, we can define evolution as changes in the <u>gene pool</u> of a species.

There are several pathways that may result in speciation. Unfortunately, the least common, phyletic speciation, is the one that most people think of - - incorrectly. The "Man evolved from monkeys" syndrome is surprisingly widespread and is based on a misconception about phyletic speciation. This is one we reinforce with misleading illustrations such as the straight-line evolution of the horse frequently pictured in textbooks. All the diverging branches, leading to extinction are omitted). In phyletic speciation, the ancestral species gradually changes until a new species is recognized and the ancestral species no longer exists. Both humans and monkey (as well as other living primates) are extant; therefore, one cannot be ancestral to the other. Instead, they have diverged from common ancestors.

Most of the remaining misconceptions are in one way or another connected to religious beliefs. The most important is to address the meaning of "theory." Creationists take advantage of the public's colloquial understanding of the word "theory," equivalent to a scientists use of "hypothesis," which is the usage familiar to most students. In science, of course, "theory" has a much more definitive meaning. The "Theory of Evolution" is as strongly supported as the "Cell Theory"! Another, though less widely held, misconception is that to accept the theory of evolution you must reject belief in God. An interesting note in this regard is that Wallace, who was an atheist, refused to accept that natural selection also applied to humans. Darwin, whose training was in theology, insisted that the human species was no different in this respect from any other species.

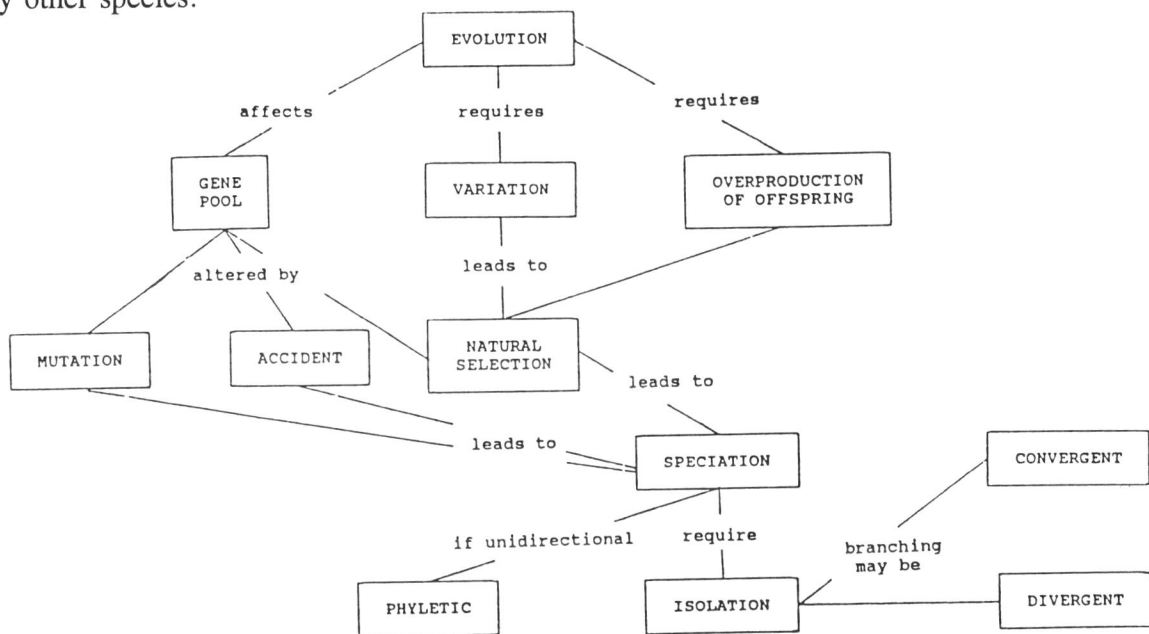

What if you happen to have a vocal fundamentalist in your class? The most important thing is NOT to ridicule or "put down" the student! In the first place, every student has a right to their opinion, and the right to express it, so long as s/he does not disrupt the class. If the person is overly persistent, your class will probably be behind you in wanting the person to quiet down. If you humiliate her/him, however, you may very likely loose a significant portion of

Chapter 17

your class. The best situation is if you can listen with sensitivity and suggest that the student come by your office to discuss points of mutual misunderstanding.

INSTRUCTIONAL AIDS

Lecture - I have a strong preference for beginning a unit on evolution by providing a historical background to the development of the mechanism of natural selection. This was an age of empire building and expansion that had a profound effect on European thinking. This is especially true in terms of the natural world. From the standpoint of biological diversity, Europe is a depauperate region so it is no wonder that the fixity of species, and origin by divine creation of those few species, was so readily accepted. European explorers brought back innumerable strange and wonderful organisms . As more "new" species were sent to Europe, especially from the tropics, the concept of fixed species became more blurred.

The biology texts cover the discoveries on the Galapagos Islands well, but equally important to the development of Darwin's thinking were two observations made earlier on the voyage. One was the discovery of giant fossils in Patagonia that were obviously related to extant species Darwin saw previously in the Amazon. The second was an earthquake and the discovery of mollusk shells in the Andes of Chile. The shells were similar to those he had been collecting in the harbor just days earlier, yet here they were high above sea level. The earthquake had just raised one side of a fault higher above sea level than the other. These, and many other observations and impressions that affected his thinking come to light in *The Voyage of the Beagle*.

Use an island situation, like the Galapagos, to illustrate the Hardy-Weinberg equilibrium. Suppose this hypothetical island had a total population of five individuals, each heterozygous for some trait, e.g., $\underline{Aa}$. There are a total of 10 genes in the island gene pool; what is the frequency of each allele? The frequency of $\underline{A}$ is 0.5 while that of $\underline{a}$ is also 0.5. Now ask what the expected frequency would be on the island if you came back in 100 yrs. Students' initial inclination will be that the frequencies must change because the genotype and phenotype frequencies may change from one generation to another. But have them look at the individual genes themselves. A gene must be either $\underline{A}$ or $\underline{a}$, so the frequency of gene $\underline{a}$ must be 1.0 - frequency of gene $\underline{A}$. This is your lead-in to the Hardy-Weinberg equations.

After working through the allele frequencies and genotype frequencies for the population that started off with all heterozygous individuals, move to another example where the initial frequencies are different, e.g., $\underline{A} = 0.8$, $\underline{a} = 0.2$. In spite of having just worked through an example where the frequencies did not change from one generation to another, a significant number of your students will still predict that there will be change in this situation. Work it through, as before, to reinforce the idea that gene frequencies will not change, in and of themselves, from one generation to another.

If the Hardy-Weinberg equilibrium prediction of no change is true, then evolution is not occurring. But in order for the equilibrium to be maintained, the population must meet several conditions. Hardy-Weinberg assumes that these conditions are met, but they rarely are. The population must be large and breeding must be completely random. There can be no migration, in or out of the population, and mutations cannot occur. Finally, there can be no selection. If any of these conditions are not met, then the gene frequencies will change from one generation to the next. This is evolution.

Pre-lecture Review - Begin by reviewing a dihybrid problem in simple Mendelian genetics. The offspring will always be similar to at least one parent, or perhaps a grandparent, and the phenotypes are distinct and predictable. There is no reason to believe that, over time, some differences will appear which could ultimately lead to the formation of a new species. This, of course, was the dilemma faced by biologists during the first decades of this century. Genetics was so logical, with testable predictions Furthermore, these predictions did not seem to support the idea of gradual change as proposed by the evolutionists.

Post-lecture Review - Concept map! There are so many misconceptions about evolution that it's a wonder so many students as do can answer questions on exams correctly. Several published studies show that frequently students will memorize the proper answer for an exam, but in follow-up discussion retain the same misconceptions they brought into the class. The best way to find out what your students are really thinking is to get them to map their thoughts

LABORATORY - EXERCISES

1. Population Growth

Darwin and Wallace realized that most organisms produced far many more offspring than will survive to maturity and reproduce. This, along with the variation between individuals, is a requirement for natural selection to be a mechanism for evolutionary change. When grown under optimal conditions, bacteria may divide by binary fission as rapidly as every 20 min. Because of this rapid rate of increase, however, optimal conditions are not maintained for very long. We can estimate increase in number of bacteria in a culture by measuring the turbidity of the culture medium over a period of time.

Each group will require two culture tubes. With sterile pipettes, transfer 5 ml of sterile nutrient broth to one tube and 5.5 ml to the other. Inoculate the first tube with 0.5 ml of stock bacterial culture and mix gently but completely. Set a spectrophotometer to 500nm, zero the transmittance, then use the second, control, tube to zero absorbance. Record the initial reading of the inoculated tube, then place both tubes in a 37C incubator. Repeat the absorbance measurements at 30 min. intervals for the rest of the laboratory period and periodically for the next 24 hr. The first few readings will show little change, but exponential growth will be evident within four hours, and within 24 hours, growth will have peaked.

Chapter 17

2. Diversity

Reassemble the same collection of materials used the first day. They were initially chosen to force students to make careful observations, but of course they were "trick" examples to demonstrate that things aren't always what they seem. These "tricks" were examples of convergence or divergence. Now the class can begin to examine the specimens more closely with some background in evolution. Divide the class into research teams and assign each team a set of three specimens. Examine both the external morphology and the anatomy (use hand sections). In each specimen, students should look for differences from a "typical" plant, and hypothesize about the kind of selection responsible for this change. In many of the examples, vegetative structures are modified to become similar is appearance. Reproductive structures, and internal structures, however, are less subject to change. The result is convergence. For instance, several families contain succulent species, but the relative proportion of cortex vs pith will vary in each.

3. Population Genetics

The Hardy-Weinberg Equilibrium is a useful tool for studying changes in the gene pool of populations. It states that, given certain conditions, the relative frequencies of the dominant and recessive alleles, represented by p and q respectively, will remain constant. Furthermore, p^2, 2pq, and q^2 will estimate the frequencies of the homozygous dominant, heterozygous and homozygous recessive genotypes in the population. We use the Hardy-Weinberg Equilibrium to make predictions about the frequency of alleles in a population, to determine the rate and direction of evolution in a population, and to direct studies towards understanding the cause of evolutionary change.

A. Estimating Allele Frequency

Determine the frequencies of alleles and genotypes in a planting of green:albino genetic corn. Albinism is a recessive trait that is inherited following simple Mendelian rules. Plant the seeds (they are F2 hybrids) in enough pots that you can distribute one to each laboratory group. Each group should count the total number of normal and albino plants in its sample. Students should then use the Hardy-Weinberg equations to calculate the allele and genotype frequencies. Albino plants must be homozygous recessive; their genotype frequency is q^2. The recessive allele frequency is the square root of this number, q, and p = 1-q.

Pool the individual group data to obtain class data and repeat the calculations. Although one or two groups may get the expected 1:2:1 genotypic ratio, most groups will not. It is probable that one or two groups will not observe any recessives in their sample!. The pooled data, however, will give a reasonable approximation. Make two points here: 1) the effect of sample size on reliability of an estimate, 2) how genes may be lost completely in small populations.

B. Genetic Drift, the Founder Principle

Genetic drift is the random change of allele frequencies in a breeding population that may have profound effects if the population is small. When a small subpopulation splinters off from a larger population, genes that were rare in the original population frequently are over represented or under represented (or even lost) in the subset. You can demonstrate this phenomenon with seeds, e.g., white soybean and green peas.

Provide each group with a dish containing 360 of one type of seed and 40 of the other. The two types of seeds represent the dominant and recessive alleles respectively. Calculate the frequency of each allele (0.9 and 0.1 respectively). Randomly select 20 seeds from the original dish and place them in a petri dish half. This is the founder population. Each group should calculate the allele frequencies in its founder population and put the results on the board. In some, if not all, of the founder populations, the allele frequencies will differ from the original population.

C. Gene Flow, Migration

Gene flow, the movement of alleles from one gene pool to another, is the result of emigration or immigration of individuals. If gene flow occurs randomly between two populations, the allele frequencies will become more and more similar.

Match student groups into paired teams. Each group will use its founder population from the above exercise. Have one person from group one randomly choose a seed from group two's dish and move it to dish one. A person from group two will select in the same way from dish one. Repeat this random transfer for a total of 10 exchanges. Recount the seeds in each dish and calculate new allele frequencies. Repeat this process three more times. Then have each group pair report its results to the class.

D. Selection

For natural selection to occur, one phenotype must have a selective advantage over another. Students tend to think that the phenotype controlled by a dominant gene must be the one with selective advantage, but this is not necessarily so.

Provide each student group with a swatch of carpet remnant, either green or white. Each group designates one person to be a "predator". The predator does not watch as the rest of the group scatters seeds from the previous exercise over the carpet. The predator is then given 5 sec. to remove as many seeds as possible, one seed at a time, and put them into a new petri dish half. After this selection event, recalculate the allele frequencies. Seeds on contrasting colors will tend to be removed at a greater rate, regardless of their original frequency.

LABORATORY - MATERIALS, EQUIPMENT, AND PREPARATION

1. Population Growth
-per class
- -37C incubator
- -100-ml stock culture of *E. coli*. The culture should be less than 24 hr old to ensure that the growth rate will be maximal.
- -one spectrophotometer for every four student groups

-per student group
- -50-ml sterile nutrient broth
- -2 sterile, capped test tubes to fit spectrophotometer (mark orientation line on each to be able to reproducibly orient tubes at each reading)
- -5-ml and 10-ml sterile pipettes (at least 2 of each)
- -pipetter or rubber bulb
- -bunsen burner to flame sterilize tube

2. Diversity
-the same plant materials used in part 1 of lab 1.

3. Population Genetics
-per student group
- -dish containing 360 seeds of one color and 40 seeds of another. The seeds should be of similar size and shape. For instance, white soybean and green pea or Great Northern bean and pinto bean.
- -petri dish, two halves will be used separately
- -calculator
- -swatch of medium pile carpet remnant, ca 1 yd.^{2} Two colors should be available, one similar to each of the seed types. Half the student groups should receive one color, the others use the second.

REFERENCES

General:
Darwin, Charles. 1962. *The Voyage of the Beagle*. edited by Leonard Engel. Doubleday and Co., New York. 524 p. An annotated version of Darwin's last (1860) revision, published with the cooperation of the American Museum of Natural History.

Dawkins, R. 1976. *The Selfish Gene.* Oxford University Press, New York. 224 p. A provocative interpretation of the genetic basis of natural selection.

Dobzhansky, T. 1973. Nothing in Biology Makes Sense Except in the Light of Evolution. *Amer. Biol. Teach.* 35:125-129. The title has become a classic statement: by one of the founders of the "modern synthesis."

Lawrence, J. and R.E. Lee. 1955. *Inherit the Wind.* The authors. Bantam Books, New York. 115 p. A play based on the Scopes "Monkey Trial," which was made into a movie.

Resources:

American Society of Zoologists. 1984. *Science as a Way of Knowing - Evolutionary Biology.* *American Zoologist* 24(2). This is the first of a series of symposium volumes dedicated to single topics of concern to biology educators. Moore's contribution of teaching evolutionary biology is particularly relevant.

Ayala, F.J. 1976. *Molecular Evolution.* Sinauer Associates, Sunderland, Mass. 277 p. A well written presentation of the most modern evidence supporting evolutionary theory.

Armstrong, J. E. 1989. *The Problems with Biology: Laboratory Exercises in Introductory Biology.* Stipes Publishing Co., Champaign, IL. This manual contains a number of innovative exercises emphasizing concept and process.

Futuyma, D. J. 1986. *Evolutionary Biology.* Sinauer Associates. Sunderland, Mass. 600 p. A good general introduction to evolutionary biology.

Stebbins, G.L. 1950. *Variation and Evolution in Plants.* Columbia University Press, New York. 643 p. The classic botanically oriented contribution to the "Modern Synthesis."

Sundberg, M.D. 1989. *Biology for Science Majors, 2nd ed.* Burgess International Group, Edina, Minn. 169 p. Contains detailed instructions for the probability exercises.

CHAPTER 18: CLASSIFICATION

Please pass the corn. To most Americans there would be no question as to what you are requesting. If you were sitting at a table of English speaking people from other parts of the world, however, you may be in for a surprise. The Scottish would pass what we call oats, people from Ireland or England might pass wheat, while in other English-speaking countries, someone might pass you whatever is the most common cereal (Heiser, 1973). Most Americans would pass you Indian corn or maize. Ask for *Zea mays*, however, and a botanist from anywhere on earth would pass you the same thing. Naming organisms was the earliest function of taxonomy, the branch of botany dealing with classification. Today, however, the major concern is to identify evolutionary relationships so the names and hierarchical classification employed reflect the natural phylogeny of plants.

OBJECTIVES

1. Describe the difference between an artificial and natural classification system; explain the advantages and disadvantages of each.
2. Explain what is meant by a hierarchical system of classification and list the levels of classification of the Plant Kingdom.
3. Describe the difference between analogy and homology.
4. Describe the kinds of information useful in generating classifications, and the advantages and disadvantages of each.

CONCEPTS

It is intuitively clear that the more alike two plants are, the more closely related they must be. If in a group of three plants, two share 90 % of all observable features, but each shares only 5% with the third, then the first two must be more closely related to each other than either is to the third. This is the basis of phylogenetic classifications. At least in theory, if you gather enough data, you should be able to form an accurate phylogenetic classification. Problems arise when similarities arise from convergence (analogies) instead of relationship (homologies). Frequently, the same data can be, and is, interpreted in different ways by different people. Thus, two botanists working on the same group, and given the same data, could generate slightly different classifications because of their different interpretations of the data. Other problems occur if a plant shares an equal number of characters with two or more different plants. Are some characters more important to consider than others? Should you weight characters to reflect their importance? Personal philosophy, thus, also plays a role in the construction of classifications. Probably the biggest conceptual problem for most undergraduates is to realize that although there is only one true phylogeny, we do not have one correct phylogenetic classification system that accurately represents that phylogeny. Some students may be troubled by this lack of consensus, but as long as you inform them that, for this course, you will accept the system proposed by Mauseth, they will be happy.

The concepts of analogy vs homology actually present few problems to students at this level. Some good examples like succulent plants and the wing of a bird vs the wing of an insect are convincing evidence for convergent evolution in response to a similar selection pressure. The difficulty with these concepts comes in actually recognizing one or the other in a group being studied.

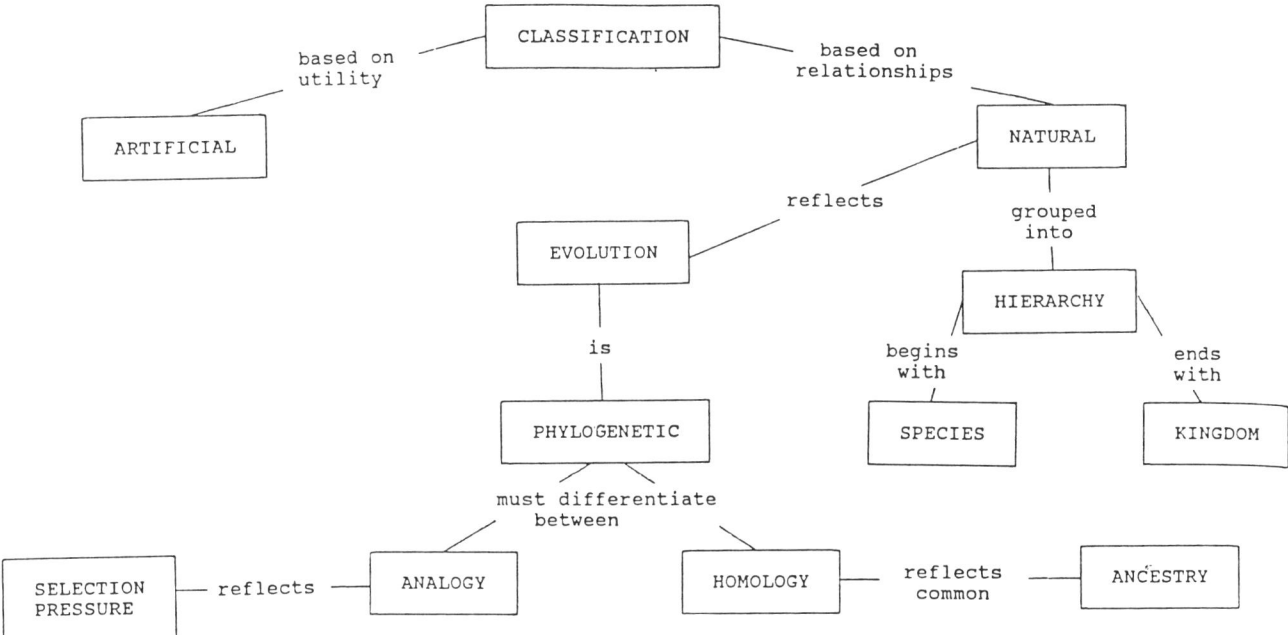

INSTRUCTIONAL AIDS

Lecture - Put an overhead containing a collection of twelve abstract geometric figures on the board. Some will be variations of squares, some of circles, and some of triangles. Some may have one or more lines extending outward, some may contain multiple shapes, some large, some small, etc. Have an identification label (e.g., letter of the alphabet) for each figure. Explain to the class that each person is responsible for assigning these figures to the proper "groups." It doesn't matter how many groups there are or how many members are in each group, but students must be prepared to justify their grouping. Once they decide on the proper grouping, students should represent this by listing the appropriate letters together in respective groups. Give them 5 min. to work on their classification. While they are doing this, circulate among the class and look for some widely discrepant classification schemes so you can ask three or four people to put their schemes on the board.

Once your student examples are on the board, take a vote of the class. How many of you had exactly the same grouping as James? As Joan?, etc. Depending on your class size, you may have ten or more people generate the same classification, or it may be just the person who put it on the board. How come you have so many different classifications? You all have exactly

Chapter 18

the same data! A brief discussion will reveal that differing philosophies concerning what factors are important also have a major influence on how one forms groups. Philosophy determines the rules for grouping into a classification.

Now go back to the most "popular" classification scheme on the board and ask the author to explain the basis for her grouping. What was the philosophy behind this grouping? Once your "expert" explains her criteria to the class, place three new figures on the board. Tell the class that they should accept the criteria just explained as the best consensus and they should use them to place the last three figures into the appropriate groups.

After a minute or two, ask the "authority" (the person who put the original scheme on the board) what the "correct" grouping is. Does anyone have something different? How can that be - you had the same data AND the same rules! The final variable is interpretation of the rules. This example illustrates exactly the problem botanists have in forming a classification of plants. Even if every systematist is given exactly the same data and sets out to devise a phylogenetic classification based on evolutionary principles, each resulting classification will vary in some ways from every other one.

You are now ready to look at classification of plants in particular. The example illustrated above is an artificial classification based on geometric similarities. The first plant classifications were also artificial, but served a useful purpose as some modern artificial classifications do. Mauseth gives some good examples. In systematics, however, we attempt to show natural relationships based on evolutionary principles - a phylogenetic system. The resulting classification should thus reflect common ancestry. This is done by using a hierarchical system of classification. Introduce the taxa and their relationship to each other. An analogy can be drawn to an extended postal address for a specific person. For instance, surname is analogous to genus, street address is to family, city is to class, state is to order, country is to division, and continent is to kingdom.

Pre-lecture Review - Systematists base phylogenetic classifications on the concept that extant plants diverged from common ancestors. Closely related species have a relatively recent common ancestor, while the common ancestor of greatly different species occurred far in the past. The idea of evolutionary divergence is the key to understanding systematic relationships. It is also one of the most misunderstood concepts in evolutionary theory.

Post-lecture Review - At the end of the text chapter, Mauseth presents a key to 39 common supermarket fruits and vegetables, including their scientific name. Assign each plant to a portion of the class, e.g., all the A's do *Daucus carota*, who must then determine the appropriate taxa names for each level of classification. Their may bring their data to lab or to the next class period where you can construct a phylogenetic tree of the edible fruits and vegetables based on shared taxa.

LABORATORY - EXERCISES

1. Artificial Classification - Constructing Keys

A key is an artificial classification presenting a series of steps that allows you to identify an unknown plant. Keys are usually dichotomous, that is, two contrasting descriptions are given and you choose the one that most closely fits your unknown. Your students may already have used a key to tree species in Chapter 6, and Mauseth presents an excellent key to common grocery store plants in the box at the end of this chapter.

Students have an easier time using a key if they have already made one of their own. Ask each student to make a dichotomous key to the other members of her/his group using characters that are easily visible. An example might be something like this:

A.	Blond hair			
	B.	Blue eyes		-Nels
	BB.	Brown eyes		-Mary
AA.	Hair not blond			
	C.	Brown hair		
		D.	Big feet	-Sam
		DD.	Small feet	-Judy
	CC.	Black hair		-Maria

After most people in each group have completed their classification, ask for two volunteers from each group. One person is the "unknown," the other provides a key. Use the student's key to assist the class in identifying the "unknown" person. In the process, you will help the students identify ambiguities in their keys, such as what is a big foot vs a small foot, and in their thinking.

2. Constructing a Cladogram

Phylogenetics, the classification of organisms to show natural relationship, depends on three assumptions: 1) evolution occurs; 2) evolution is monophyletic, lineages derive from a common ancestor; and 3) characters passed from generation to generation are either modified or not. Although phylogenetics concerns genealogical relationships, the latter cannot be observed. Rather, we must inferred these relationships from observable characters such as morphology, physiology, or biochemistry. Cladistics is a quantitative method that attempts to recover evolutionary relationships, based on observable characters, and present the resulting phylogeny in the form of a tree called a cladogram.

Chapter 18

Present each student group with a copy of the hypothetical organisms sketched below. Ask them to make a list of as many different morphological characters (character states) as they can identify.

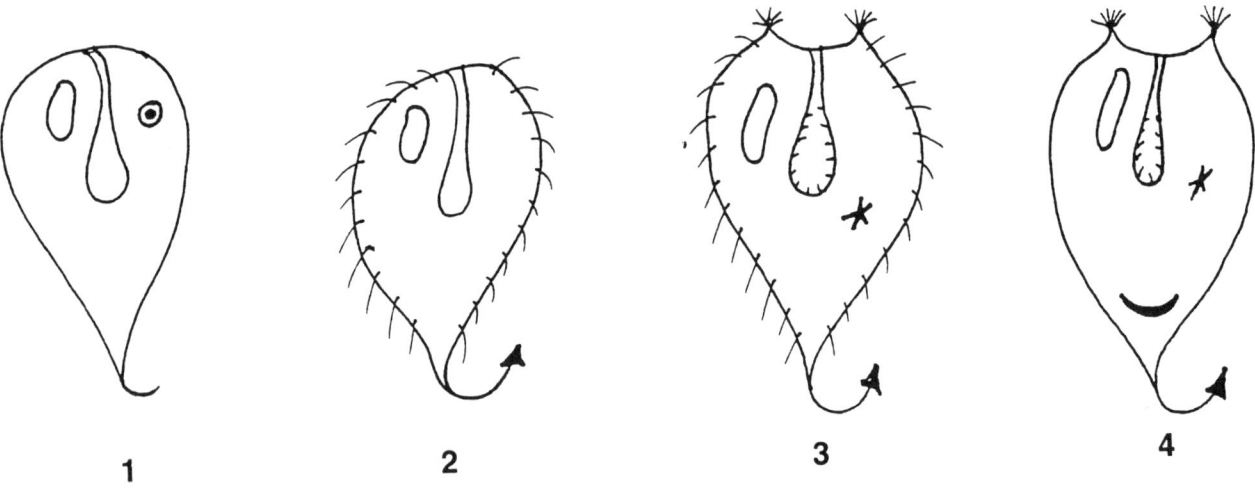

Once they have made their list, they will be ready to construct a data matrix. On one axis, list the identity of the organisms; on the other axis, list the different characters. To fill in the data matrix, students will have to assign a numerical value to each character for each organism. Begin by assigning a score of 0 to each character state of the outgroup species, that species that seems "outside of" the natural group to be classified but that appears to have more relictual features. One character at a time, compare the character state of each organism with that of the outgroup. If there has been no change, the character also scores as 0. The first time that character changes in a new species, score the new state 1. Score all other species with the same character state 1 also. If a species has a second change (a third state for that character), score the new trait 2. Repeat this procedure for each character in the matrix.

Character State	Taxon 1	Taxon 2	Taxon 3	Taxon 4
	0			
	0			
	0			
	0			
	0			

A simple approach to analyzing the data to construct a tree is to use the Distance Wagner Tree Method. Like any good tree, a cladogram must have a root, the origin from which the branches of the tree grow. We use the outgroup to root the cladogram. Add taxa sequentially to provide the most parsimonious (simplest) tree. To begin, connect the outgroup to the next two taxa as shown below. In parentheses, put the character state values for that species from the data matrix.

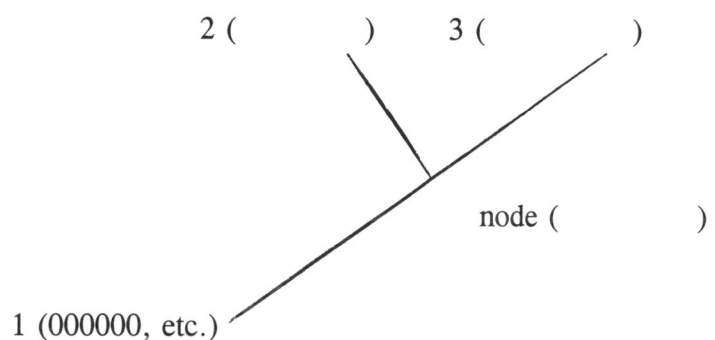

We call the three-taxon statement a "Wagner Neighborhood" with the taxa joined at a node. This node represents the condition of a hypothetical ancestor of both branches. Determine the character states for the node. If there are only two character states for a character (it is binary), the node characteristic is defined as the majority state of the three taxa in the neighborhood. For instance, if the outgroup and species 2 are both scored 0 for the first character, but species 3 is scored 1, then the node must be 0. If a character is multi-state, the node characteristic is the median state of the neighborhood. For instance, if the first character was 0, 1, 1 for the first character, the nodal character state would be 1.

There are three positions where the fourth species could be added to this tree, indicated by the three? in the diagram above. Construct three new cladograms adding the new species in each of the possible positions. Compute the nodal characteristics for each possibility. Examine

Chapter 18

the three possible cladograms. The preferred one provides character state values for a unique node and a minimum number of changes between the new node and the added species.

Add additional species in the same way, but for each new species added, the number of possible trees that must be evaluated increases rapidly. Similarly, with many character states to examine, the task quickly becomes too laborious to work by hand. Widespread application of numerical taxonomic techniques did not occur until computers became generally available.

Like the geometric-shapes exercise suggested for lecture, in this exercise you may get more than one "correct" phylogeny. The discrepancies will again be due to differences in philosophy and interpretation. For instance, are the "cilia" coming off the two horns in "4" a new trait, or simply a different character state of the "cilia" covering the body of "3"?

3. Constructing a Fern Phylogeny

You can apply the same procedure as above directly to studies of plant taxa. Ferns provide a readily available taxon with enough easily observed traits to produce a meaningful analysis during the course of a laboratory period. Obtain samples of four different fern species with sori present. Divide the class into four groups and assign each group one species to examine in detail. Ask each group to set up a demonstration showing each of the character states examined and to summarize the condition expressed for each character on a tally card left with the demonstration. Have students examine the following characters: shoot symmetry (radial or dorsiventral); stipe hairs (simple, branched, or scale-like); stele type (protostele, solenostele, or dictyostele); leaf traces (single, two, many); leaf venation (dichotomous, reticulate); sorus arrangement (discrete, linear, massed); sorus protection (naked, marginal [false indusium], indusiate); sequence of sporangium development (simultaneous, basipetal, random); sporangia per sorus (few, many); and sporangial stalk (short and thick, long and thin).

After each group has completed its character analysis and set up demonstrations, along with the tally card, each group rotates to every other demonstration. They should check the demonstrations against the information on the tally card to see if they agree with the authority group's interpretation. Students use the information gathered at each station to construct a data matrix for each of the characters of each species. Once students complete the data matrix, construct a "Wagner Tree" as outlined in part 2 above. Details for this procedure are cited in the references.

LABORATORY - MATERIALS, EQUIPMENT, AND PREPARATION

1. Artificial Classification - Constructing Keys
-no special equipment or supplies are required to perform this exercise. A hand lens will be useful when applying keys such as the tree keys in the appendix or Mauseth's key to grocery store plants.

2. Constructing a Cladogram
-each student should be given a copy of the organisms sketched above.

3. Constructing a Fern Phylogeny
for each student group
-compound and dissecting microscopes
-dissecting needles
-razor blades
-microscope slides and cover glasses
-phloroglucinol (to differentially stain xylem strands in vascular bundles)
-note cards
-fern specimen (at least one from each of the following groups)
- *Adiantum, Lygodium, Pteris*
- *Cyrtomium, Dryopteris, Nephrolepis, Phyllitis, Pteridium*
- *Polypodium, Platycerium*

REFERENCES

General:
Bailey, L.H. 1963. *How Plants Get Their Names*. Dover Publications, Inc. New York. 181 p. A delightful introduction to the logic behind the nomenclatural system used in botany, reprinted from the original 1933 edition.

Heiser, Charles B., Jr. 1973. *Seed to Civilization: The Story of Man's Food*. W.H. Freeman, San Francisco. 243 p. A concise description of the origins of agriculture and our attempts at providing enough food to sustain human population growth.

Heslop-Harrison, J. 1967. *New Concepts in Flowering-Plant Taxonomy*. Harvard University Press, Cambridge. 134 p. A concise statement of the influence of genetics, ecology, phytogeography, cytology, and experimental studies on taxonomy.

Sokal, R.R. 1966. Numerical Taxonomy. *Scientific American* 106:

Resources:
Brooks, D., J. Caira, T. Platt, and M Pritchard. 1984. *Principles and methods of phylogenetic systematics: a cladistics workbook*. University of Kansas Press, Lawrence, Kansas. 92 p. The basics of setting up and working a cladistic analysis.

Cronquist, A. 1968. *The Evolution and Classification of Flowering Plants*. Houghton Mifflin, Boston. 396 p. This is an earlier synopsis of the much larger work cited by Mauseth in the text, which is much more undergraduate friendly.

Chapter 18

Sundberg, M.D. 1984. An Evolutionary Approach to Teaching About Ferns in a Plant Kingdom Course. in: C. Leon Harris (ed.) *Tested Studies for Laboratory Teaching, Vol 4*. Kendall Hunt, Dubuque, Iowa. pp43-52. A non-cladistic approach to fern phylogeny which deals with many of the same concepts.

Sundberg, M.D. 1989. *Biology for Science Majors, 2nd ed*. Burgess International Group. Edina, Minn. 169 p. Contains detailed instructions for the cladistic analysis of fern phylogeny.

Wiley, E.O. 1981. *Phylogenetics: The Theory and Practice of Phylogenetic Systematics*. Wiley - Interscience, New York. 439 p. A technical treatment of the subject.

CHAPTER 19: KINGDOM MONERA

Inclusion of bacteria, and fungi and algae as well, in botany courses is a reflection of the tradition deriving from a two-kingdom classification. If an organism is not an animal, then it must be a plant. This classification has been obsolete for decades, yet the tradition remains. Including the Kingdom Monera in a botanical survey is especially striking because of the fundamental difference between bacteria and all other forms of life. The single word prokaryotic implies a host of characteristics separating this kingdom from all others.

OBJECTIVES

1. Describe the fundamental differences between prokaryotic and eukaryotic cells.
2. Describe the method of asexual reproduction and the mechanism of genetic information transfer in bacterial cells.
3. Distinguish between cyanobacteria and the other eubacteria on the basis of structure and physiology.
4. Describe the ecological and evolutionary significance of eubacteria, including cyanobacteria.

CONCEPTS

The major concept in this chapter is that of a prokaryotic cell. Eukaryotic cell structure is so well ingrained in students' minds that it is sometimes difficult for them to accept that there are viable alternatives. As the name pro- (before) karyo (nucleus) suggests, the obvious feature is lack of a membrane bound nucleus. With no nucleus there can be neither mitosis nor meiosis, two types of nuclear division. As alternative mechanisms, binary fission and conjugation are relatively straightforward and easily understood by students. Transformation and transduction, however, are more difficult, perhaps because they involve only segments of DNA, or perhaps because there is no physical connection between the cells involved. Lack of a nucleus correlates with small size and lack of other membrane-bound organelles.

The blue-greens, as a group, have evolved several features suggesting major trends in evolution. Cells tend to remain joined in groups with specialization of some cells occurring in many taxa. This is a step toward multicellularity. Cell size and internal structure suggests a trend toward the eukaryotic condition. Heterocysts, and especially akinetes, approach the dimensions of eukaryotic cells, and the photosynthetic pigments occur on an internal thylakoid membrane, rather than in the plasma membrane as in other photosynthetic bacteria.

Probably the most difficult concept to get across to students is that the vast majority of bacteria are beneficial. All of your students will know that bacteria cause illness and disease - at least in animals. Many will not realize there are also bacterial pathogens of plants. They will have no idea of the ubiquity of the group nor of the essential nature of bacteria in decomposition, nitrogen fixation, or nitrogen cycling.

Chapter 19

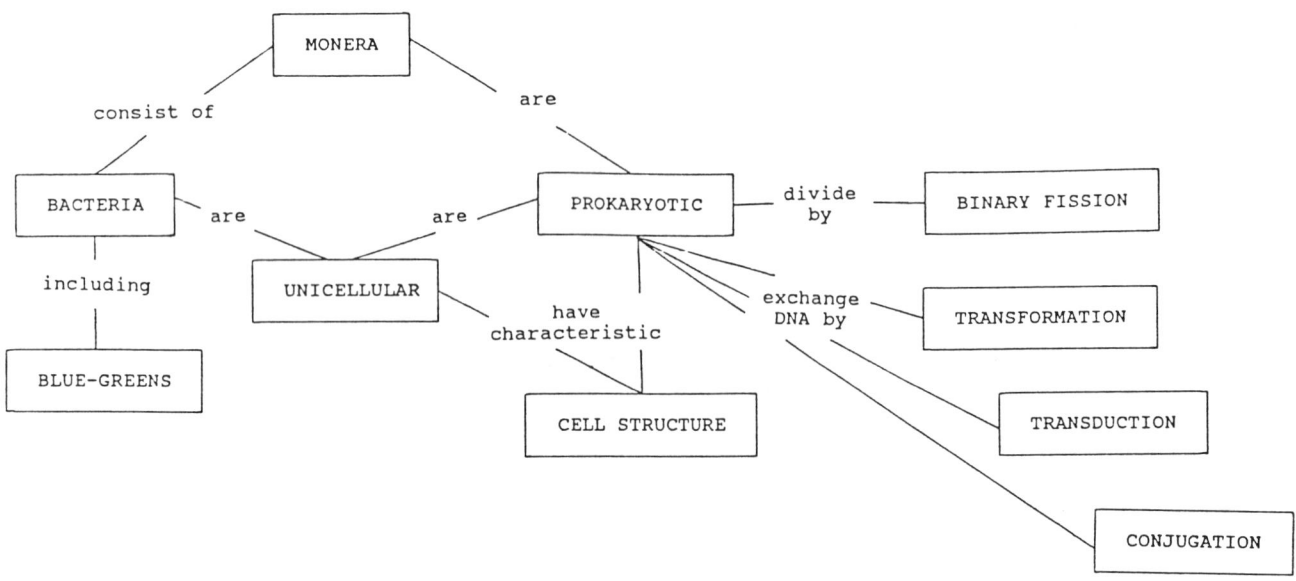

INSTRUCTIONAL AIDS

Lecture - To pique interest, begin with a series of "believe it or not" statements such as: living bacteria have been isolated from 10,000-year-old ice, from near boiling hot springs, from beneath the soil in Death Valley, and from the air more than 5 miles up in the atmosphere; the mass of bacteria on earth is more than that of ALL other organisms combined; your skin and hair is virtually covered with a layer of bacteria, etc. This will emphasize the ubiquity of bacteria and allow you to immediately address the misconception that bacteria are exclusively disease-causing organisms.

The most distinguishing characteristic of bacteria is their prokaryotic nature. Have your students recall that one of the characteristics of life is that living things are composed of cells. Then ask for recognizable features or characteristics of cells and list these in a column on the board. Be sure that processes such as mitosis, meiosis, respiration, and reproduction are listed as well as physical structures. Then, in a parallel column list the corresponding features of prokaryotic cells. This list can then serve as an outline for the remainder of your lecture(s) on the Kingdom Monera. If a particular structure or process is not necessary for a cell to be alive, but the function usually associated with that structure or process is essential, then the bacterial cell must have some other mechanism for carrying out the function.

Lack of a nucleus is particularly striking and is associated with a number of other characteristics. Students will already be familiar with the bacterial chromosome from Chapter 15, but it will be worthwhile to review this structure and the mechanism of replication in association with asexual reproduction by binary fission. Similarly, previous coverage of the techniques of genetic engineering will help students understand and differentiate between the various means of "sexual" genetic transfer in bacteria: transformation, conjugation, and transduction. A set of overheads illustrating various stages of each of these processes will be extremely useful.

Morphology and anatomy of bacterial cells is straight-forward and does not pose major problems for students. Do not worry about the details of why gram + and gram - bacteria stain the way they do, but be sure your students understand there are consistent differences between these groups. These differences turn out to be indispensable for bacterial identification and thus control via antibiotics. Similarly, the details of metabolism are "trees in the forest."

You will want to spend some time differentiating between cyanobacteria and the other eubacteria. Evolutionarily, this group is important to illustrate several trends of evolution, including cell specialization and the formation of colonies. They are also the first group to produce O_2 as a byproduct of photosynthesis. Botanically, they are important because of their symbiotic associations with certain bryophytes, ferns, and cycads (as well as their role in the lichen symbiosis). Ecologically, they are important because of their ability to fix nitrogen and their association with algal blooms.

Pre-lecture Review - Briefly review the factors that systematists consider when formulating a classification: facts, philosophy, and interpretation. In general, the philosophy employed by botanists in trying to establish phylogenetic classifications is evolutionary theory. Ask your class to list some of the philosophical guidelines that scientists might use. One of the first that they will mention is that evolution moves from simple to complex. Of course this is not always the case. You can give some examples of exceptions, such as the reduction of the plant body in parasites, e.g., mistletoe, or reduction in flowers of many groups. Nevertheless, as a general trend, this guideline of increasing complexity is useful. It serves as the major guideline in surveying groups of organisms. A second philosophical guideline is parsimony. Given a choice between two pathways, one involving two steps and the other involving ten, the shorter one is more likely to have evolved. A third guideline is that evolution must build upon preexisting structures of processes.

Post-lecture Review - One of the most interesting aspects of the survey portion of a course is the economic importance different taxa. Unfortunately, this information is frequently omitted from lecture for lack of time. An alternative approach is to assign outside reports, focusing on the economic importance of a group, to the class. Make assignments a month or so in advance for presentation during the week you discuss a particular taxon. Alternatively, make the assignments at the time you cover a taxon, with presentations scheduled for later in the semester. Give students at least two weeks to research their topics and prepare presentations. Schedule presentations outside of class, with every student required to attend at least one or two presentations besides the one they will be giving. Most of these group presentations will be straightforward student reports, but if you encourage innovation, students will come up with a

Chapter 19

variety of interesting presentation formats. These may include: dinners, skits, broadcasts on campus radio, games, field trips, etc.

LABORATORY - EXERCISES

1. Bacterial Cell Shape

Prepared slides of bacteria to illustrate cocci, bacilli, and spirilla are readily available; however, it is a straightforward procedure to prepare fresh slides from pure cultures. Thoroughly clean a microscope slide with soap and water and allow it to dry (passing the slide quickly through a flame will speed this process). Using sterile technique, place a loop full of inoculant from a culture tube on the slide and spread it around in an area about the size of a dime. Allow the slide to air dry, then fix the bacteria by passing the slide through a flame, smear side up, several times. Cover the fixed smear with one of the following stains, crystal violet (30-60 sec.), methylene blue (2-3 min.), or safranin (1-2 min.), then gently rinse off the stain with water and blot dry with a paper towel. The slide is now ready to examine - no coverglass is necessary. Different members of each group can prepare slides of different organisms; share the slides so all students see examples of each characteristic type.

2. Motility in Bacteria

Many bacteria contain flagella and are motile, but this is difficult to determine from fixed material without special staining and oil immersion objectives. The simplest, and most dramatic, method is to observe a drop of sample using dark-field microscopy.

Add a drop of hay infusion, or other rich bacterial source, to a clean microscope slide and carefully add a cover slip. Obtain an initial focus using bright field, then insert a field stop (circular disk with an opaque center) in the filter holder and open both the iris and field diaphragms wide. Motile bacteria will be bright moving "specks" in a dark field.

3. Serial Dilution and Estimating Number of Bacteria

Many diseases transmitted by food or water are the result of fecal contamination. Among these are typhoid fever, cholera, dysentery, and food poisoning. The responsible bacteria enter the water and soon become diluted. Consequently they are difficult to detect. By sampling a water source, and spreading the sample on a place of nutrient medium under optimal conditions, each bacterium in the sample will grow to form a colony visible to the naked eye. If bacteria are abundant in the sample, however, the resulting colonies will overlap. To avoid this, have students make a series of dilution plates so that accurate counts may be made. Multiply the count by the dilution factor to determine the initial number of bacteria per unit volume of sample.

Obtain three uncontaminated nutrient agar plates and three sterile test tubes. Label the tubes undiluted, 1/10, 1/100 and the plates 1/10, 1/100 and 1/1000. Half fill the undiluted tube with sample water. With a sterile 10-ml pipette, transfer 9 ml of sterile distilled H_2O into the other two tubes. With a sterile 1-ml pipette, transfer 1 ml of undiluted solution to the 1/10 tube and mix. With a new sterile pipette transfer 1 ml of the 1/10 dilution to the 1/100 tube and mix. Using a clean sterile pipette each time, transfer: 0.1 ml from the undiluted tube to the 1/10 plate; 0.1 ml of the 1/10 tube to the 1/100 plate; and 0.1 ml of the 1/100 tube to the 1/1000 plate.

Dip the end of a bent glass rod into a small beaker of alcohol, then pass it through a flame and let the alcohol burn off. Allow the rod to cool, then spread the drop of inoculant around the surface of one plate. Resterilize the rod and repeat for the other two plates. Place the plates, upside down, in an incubator overnight or for two days at room temperature, then count the number of colonies visible on each plate. If more than 300 colonies are present, simply record the plate as solid. From the number of colonies at a known dilution, calculate the number of bacteria in the undiluted sample by multiplying the count by the appropriate dilution factor. Students should also estimate the number of different types of bacteria by the number of different colony types. Color, size, shape, and texture are distinguishing features. More than 200 bacteria / ml is considered potentially unsafe.

4. Cyanobacteria

Cyanobacteria have several easily observable characteristics. *Oscillatoria* exhibits a characteristic waving movement in fresh cultures and forms separation discs that divide the filamentous colonies into hormogonia. *Nostoc* forms bead-like filaments within a gelatinous matrix that often forms macroscopic balls - sometimes the size of marbles. Heterocysts are obvious in these filaments. Symbiotic blue-greens live in the thalli of *Azolla* and hornworts, as well as in the cortex of corraloid roots of cycads.

5. Optimum Growth Medium

A. *Streptococcus thermophilus* (Yogurt)

Bring 1 qt. milk to a boil, then set aside to cool to touch (ca. 37° C). Pour all but one cup into a clean quart jar; mix 1 tbsp. plain yogurt into the remaining cup of milk, then pour the mixture into the jar as well. Cover with foil and place in a 37°C incubator for 6 or more hours (longer incubations produce a more sour culture). The culture may be eaten plain or with fruit according to the taste of the individual.

B. *Leuconostoc* spp. and others.

Chapter 19

    Coarsely grate *Brassica oleraceae* var. capitata (cabbage) and mix well with 2 tsp. salt per pound of cabbage. Firmly pack mixture into a wide-mouth container, cover with a clean cloth, and add a weight to keep the plant material below the liquid surface. Set aside in a cool corner of the room (ca. 15°C) for at least 4 weeks. Periodically skim the surface of the medium. After fermentation period, boil mixture, uncovered, for 15 minutes and transfer to clean plates. You may want to add celery seed to boiling medium before plating. Many people prefer to add bratwurst or franks to the plates.

LABORATORY - MATERIALS, EQUIPMENT AND PREPARATION

1. Bacterial Cell Shape
    -compound student microscopes
    -demonstration microscopes with oil immersion lens (optional)
    -prepared slides of various bacteria (optional)
    -glass slides and coverslip
    -soap to wash slides
    -bacterial cultures, eg. *Escherichia coli, Bacillus megatherium, Streptococcus lactis,* and *Rhodospirillum rubrum.*
    -bunsen burner and igniter
    -transfer loop
    -0.5% solutions of: Crystal violet, methylene blue, safranin in dropper bottles
    -paper towels

2. Motility in Bacteria
    -compound student microscopes
    -hay infusion (boil
    -field stop (cut a clear plastic circle to fit the filter holder in a microscope. Remove the eyepiece from a microscope and observe the image of the field of view in the tube with the high power lens in place. Open the iris diaphragm until the field just fills the tube. Measure the diameter of the opening in the diameter - this is the correct diameter for your field stop. Make an opaque circle of this size in the very center of your clear plastic circle. To obtain dark field illumination, simply insert this field stop filter in the filter holder and open the iris diaphragm to its fullest.)
    -droppers

3. Serial Dilution and Estimating Number of Bacteria
per class
    -sample of river or pond water
    -37° incubator
per student group
    -3 sterile test tubes with caps
    -3 petri dishes with nutrient agar

-sterile pipettes (1 10-ml, 5 1-ml)
-bunsen burner
-bent glass rod
-covered beaker of alcohol
-wax pencils or marking pens

4. Cyanobacteria
-compound student microscopes
-cultures of various taxa, e.g., *Oscillatoria, Nostoc, Anabaena, Gleocapsa, Gleotrichia*, etc. (or prepared slides of various taxa)
-clean microscope slides and cover glasses

5. Optimum Growth Medium
A. *Streptococcus thermophilus*
-1 quart jar with lid (preferably wide-mouth)
-1 qt. milk
-1 tbsp. plain yogurt
-powdered milk (optional)
-37°C incubator

B. *Leuconostoc* spp. and others
-1 wide-mouth container, preferably crockery
-1 or more shoots of *Brassica oleraceae* var. capitata
-NaCl
-food grater

REFERENCES

General:

Andrews, M. 1977. *The Life that Lives on Man*. Taplinger Publishing Co., New York. 183 p. A popular account of the bacteria, viruses, fungi and animals that live on human skin.

de Kruif, Paul. 1954. *The Microbe Hunters*. Harcourt Brace Jovanovich, San Diego. 337 p. Originally published in 1926 this is a narrative history of the foundation of the science of microbiology.

Zinsser, H. 1934. *Rats, Lice and History*. Little, Brown and Co., Boston. 228 p. An informal history of typhus, its causes, vectors and social impact.

Resources:

The American Type Culture Collection Catalogue of Strains I, 14th ed. 1980. Rockville, MD. 648 p. A listing of all organisms in the collection, including algae, fungi, protozoa, viruses, and bacteria, with the media of choice for optimal growth.

Chapter 19

Difco Manual of Dehydrated Culture Media and Reagents for Microbiological and Clinical Laboratory Procedures. 9th ed. 1953. Difco Laboratories, Detroit. 350 p. An invaluable resource for media recipes and general information for growing bacteria and fungi.

Stanier, Roger Y, M. Doudoroff, and E.A. Adelberg. 1976. *The Microbial World, 4th ed.* Prentice-Hall, Englewood Cliffs, N.J. A classic text.

Sundberg, M.D. 1988. *General Botany, BOTY 1202.* Burgess International Group, Edina, Minn. 106. Introductory laboratory manual including more elaborate instructions for the suggested laboratory procedures.

CHAPTER 20: FUNGI

Contrary to student opinion, studying fungi can indeed be FUN. You can obtain fungi from just about anywhere: your refrigerator, shower, the woods, the grocery store, between your toes, etc. Unfortunately for students, the terminology used to describe fungi is enormous, and the structure and life cycles of typical (whatever that is) fungi are extremely variable and frequently complex. As a result, textbooks and chapters on fungi tend to be intimidating and turn students off. Yet fungi are extremely important for the world economy and in terms of social and historical change.

Fungi are wonderful experimental organisms for laboratory investigation. Specimens can be "trapped" in their environment or "isolated" from mixed cultures with comparative ease. Experimental conditions are manipulated easily to examine their effects on growth. The major disadvantage is the difficulty of forcing many species to form reproductive structures - the key characteristic for classifying to subdivision. Luckily there are other microscopic characters that correlate well and permit a good "educated guess."

OBJECTIVES

1. Describe the shared characteristics that unite the fungi, albeit in an artificial way, and separate them from the other kingdoms of living things.
2. Describe the shared characteristics of each of the five subdivisions of fungi and explain how you can recognize members of each subdivision on sight.
3. Describe the lichen symbiosis; describe the mycorrhizal association.
4. Explain the general ecological importance of fungi and give an example of the economic importance of one member of each subdivision.

CONCEPTS

There are several difficult concepts concerning fungi that you should address The first is more of a misconception that we reinforce merely by including fungi in the course - a problem addressed by Mauseth in the text. Fungal cells have walls and fungi produce spores, but these are not the strong lines of homology with plants that we once thought they were. As Mauseth points out, as we look ever more closely at the members of this group, it becomes clearer that the traditional fungi are an artificial grouping of organisms that have undergone convergent evolution.

The concept of extracellular digestion is foreign to most students and will require some explanation. They will be familiar with animal digestion, so use this as an analogy. Related to extracellular digestion is the method of growth. Tip growth is characteristic of fungal hyphae. When combined with extracellular digestion ahead of the growing tip, the hypha can penetrate through the substrate.

Chapter 20

Even more confusing will be the cellular organization of fungi. Prokaryotic cells are unusual, but unusual enough that this character alone can define an entire kingdom. The fungi, however, are eukaryotic, so students assume fungal cells must be like those of plants and animals. They must contain a large nucleus that undergoes either mitosis or meiosis. Unfortunately for students, this is not the case. First, students will encounter coenocytic cells, then dikaryotic cells. Even mitosis and meiosis are not typical. Even when cells have a single nucleus, it is usually haploid, not diploid, and the septum between cells in a hypha are perforate.

A final problem will be the various, and characteristic, ways that different groups of fungi eventually achieve syngamy. The structures involved: oogonia, zygospores, asci, and basidia, are so diagnostic that the subdivisions are named after them. If syngamy is absent in a group, or at' least has not been observed, then such absence is diagnostic as well and the fungus is classified as imperfect.

The final concepts deal with associations between various fungi and other groups of organisms, particularly photosynthetic ones. Lichens are the classic example of mutualistic symbiosis. Both the fungus and its algal partner benefit from the arrangement. Mycorrhizae are a much more cryptic, although ecologically and economically more important, arrangement whose widespread occurrence has only begun to be realized recently.

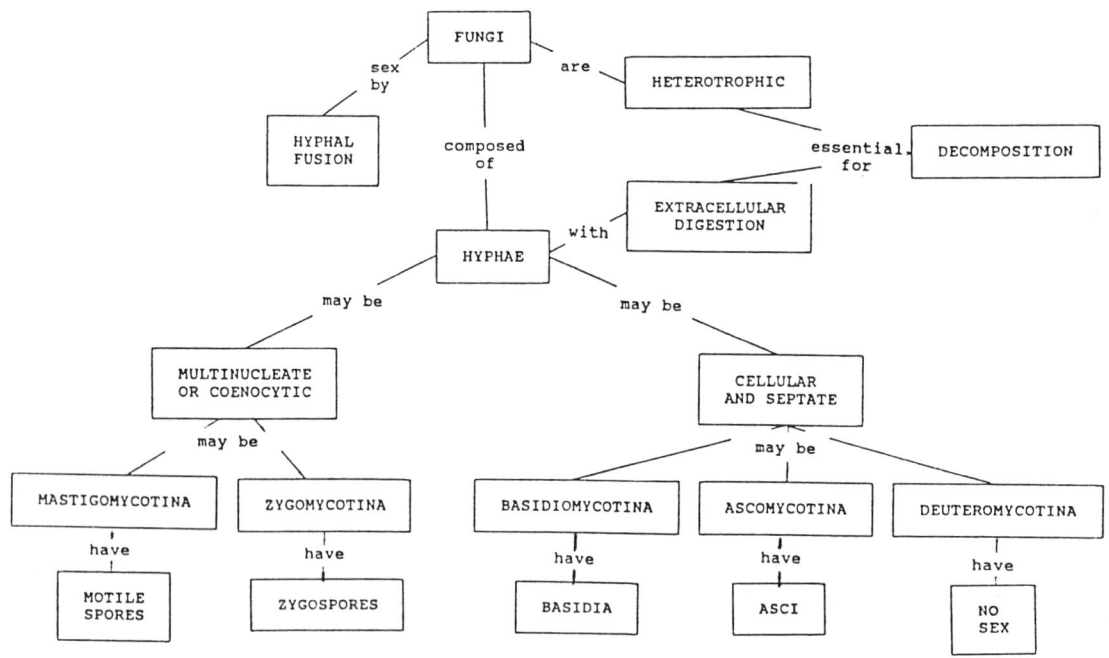

INSTRUCTIONAL AIDS

Fungi

Lecture - Begin by asking students how they can distinguish members of Kingdom Fungi from members of the other kingdoms. The easy discrimination will be to separate fungi from members of the Kingdom Monera, which you just finished studying. Monerans have prokaryotic cells, fungi will have eukaryotic cells. Great, but the same is true for plants, protists, and animals. How can you distinguish fungi from these? Some students may note the presence of cell walls distinguish fungi from animals and nonphotosynthetic distinguishes fungi from plants. A multicellular body separates fungi from protists. These are all fine, although there are exceptions to each, but they are all negative - a means of identifying fungi by process of elimination. Are there any positive characters that, if seen, would allow you to identify an unknown eukaryotic organism as a fungus?

One such character is extracellular digestion. This is a concept that is not immediately obvious, so begin with a common example. In animals the cells lining certain portions of the gut secrete digestive enzymes to hydrolyze ingested food. If those cells instead lined the surface of the body, near the mouth, it would be a situation similar to what is found in fungi. The enzymes would be released to the outside where they could hydrolyze any food materials present. The already digested "soup" could then be ingested. The major difference is that rather than having a multicellular organ involved, digestion is accomplished by excretion of enzymes by the single cell at the tip of a filament, the hyphal tip. This process also permits the growing hyphal tip to penetrate through otherwise firm substrate, e.g., wood. An effective illustration of the efficiency of this process is to bring in a piece of rotted log. Its weight, and strength, will be a fraction of what it was fresh, as fungi will have digested away significant amounts of biomass.

Filamentous growth, of sorts, was already seen in the cyanobacteria, so this should not be a problem. The problem will be in the kinds of cells making up these hyphal filaments. Although in most eukaryotic organisms one nucleus is usually sufficient per cell, fungi frequently exhibit other organizations. Coenocytic (multinucleate) hyphae characterize two primitive groups, the Oomycetes and Zygomycotina. Challenge the class to describe how such a situation could arise during the life cycle of a fungus. The solution is that mitosis occurs repeatedly without cytokinesis. There is obviously an advantage for cytokinesis and mitosis to occur in synchrony, but there is no reason that they must do so.

Likewise, when the cytoplasms of two cells fuse, there is no reason that the corresponding nuclei also must fuse. Cell fusion will automatically imply the process of fertilization of egg by sperm, where the two nuclei do fuse shortly after cytoplasmic fusion. In both the ascomycetes and basidiomycetes, there is a protracted period of dikaryotic hyphal growth where a binucleate cell will undergo a specialized form of cytokinesis with simultaneous mitosis by both nuclei. This tendency is much more pronounced in basidiomycetes where dikaryotic hyphae compose the entire fruiting body of a mushroom.

Chapter 20

Even when there is but a single nucleus, it is usually unlike that found in animals and most plants. Haploid, not diploid, cells make up the uninucleate hyphae in both the ascomycetes and basidiomycetes. If students learn now that it is equally possible for haploid cells or diploid cells to divide mitotically, thus increasing the number of cells in a tissue or population, life cycles will be much less confusing.

A careful examination of vegetative hyphae will often provide a clue about the subdivision to which an unknown fungus belongs. Coenocytic hyphae occur in oomycetes and zygomycetes; the former are mostly aquatic, the latter are mostly terrestrial. Septate hyphae characterize ascomycetes, basidiomycetes, and imperfect fungi. Most basidiomycetes have a complex pore structure associated with the septa which appears like a small circular dot in the light microscope. You can frequently identify the dikaryotic hyphae of basidiomycetes by a small bulge in the wall on one side of the septum, the clamp connection. The surest way to make a positive identification, however, is to examine reproductive features.

The oogonium is a spherical gametangium located laterally along the length of a vegetative hypha. Meiosis occurs within the oogonium to produce egg cells, which when fertilized may germinate to form coenocytic hyphae with diploid nuclei. Oogonia and cells with diploid nuclei are characteristic of oomycetes. In zygomycetes, hyphal tips of two compatible strains act directly as gametangia and fuse with each other. Crosswalls develop behind the fusing tips forming a coenocytic cell containing nuclei from two different strains. Syngamy occurs between pairs of complementary nuclei and the cell enlarges, ultimately forming a thick protective wall. This zygospore, characteristic of the zygomycetes, is large enough to be seen with the naked eye. Meiosis occurs at the time of germination so the hyphae contain haploid nuclei.

Karyogamy and plasmogamy are temporally separated in the ascomycetes and basidiomyctes. In the former, specialized cells develop in the base of a fruiting body to facilitate cytoplasmic fusion. These structures become multinucleate; following plasmogamy, compatible nuclei pair up but karyogamy does not occur. Instead, new dikaryotic hyphae grow out through the fruiting body. Ultimately, nuclear fusion will occur in the tip cell of each dikaryotic hypha forming a diploid ascus. Meiosis occurs (sometimes followed by a mitotic division) to produce four (or eight) ascospores within the ascus wall. In basidiomycetes, dikaryotic hyphae form directly from fusion of pairs of compatible haploid hyphae. The resulting dikaryotic cells proliferate to form a dikaryotic mycelium that ultimately produces a fruiting body. Again, the nuclei at the tips of dikaryotic hyphae fuse to produce a diploid cell, the basidium. Following meiosis, the nuclei migrate to outgrowths forming from the surface of the basidium. Thus, basidiospores form on the outside of the basidium. The absence of sexual structures is characteristic of imperfect fungi.

We now understand the concept of symbiosis to include a variety of associations between two organisms. The meaning understood by most students, however, is that classically observed in lichens. Although we can dissociate some lichens in the laboratory to produce pure cultures of either mycobiont (fungus) or phycobiont (alga), few are found in nature. Most occur only as lichen symbionts. Furthermore, lichens produce a variety of chemicals, lichen substances, that are characteristic of that "species" of lichen, but cannot be produced by either partner alone. Mycorrhizae are considerably more widespread than we thought. Virtually all orchids, oaks, and pines, as well as many other families, depend on mycorrhizal fungi to provide essential nutrients or assist in their uptake. The take-home message is that many, if not most, vascular plants depend upon mycorrhizal associations for nutrient uptake. If you collect seeds from an unusual plant on your vacation that you would like to grow at home, you also should collect some of the soil from that location to ensure that you have the mycorrhizal fungi as well!

Pre-lecture Review - Ask the class to list the corresponding features of prokaryotic and dikaryotic cells, then single out those characters related directly to the nucleus - presence of a nucleus, mitosis and meiosis. Sketch an outline of a cell with a nucleus containing a single pair of chromosomes on the board and challenge someone to diagram what that cell will look like following mitotic division. Invariably you will get two daughter cells, each with a single nucleus. Each nucleus may have a pair of chromosomes (if the volunteer understands mitosis) or a single chromosome (if the difference between mitosis and meiosis are understood). Depending on the results of this first example, either relabel it meiosis and have another student come up to put mitosis on the board, or ask for a volunteer to do likewise for meiosis. Now you diagram another alternative where no cytokinesis occurs for either mitosis or meiosis -then ask if this isn't equally possible? Students understand mitosis and meiosis to by synonymous with cell division - either asexual or sexual. We seldom make a clear distinction between karyokinesis and cytokinesis. It is essential to understand this distinction when dealing with fungi. All four of the situations you now have diagrammed on the board occur in the fungi!

Post-lecture Review - The fungi are the most FUN group for outside assignments on economic and historical importance. History majors can research the role of fungi in the Irish immigration to the U.S. or the probable role of fungal- contaminated flour on the Salem witch trials. Political scientists can examine the effect of fungi on the repeal of the Corn Laws and development of social welfare. Anthropologists can examine the role of fungi in religion and sociologists can look at the psychedelic drug culture of the 60s and 70s. English majors can analyze why Dostoyevsky had peasants standing outside the rear door of the house when aristocrats were having a party (according to Oliver Goldsmith, the intoxicating principle of certain wild mushrooms can pass through the bladders of several people without loosing its potency!). Of course there is also the billions of dollars worth of damage done annually by plant pathogens, the interesting human pathogens (and not just athlete's foot!), and other medicinal aspects, etc.

LABORATORY - EXERCISES

1. Fishing For Water Molds

Chapter 20

Water molds, oomycetes, can be isolated from almost any body of water. The surest method is to place one or two hemp seeds, cut in half, into a small container of water from a natural source. Within a week there should be a cottony mycelial mass surrounding each seed half. Swollen sporangia form at the tips of each hypha. These produce numerous flagellated spores. In a good culture you will be able to find sporangia releasing spores and the little "beasts" swimming all over. In another week spherical oogonia will be visible in older regions of the same hyphae.

2. Hunting Fungi

A standard medium of potato dextrose agar (PDA) is useful for growing many different types of fungi; the fact that they grow from the tips makes it relatively easy to isolate individual fungi from mixed cultures. Have students obtain their own sources of inoculum. Rotting wood and decaying leaves will frequently harbor slime molds as well as filamentous fungi. Spoiling fruit and vegetables exhibit symptomatic lesions and refrigerators frequently house vigorous fungal cultures. Place a small bit of material in the center of a petri dish of PDA and watch for hyphal or plasmodial growth. Several different organisms will probably grow out. To establish a pure culture, cut off the tip of a growing hypha with a sterile blade and transfer it, along with the chunk of agar it was growing on, to a new dish of PDA. Students repeat this procedure until they have a pure culture.

Squash a small piece of hypha and agar beneath a coverslip on a slide to provide some clues about the nature of the isolated unknown. A drop of Cotton Blue will stain the fungal cell walls. In many cases students can observe enough detail directly from the culture, without removing the lid, by placing the plate on the stage of the microscope. To use medium power it may be necessary to turn the plate over and focus through the agar. High power will not work. A typical class will isolate several different zygomycetes, ascomycetes and imperfect fungi. Basidiomycetes are rarely found unless a portion of fruiting body was used as the initial inoculum.

3. Fungal Growth

Each group should choose one of its cultures, or be provided with a standard culture, such as *Rhizopus*, to examine the effects of various environmental conditions on fungal growth. Temperature and light are the easiest variables to manipulate, but students can also control moisture and nutrient medium. The experiment is to determine the effect of some parameter on growth. Each student group will have to agree on the conditions they want to test and determine how the experiment should be set up. In all cases, the control can be room temperature, under ambient lighting, and on PDA. Inoculate each plate by placing a small piece of mycelium (or spores) in the center of the dish. Mark the center of the inoculum and seal with parafilm. Record the date and time. At periodic intervals over the next two or three days, measure and record the radial distance from the inoculation point to the edge of the mycelium. They should also note any qualitative differences in appearance. At the end of the experiment, data can be graphed and growth rates determined. There may also be qualitative differences such as color of the culture (usually associated with production of asexual spores), discoloration of the medium (due to build up of metabolic products) or periodic growth patterns.

4. Ecological Succession on Dung

Horse manure, if collected fresh (the fresher the better), frequently provides a succession of fungi in perfect order for study. Bring fresh manure to lab and placed in a humid environment - a bell jar with wet paper towels works well. In about a week, a thin mycelium should be visible covering the mass. Cover the bell jar with foil, except for a small hole to the top on one side. The first fungus to appear is *Pilobolus*, a zygomycete that is phototropic and has forceful spore discharge. The small, translucent sporangiophores have a swollen yellow tip beneath a black sporangium. When mature, the spores are shot towards the opening in the foil. You can judge the accuracy of this "artillery" when the foil is removed. The black spores will be sticking to the inside of the bell jar in the vicinity of the light opening. An alternative is to leave the bell jar uncovered but provide a unidirectional source of light. When spore release begins, remove the bell jar and place a sheet of white paper between the light source and the fungi (at least four feet of paper). The spores shoot in the direction of the light for various distances.

About a week after the *Pilobolus* show, cup-shaped apothecia of ascomycetes will appear. The velvet-like lining of these ascocarps contain asci which students may examine microscopically. In good cultures, a crop of delicate white mushrooms follows the ascomycetes by about one week.

5. Mushrooms

Mushrooms are one of the few fungal types readily recognized by most students. Depending on the season of the year, you can ask students to collect their own samples to bring into lab - otherwise you will have to depend on fresh mushrooms from the store and preserved specimens. Have students examine such external features as: general size, shape and color. On the cap there may or may not be scales or hairs. Portions of a membrane, the partial veil, which once covered the gills may remain attached to the edge of the cap and/or form a ring or annulus around the stalk or stipe. At the base of the stalk, there may or may not be a cup, or vulva. Gills or pores will be visible on the under surface of the cap. To make a spore print, remove the stalk and place the cap on a clean sheet of paper. If students place a clean microscope slide between the cap and the paper, they will be able to examine spores microscopically once the spore print is made. Place a small drop of water on top of the cap, and cover the cap and paper with a beaker or glass to maintain high humidity. After an hour or two (or preferably overnight), you will have a spore print outlining the gills and indicating

Chapter 20

the color of the spores. Students can make hand sections through the gills and examined microscopically to see basidia with basidiospores borne on tiny projections, the sterigmata.

LABORATORY - MATERIALS, EQUIPMENT, AND PREPARATION

1.   Fishing for Water Molds

-fresh water from a natural source
-hemp seeds, cut in half. You can purchase non-viable hemp seeds from
biological supply houses. Radish seed is a poor second choice.
-petri dishes or small jars
-microscope, slides and coverslip
-cultures should be "baited" with seeds one and two weeks prior to laboratory in order to see both asexual and sexual stages.

2. and 3.   Hunting Fungi and Fungal Growth
-sterile petri dishes containing potato dextrose agar
-metal scalpel
-covered test tube of alcohol and bunsen burner to flame sterilize scalpel for isolating and transferring growing fungal tips
-wax marker or marking pen
-ruler
-microscope, slides and cover glasses
-0.5% Cotton Blue or Evans Blue to stain hyphae

4.   Ecological Succession on Dung
-Fresh horse manure (must be fresh)
-container to maintain high humidity, e.g., bell jar
-aluminum foil to cover jar and or lamp to provide unidirectional light source

5.   Mushrooms
-fresh mushrooms, preferably field collected. The best way to preserve basidiocarps is to freeze-dry them. If a freeze drier is not available, place uncovered mushrooms in a freezer. Once frozen, sublimation will occur gradually. The time required to dry specimens will depend on their size - larger specimens collected this year may be ready next year. Smaller specimens will take a month or two.
-white paper for spore print (half-black, half-white will aid examination of white-spored prints. A clean microscope slide between the cap and paper will allow direct examination of spores from the print on glass.
-razor blade, microscope, slide and coverglass
-dropper bottle of water

REFERENCES

General:

Christensen, C.L. 1965. *The Molds and Man: An Introduction to the Fungi, 3rd ed*. University of Minnesota Press, Mpls, MN. 284 p. A popular treatment of mycology emphasizing their impact on humankind.

Large, E.C. 1962. *Advance of the Fungi*. Dover Publications, New York. 488 p. A popular history of plant pathology which provides a good introduction to the lives of fungi.

Miller, O.K., Jr. no date. *Mushrooms of North America*. E.P. Dutton and Co., New York. 360 p. A beautifully illustrated guide to common mushrooms.

Smith, A.H. 1973. *The Mushroom Hunter's Field Guide*. University of Michigan Press, Ann Arbor, MI. 264 p, 187 plates. Smith's is perhaps the best known guide to common fungi.

Resources:

Ahmadjian, V. 1967. *The Lichen Symbiosis*. Blaisdell publishing Co., Waltham, Mass. 152 p. The classic study of the symbiotic association between fungi and algae.

Alexopoulos, C.J. and C.W. Mims. 1979. *Introductory Mycology, 3rd ed.* John Wiley and Sons, New York. 632 p. This is the classic textbook of mycology with emphasis on taxonomic relationships.

Gray, W.D. 1959. *The Relation of Fungi to Human Affairs*. Henry Holt and Co., Inc. New York. 510 p. A wealth of information on both the beneficial and harmful activities of fungi.

CHAPTER 21: ALGAE

Linnaeus first used the term "algae" for one of his orders in the *Genera Plantarum*, 1754. The organisms he included in this group were mostly, liverworts, which we classify today as classified true plants. The first treatment of algae as aquatic plants lacking stems and leaves was by de Jussieu in 1789. In this century it has become increasingly clear that the anatomy and physiology of individual cells, not the organization of the body as a whole, should be the basis of classification. The term "algae" has now been relegated to the layperson's list of terms designating simple, aquatic plant-like organisms. The unicellular forms were among the first organisms included in the Kingdom Protista when this classification was first proposed. Today, many workers include even the macroscopic forms in this kingdom.

OBJECTIVES

1. Describe the endosymbiotic theory of the origin of the eukaryotic cell.
2. Describe the characteristic features of each of the algal divisions.
3. Explain the evolutionary significance of the Division Chlorophyta with respect of the origin of land plants.
4. Describe the ecological and economic importance of each of the algal divisions.

CONCEPTS

Based on the coverage of evolutionary theory in Chapters 17 and 18, it is logical to suppose that eukaryotic cells arose through gradual modification of prokaryotic cells in a step-by-step fashion. Indeed, as recently as the early 1970s, most botanists accepted this autogenous theory. The endosymbiotic theory, when first proposed, seemed preposterous to many, but the supporting evidence quickly converted most skeptics. Your students will also be skeptical! Be prepared to meet this skepticism with a battery of supporting evidence.

With six divisions to cover, there is a tendency simply to present information, catalog style. Use several unifying concepts to organize your presentation and facilitate comparisons between taxa. General body plan varies from consistently unicellular to consistently multicellular with every stage in between. Certain cytological and physiological characteristics, cell division, pigments, storage products, wall composition, are extremely useful in algal classification. Conceptually, the most difficult comparison will be life cycles. Chapter 9 covered the basic plant life cycle. It will be useful to review this before moving on. Algae exhibit virtually every variation of the generalized quadripartite life cycle, most of which are "foreign" to students experience and understanding.

In the next chapter, Mauseth covers the transition to land. Certain features are characteristic of all land plants and were, therefore, probably present in the common ancestor. This ancestor was a green alga. Be conscious of this connection and lay the ground work when discussing the features of the Division Chlorophyta.

Algae

We frequently overlook the ecological significance of algae. Over half of the total photosynthesis on earth occurs in the seas; the single most important division of photosynthetic organisms is the Chryosphyta! These algae not only form the base of aquatic food webs, but are also significant for CO_2 and O_2 cycling. Similarly, most people under appreciate the direct economic importance of algae. Billion dollar industries depend on the phycocolloids agar, algin, and carrageenin, extracted from members of the Divisions Phaeophyta and Rhodophyta.

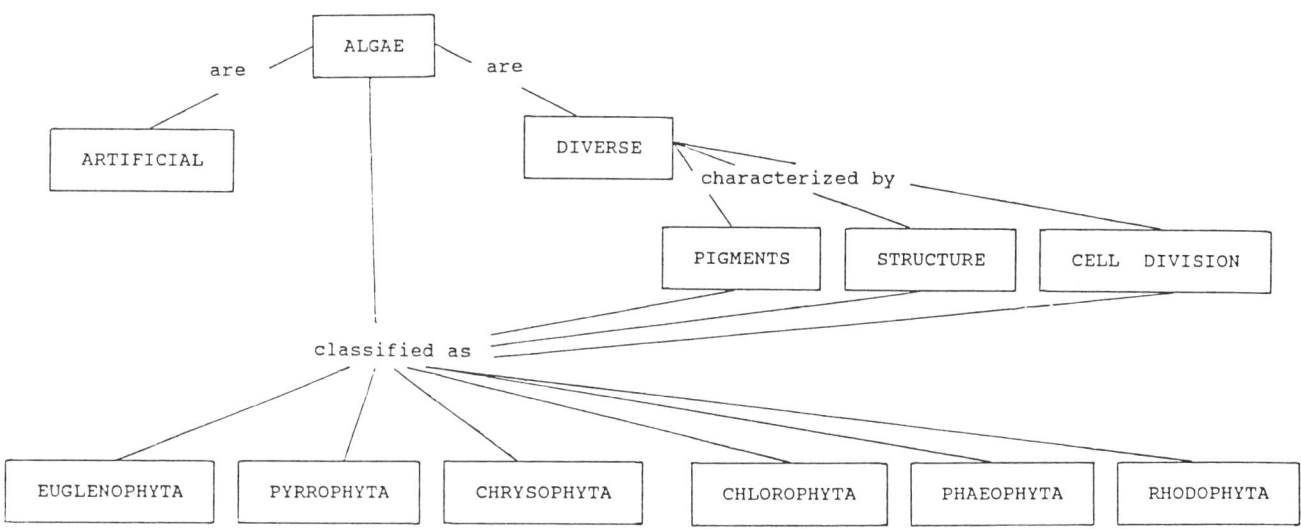

INSTRUCTIONAL AIDS

Lecture - Begin with a list comparing the features of prokaryotic and eukaryotic cells. Next, ask students to recall the structure of two of the obvious eukaryotic cell organelles, the mitochondrion and chloroplast. Make a sketch of each on the board as they are described. Be sure to emphasize the double membrane bounding each organelle as well as its unique genome. Now sketch a large, undistinguished prokaryotic cell on the board and pose the question, "How could this cell have evolved into a eukaryotic cell containing the kinds of organelles such as chloroplasts and mitochondria that we illustrated?" Discussion will probably follow along the lines of a step-wise series of mutations, etc., which eventually produced a cell with mitochondria, then chloroplasts... Diagram this possibility and explain that it is certainly a valid hypothesis and was, in fact, supported by most botanists as recently as 20 years ago.

Now ask if anyone remembers from their high-school biology how one-celled organisms, like *Amoeba*, eats? Now suppose that a large bacterium with anaerobic respiration (glycolysis), such as the one diagrammed on the board, phagocytized a smaller aerobic bacterium. Diagram this process. The aerobic bacterium now has access to a much larger supply of pyruvate, while the anaerobic bacterium has access to additional ATP so the situation would be mutually beneficial if it could be maintained. How would the structure of this association compare to that of a mitochondrion? Double membrane, one from the engulfing bacterium, one from the engulfed (the inner and outer mitochondrial membranes are biochemically different), two distinct genomes, separation of glycolysis from Kreb's Cycle and electron transport, etc. "Can the same be done in the case of chloroplasts and photosynthetic bacteria?" Repeat this blackboard scenario with the now aerobic symbiont and a smaller photosynthetic bacterium. Some of the most convincing evidence supporting this endosymbiotic theory are examples of extant symbiotic associations between protists and blue-green algae.

The most useful thing you can do for students as you cover the characteristics of the various divisions is to cover the information in the same order for each group. For instance, you might first cover the typical habitat, marine or freshwater, benthic or planktonic, shallow water or deep water, etc., then examine the morphology or general body plan characteristic of that taxon. The first time a new character state occurs, elaborate on its characteristics and how it differs from, or is similar to, the states previously observed. For example, discuss the trend from unicellular, to colonial, to multicellular body plans. Presenting material in the same order will facilitate student note-taking and allow you to emphasize similarities and point out key differences. Again, there is a problem of forest and trees. Emphasize a few key features in lecture. Students can gather details from the reading.

Pay particular attention to those features shared with the land plants when you are discussing the Division Chlorophyta. This will make the transition to the Kingdom Plantae much smoother in the next chapter.

In addition to habitat, there are other important ecological and economic aspects of certain groups that you should stress. Blooms of dinoflagellates not only are a major perturbation of the local environment causing massive fish kills, but they are economically important, especially to shell fishermen. The red tide toxins do not harm shellfish, but instead are concentrated in their bodies making them potentially lethal to vertebrates, including humans. A bloom in the Gulf of Mexico may close oyster beds and limit shrimping for weeks at a time, causing severe economic hardship to the fishermen. The same group also includes important endosymbionts of coral animals. Without zooanthellae, the coral animals are unable to fix calcium and the reefs die. Similarly, over half of the calcium accumulated on coral reefs is due to coralline algae (including green algae as well as red algae). The white, sandy, tropical beaches are, for the most part, the broken encrustations of algae! The single most important photosynthetic organisms on earth belong to the Division Chrysophyta, one of whose classes, the diatoms, have numerous industrial applications. The brown and red algae, in addition to having a number of commercially farmed edible species, are the sources of phycocolloids that are indispensable in food processing and other industries.

Pre-lecture Review - Review two topics, either before or during lecture. One is the characteristics of prokaryotic vs eukaryotic cells, Ch. 19, and the other is the generalized life cycle, Ch. 9. In the latter case, modify the quadripartite scheme to illustrate the relative importance of sporophyte and gametophyte in each group studied (or to illustrate the diversity of life cycles within a division.)

Post-lecture Review - The best way for students to pull together the information from this chapter is to construct a table of characteristic features for each of the divisions examined. This will force students to go back over their notes (again it will help if you presented the relevant information in a consistent order during lecture) and/or to organize material from the text. It also forces them to create a useful study guide for their next examination.

LABORATORY - EXERCISES

1. Water Sample

Obtain a sample of water from a pond, lake or slow-moving stream. Allow the sample to sit for a day in diffuse light, then sample at various levels in the water column, at the edge near the glass and in the middle, and from the sediments on the bottom. If possible, bring an algal-encrusted rock as well. A slimy rock will be coated with diatoms while free-swimming plankton accumulate near the glass and near the surface. Darkfield microscopy (Ch 3) produces spectacular results! The only disadvantage is that any animals in the field, and there will be some, will "steal the show." But at least students will be fascinated and enjoy their time peering through the microscope.

An interesting variation is to collect water samples from the same location at weekly intervals over the period of a month, then twice during the last week. You will have a representative sample of succession in a microenvironment with a few organisms carrying over between successive collections but a variety of "new" organisms appearing. Algae will dominate early samples, later samples will have many fewer algae but a variety of worms, larvae and other "beasts."

2. Evolution within the Green Algae

The volvocine line of green algae provides a simple example of evolutionary trends that students may either examine qualitatively or use to construct a data matrix and cladogram (Ch. 18). Beginning with a unicell like *Chlamydomonas*, there are tendencies toward increasing cell number, colonization and, ultimately, multicellularity. A simple exercise is to arrange the genera in order of increasing cell number and note evidence of specialization such as polarity in the colony, or any cell specialization. Darkfield also works well here.

Chapter 21

To construct a cladogram of the group, include at least one additional filamentous type, such as *Ulothrix*, to serve as an outgroup. Students will have to examine the specimens carefully to identify a sufficient number of character states for their data matrix to make this a meaningful exercise.

3. Comparative Morphology

Most botany laboratory manuals contain a fairly extensive exercise on structure and organization of various algal divisions. This is the traditional approach for studying algae in the laboratory. The tremendous variety of algae makes this an interesting approach, provided it does not become a memorization marathon. Students appreciate having living material whenever possible, and even pickled specimens of macroalgae, over straight slide material. Most large specimens preserved in formalin may be handled if first soaked overnight in fresh water. Encourage students to handle specimens carefully and experience the tactile sensation of the rubber-like consistency of phycocolloid embedded walls.

4. Savory Seaweed

Several edible seaweeds are stocked in oriental markets and the deli section of some larger food stores. At the very least, some sample packets of Nori (Laver in Britain), Kombu, or Agar should be available for students to examine. A preferable alternative is to prepare, or have students prepare, some simple seafood dishes for the class to sample.

LABORATORY - MATERIALS, EQUIPMENT, AND PREPARATION

1. **Water Sample**
    -sample of water from a natural source
    -compound microscope, slides and cover glasses
    -droppers
    -darkfield stops

2. **Evolution within the Green Algae**
    -compound microscope, slides and cover glasses
    -samples of as many of the following genera as possible: *Chlamydomonas, Carteria, Gonium, Pandorina, Eudorina, Pleodorina, Volvox.*
    -for cladistic analysis also add: *Ulothrix*, and perhaps *Stigeoclonium, Spirogyra,* or *Oedogonium*

3. Comparative Morphology
    -compound microscope, slides, cover glasses
    -variety of algal cultures or prepared slides
    -preserved specimens of larger algae

4. Savory Seaweed.

    a. Seaweed Soup
1 tbsp. salad oil
1 small onion, thinly sliced
1/2 cup thinly sliced green beans
1/4 in. slice of fresh ginger, minced
1 tbsp soy sauce
1 can (14 oz.) regular strength beef broth
1 cup water

3 strips (2x5 in. each) roasted nori, crumbled. (roast nori by spearing it on a fork and holding it over a stove burner - it will turn yellowish.)

Heat oil in a saucepan over medium heat and cook onion and beans until onion is translucent. Stir in ginger, soy sauce, broth, water, and crumbled seaweed. Bring to boil, reduce heat and simmer covered for 5 min. Makes 3-4 servings.

    b. Seaweed Aspic Salad
1 can (24 oz.) V8 juice
1 cup *Chondrus crispus*
1/4 cup chopped celery
1 tbsp. finely chopped parsley

Pour V8 juice into top of a double boiler. Put *Chondrus* into a cheese-cloth bag and tie. Place in V8 juice and steep over boiling water for 30 min., give bag a final squeeze. Add celery and parsley to juice. Pour into a mold and allow to set.

REFERENCES

General:

Hillson, C.J. 1982. *Seaweeds: a Color-coded, Illustrated Guide to Common Marine Plants of the East Coast of the United States.* Keystone Books, The Pennsylvania State University Press, University Park. 194 p. A layman's key to common algae of the east coast.

Prescott, G.W. 1978. *How to Know the Freshwater Algae, 3rd ed.* William C. Brown, Dubuque, IA. 293 p. A popular key to common freshwater algae

Chapter 21

Waaland, J.R., 1977. *Common Seaweeds of the Pacific Coast*. Pacific Search Press, Seattle. 120 p. A laymans guide to common algae.

Resources:

Bold, H.C. and Wynne, M.J. 1978. *Introduction to the Algae: Structure and reproduction*. Prentice-Hall, Englewood Cliffs, N.J. 706. The standard textbook of phycology.

Chapman, V.J. 1980. *Seaweeds and Their Uses*. Methuen, New York. 334 p. The economic aspects of algae; includes an extensive bibliography.

Scagel, R.F., R.J. Bandoni, J.R. Maze, G.E. Rouse, W.B. Schofield, and J.R. Stein. 1984. *Plants: An Evolutionary Survey*. Wadsworth Publishing Co., Belmont, CA. 757 p. A survey text of all of the groups traditionally studied in   botany courses.

Taylor, W.R., 1957. *Marine Algae of the Northeastern Coast of North America, 2nd ed*. University of Michigan Press, Ann Arbor, Mich. 520 p. The standard manual for algae of this region.

CHAPTER 22: NONVASCULAR PLANTS

This chapter begins an examination of the land plants, Kingdom Plantae, in the strict sense. There are several common problems that confronted organisms making the transition to the land habitat. Mostly these involved obtaining and retaining water. A shared feature of all plants is multicellular reproductive structures with one or more layers of sterile cells surrounding and protecting the spores or gametes. Beyond this, the nonvascular and vascular plants have adopted divergent means of dealing with the problems associated with life on land.

Mauseth points out that many students will have an intuitive feeling for what a moss is, but frequently this preconception will be misleading. For instance, the Spanish moss familiar to students in the Southeast is not a moss at all. Neither is the reindeer moss of the Northern states. In addition, two of the groups included with mosses in the Division Bryophyta will most likely be unfamiliar to most students. Not only are these organisms small and inconspicuous, but they are also much more restricted by their environmental requirements and, therefore, more limited in their distribution.

OBJECTIVES

1. Describe the structural adaptations of bryophytes associated with the transition to the terrestrial environment.
2. Explain alternation of generations as it relates to the typical bryophyte life cycle.
3. Compare the morphological features of mosses, liverworts and hornworts.
4. Discuss the ecological and economic importance of bryophytes.

CONCEPTS

Several problems associated with the terrestrial environment had to be solved for plants to successfully colonize this new habitat. The most serious involved obtaining, distributing and retaining water. Many of the adaptations evolved by bryophytes limit their exposure to desiccation and thus bryophytes tend to avoid these problems. Bryophytes usually grow in damp places where water stress is less severe. They grow in clumps so no individual is completely exposed to the environment. They stay small, thereby minimizing H_2O requirements and loss by evaporation.

A major requirement for successful colonization of land is to be able to reproduce. Multicellular structures, sporangia, antheridia, and archegonia, provide maximum protection to developing reproductive cells, but water is still necessary for fertilization to occur. It will be a surprise to most students that except for conifers and flowering plants, plants have swimming sperm that must have a pathway of liquid in order to swim to the egg for fertilization.

Chapter 22

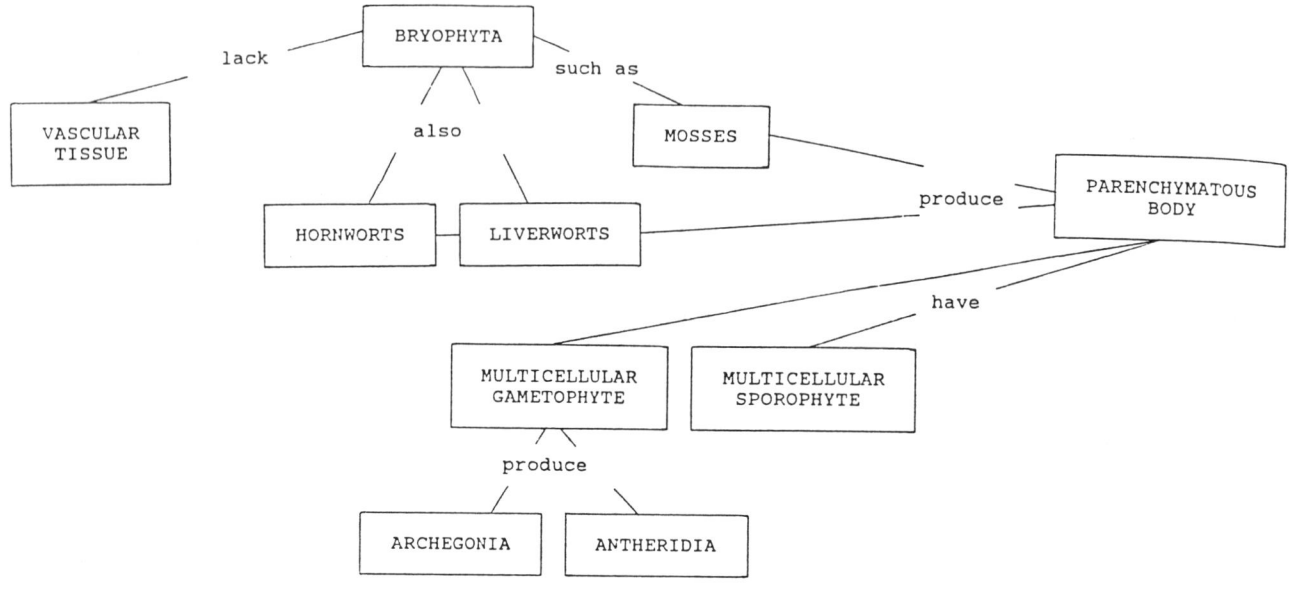

INSTRUCTIONAL AIDS

Lecture - Many of the characteristics of major taxa of land plants associate with mechanisms they adapted to live in the terrestrial environment. Begin by asking your students to list some of the problems associated with a terrestrial environment - write these on the board. Your list will likely include: obtaining water, retaining water, moving water, providing support, and reproduction. Now take one problem at a time and ask how it could be solved? What kinds of adaptations would have a selective advantage given these conditions? For obtaining water, students will immediately respond - develop roots. What are some other ways? For example, trapping precipitation (tank bromeliads), absorption through leaf surfaces (Spanish moss), or simply avoiding the problem by staying small and living in a constantly moist environment. To retain water some solutions include: develop protective coverings such as bark and cuticle; develop structures to regulate water loss through specific channels (stomata); or avoid the problem by staying small, growing in clumps, and living under constantly humid locations. Evolution of vascular tissue facilitates water movement. Alternatively, plants could avoid the problem by staying small enough that diffusion is sufficient. Support is necessary to gain a height advantage in competition for sunlight and to increase in size (in water, such support is not necessary because of natural buoyancy). On land vascular tissue, particularly xylem, provides this additional function. Alternatively, plants would avoid the problem by staying small and growing in clumps where one organism supports the other. In order to reproduce on land there must either be a way to protect the gametes as they are brought to one another, or reproduction must remain dependent on free water to prevent desiccation of gametes and to provide a transport medium. Pollen and the various pollination syndromes are one solution to the problem. An avoidance mechanism is to produce gametes only in wet environments or during a wet season. Your further discussion of the structure, function, and life cycle and the various bryophyte taxa can relate back to the solutions employed to overcome these problems.

Mosses are the most familiar of the bryophytes, but this familiarity does not go much beyond a small bunch of green stuff growing on the ground. The more visual you can make your presentation, with living specimens to pass around, or photographs, or at least illustrations, the better off you will be. I like to work a discussion of structure/function around the basic life cycle, starting with the upright gametophytic moss with which students are most likely to be familiar. At each panel of the life cycle diagram you can digress to a close-up look at that structure or process. Where is it happening? What adaptations are involved? How does it relate to the rest of the plant, etc.? In this way, students learn the processes involved and put them into the perspective of the life of the plant, without dryly memorizing the life cycles.

Concentrate on the mosses. Then, depending on time and your emphasis, you can either repeat this analysis of life cycles approach with the liverworts and hornworts, or simply cover some of the distinguishing characteristics that separate the groups. My preference is to concentrate on mosses as typical representatives of the bryophytes, then introduce enough information about the other two groups that students will be able to recognize them if they come across them in the field. Be sure to check your campus to see if liverworts and/or hornworts are growing nearby. Liverworts prefer moister conditions, hornworts may grow in full sun).

Although the ecological and economic importance of bryophytes are relatively minor, compared to other groups, in local areas it is considerable. Bryophytes are frequently pioneer species, along with lichens, especially in colder climates. *Sphagnum* is a dominant species in many bog communities and has direct economic importance in the horticultural industry both in its natural form, and as partially decomposed peat moss. The latter is also important in some areas as a fuel source.

Pre-lecture Review - Before beginning a lecture on the bryophytes it is useful to review the characteristics of the green algae which suggest their relationship to the land plants. After completing a list of common features, such as pigments, storage products, wall materials, method of cell division, etc., you can ask what additional adaptations were necessary to make the transition from water to land. This will lead directly to the discussion of problems and solutions associated with the terrestrial environment.

Post-lecture Review - It may or may not be obvious to students by the end of this chapter that for virtually all of the problems associated with the land habitat, bryophytes have evolved adaptations that tend to avoid the problems. Point out that this is a very efficient solution in terms of energy outlay, but it is very limiting in terms of how much of the available land habitat bryophyes may successfully colonize.

Chapter 22

LABORATORY - EXERCISES

1. Scavenger Hunt

At this point in the course, students will be familiar with many of the less conspicuous "plants" in the environment, so it is a good time to allow them to go out and collect what they can. Divide the class into groups and give them a set amount of time to collect as many different specimens of fungi, algae, and non-vascular plants as possible. Of course, it is not enough for them simply to collect specimens. Students must correctly identify them to the proper division. This will be a good review of the groups studied earlier and also will provide bryophyte specimens for closer examination.

2. Bryophyte Morphology and Anatomy

Most biology and botany laboratory manuals have detailed directions for examining prepared slides of various bryophyte structures. However, this group is particularly well adapted for study of fresh material. A dissecting microscope is invaluable for examining whole mounts or fresh plants. "Leaves" are only one or a few cells thick. Have students examine them microscopically in simple wet mounts. Students can cut "stems" and the thalli of liverworts and hornworts in cross-section with a razor blade, then mounted in water with a cover glass. Such preparations are not as aesthetically pleasing as prepared slides, but students have no doubt about how to relate what they see in the microscope to the intact structure because they made the preparation themselves. Polarizers are useful, if sections are thin enough, to outline cell walls, wall thickenings in *Sphagnum*, elaters in liverwort capsules, etc.

3. Tissue Culture

A. Spores and Protonema

If ripe sporophytes are available (this depends on the season) collect spores and germinate them *in vitro* to examine protonema development, budding and the initiation of leafy sporophytes. Collect ripe capsules and remove the calyptra if present. Holding the sporophyte by the stalk with a forceps, surface sterilize the tissue by dipping it in 10% bleach for a minute or two. Then rinse it in two changes of sterile water. With a sterile scalpel or dissecting needle, remove the operculum and tease spores out of the capsule onto a plate of agar. Sterilization is not necessary to get spores to germinate and form viable protonema, however, it reduces fungal contamination that may make it more difficult to make observations. Maintain the plates at room temperature under dim light conditions. In a few days, a green filamentous protonema should be visible. Within 10 days to two weeks, buds should be visible which will give rise to leafy sporophytes.

Nonvascular Plants

B. Sporophyte Culture

While still green, sporophyte stalks may be surface sterilized and cultured *in vitro*. Be careful not to leave the material in bleach too long and to rinse the specimens thoroughly before transferring to culture medium. Over sterilization will kill the moss tissue as well. You will recognize this if the tissue turns white in a day or two). Successful explants will form rhizoids and diploid protonemae within a few days. If maintained, the resulting diploid gametophytes will form tetraploid sporophytes! Similar culture may be done with the asexual propagules, gemmae, of liverworts.

4. Water Holding Capacity of *Sphagnum*

Sphagnum is of considerable economic importance in horticulture due to its water holding capacity. Ground up and added to soil mix it improves the water holding capacity of potting soil. Wrapped around plants in shipment or in arrangements of plants it decreases water loss, thus, helps preserve the plants. Students can measure this capacity by determining the mass of a volume of dried *Sphagnum*, then thoroughly wetting the material and reweighing. This is difficult to do. Hold the plant material under water and squeeze and release like a sponge. Thoroughly drain off the excess water, and reweigh. *Sphagnum* will hold up to ten times its weight in water.

LABORATORY, MATERIALS, EQUIPMENT, AND PREPARATION

1. Scavenger Hunt
-hand lens and/or dissecting microscope

2. Bryophyte Morphology and Anatomy
-dissecting microscope
-compound microscope
-polarizing filters
-slides, cover glasses
-razor blades or scalpel, dissecting needles, forceps

3. Tissue Culture
-plates of "Hyponex" medium
-0.5 to 1.0 g "Hyponex" or other complete fertilizer
-10 g agar
-1000 ml H_2O (no pH adjustment is needed)
-autoclave in flask with capacity of at least twice the liquid volume for 15 min., pour into sterile petri dishes. 100 ml makes about 4 plates.
-green moss sporophytes for seta culture, ripe sporophyte capsules for spore culture, or liverwort gemma.

Chapter 22

      -forceps, razor blades or scalpel
      -covered test tube of alcohol (to flame instruments)
      -bunsen burner or alcohol lamp

4. **Water Holding Capacity of** *Sphagnum*
      -dried *Sphagnum* moss
      -beaker of water
      -balance

REFERENCES

General:
Conard, H.S. 1979. *How to Know the Mosses and Liverworts, 2nd ed.* W.C. Brown Co., Dubuque, IA. 302 p. A pictured key to many of the common bryophytes of North America.

Resources:
Bold, H.C., C.J. Alexopoulos, and T. Delevoryas. 1980. *Morphology of Plants and Fungi, 4th ed.* Harper and Row., New York. 658 p. An excellent survey of all organisms covered in traditional botany surveys with an emphasis of the relationship of structure to function.

CHAPTER 23: SEEDLESS VASCULAR PLANTS

The definitive characteristic of the remainder of the plant kingdom is the presence of vascular tissue in the plant body. Vascular tissue arose early in the history of land plants. the We can recognize the basic patterns associated with extant species in the fossil forms that are now extinct. We can find nearly all the modifications adapting plants to the terrestrial environment in members of this group. The one notable exception is the continued dependence on free water for fertilization. Although some fossil lycopods formed reproductive structures analogous in many ways to modern seeds, they continued to depend on free water for sperm to swim to and fertilize an egg.

OBJECTIVES

1. Describe the adaptations to the terrestrial environment characteristic of the vascular plants.
2. Explain the differences between: dichotomous, pseudomonopodial and monopodial branching; enations, microphylls, and megaphylls; protosteles and siphonosteles; homospory and heterospory.
3. Describe the characteristic features of the extant divisions of seedless vascular plants.
4. Describe a typical life cycle of a seedless vascular plant.
5. Describe the ecological and economic importance of the non-seed vascular plants.

CONCEPTS

In the last chapter we listed several problems relating to the land habitat. We also discussed some possible solutions that would have selective advantage under certain circumstances. In general, non-vascular plants adopted avoidance solutions.. Vascular plants, on the other hand, accumulated modifications that allow them to be successful in all but the most severe environments.

Vascular tissue itself is the most important new development in this group of plants. Not only does it provide an effective means of transporting water and nutrients, thereby allowing for increase in size and specialization of organs for absorption and regulation of loss, but it simultaneously provides the structural strength necessary to support this increased size.

Mauseth's phylogenetic survey of fossil taxa emphasizes the evolutionary trends in various structures that led to the characteristic morphologies exhibited in extant plants. Earlier in the text we studied stems, roots and leaves as distinct vegetative organs (Chapters 5-8). In this chapter, it becomes clear that both roots and leaves, as distinct organs with specialized structure and function, derived from primitive stems. An examination of the psylophytes, *Psilotum* and *Tmesipteris*, reinforces this concept. The former, like many of the earliest fossil forms, lacks both roots and leaves, while the latter has a flattened fern-like leaf but remains rootless. Similarly, we still see the early trends in branching pattern, leaf type, stelar

Chapter 23

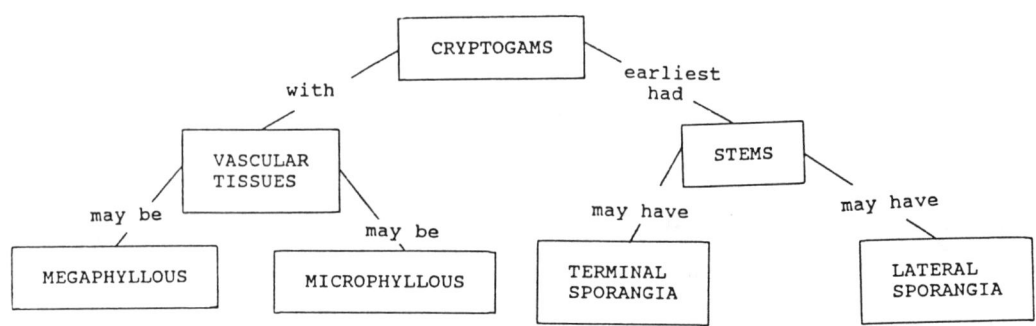

arrangement (pattern of xylem and phloem in the plant axis) and gametophyte differentiation in living representatives of the various taxa.

The life cycles of each taxon covered in this chapter are remarkably similar. It is partially because of this that we frequently refer to the seedless vascular plants as the ferns and fern allies. The major variation we find is heterospory in some Lycophyta and Pteridophyta. In an evolutionary sense, this is a significant modification because such differentiation of microspores and megaspores, giving rise to separate microgametophytes and megagametophytes respectively, was a necessary step leading to the evolution of a seed.

INSTRUCTIONAL AIDS

Lecture - Begin by asking students to recall the basic problems associated with the land habitat enumerated last time. Vascular plants developed, to a greater or lesser degree, the alternative solutions avoided by the bryophytes. Examine each of these adaptations, one at a time, with examples given either from the fossil record, as Mauseth has done so well, or from extant plants. For instance, a simple solution to avoid water loss is to maintain a low surface to volume ratio. This is done by retaining a cylindrical body plan with few elaborations. Thus *Rhynia* and *Cooksonia*, like *Psilotum* today, were simple systems of cylindrical, dichotomously branching stems.

Whether or not you concentrate on fossil forms, it will be essential to have plenty of illustrations, or living examples, available to illustrate the various conditions you are talking about. Except for ferns, students will be unfamiliar with most of the taxa covered in this chapter. Even with the ferns, you will have misconceptions to overcome. For instance, we covered leaves in chapter 6 but these were all megaphylls. Of the plants examined in this chapter, only ferns and horsetails have megaphylls, but they are very different from each other. Furthermore, most students will suspect that the stipe of the fern frond is actually the stem and that the rhizome, if they notice it at all, is the root. Intact specimens, or photographs of whole plants, along with micrographs of appropriate structures, such as cross-sections of rhizome and stipe, will greatly assist your attempt to convey the correct interpretation of these organs.

Seedless Vascular Plants

Understanding changes in the vascular system is perhaps the most difficult challenge for students. The vascular system is hidden from normal view, thus the only frame of reference students will have is what they learned in previous chapters. Again, fossil forms may be used to illustrate evolutionary trends, but it will be more real for students to concentrate on examples from extant taxa. The simplest case, an exarch protostele, remains characteristic of dicot roots as well as the roots and stems of many taxa of non-seed vascular plants. Students can observer various elaborations of this basic plan, from ridges of xylem in psilophytes to plates of xylem in many lycopods and some ferns.

Illustrate the typical life cycle of a vascular cryptogam with a fern. As with the mosses, use the life cycle as a vehicle to introduce the characteristic structure and function of the various parts of the sporophyte and gametophyte body.

Except for ferns, vascular cryptogams are a minor component of the modern landscape and living representatives have relatively little economic value. The notable exceptions are those ferns used ornamentally in horticulture and landscaping. If you include fossil members of the group, however, their economic value is enormous. As Mauseth mentions, some of the fossil lycopods and sphenopsids developed secondary growth, formed wood, and were the dominant components of the carboniferous landscape, forming forests over large areas of land. It is these forests that we mine today in the coal seams of the eastern and central U.S. In fact, we find some of the best fossils of this era in the mine spoils beside pits.

Pre-lecture Review - Vascular tissue characterizes this group, especially the arrangement of xylem and the function of the vascular cambium, if present. Before beginning this new material, review xylem structure and its pattern of differentiation (Ch. 2). You should also briefly review the function of a vascular cambium in producing secondary xylem and phloem in order to have a frame of reference to compare with the anomalous types of secondary growth found in extinct forms and in *Isoetes* and *Stylites*.

Post-lecture Review - The details of differences between taxa examined in this chapter may easily overwhelm students unless you encourage them to organize their information. The best way to do this is to have students make a table of character states as was done for the algae. In this way, similarities and characteristic differences will become immediately obvious and much easier to study.

LABORATORY - EXERCISES

1. Comparative Morphology and Anatomy

Most laboratory manuals will have a series of exercises asking students to examine specimens, or more likely prepared slides, to observe characteristic features of representative taxa. A more effective technique is to provide as many representative specimens as possible for

Chapter 23

students to observe, dissect, and prepare their own sections. From intact specimens, students can observe differences in branching patterns, size and vasculature of leaves, and position of sporangia. Hand sections of stems and leaves will permit students to compare arrangements of vascular tissues. Students will be able to associate structure with intact plants and with particular taxa. Judicious selection of material will allow the class to examine virtually all of the evolutionary trends within this group.

2. **Fern Evolution**

The exercise suggested in Chapter 18 to demonstrate cladistic analysis using various fern specimens may be used here as well. Each group will observe in detail a different set of characteristics, depending on the specimen they examine. In rotating through the demonstations set up by other groups, they will have demonstrated some of the variety present within the Pteridophyta. Once students complete their cladistic analysis, they can compare the evolutionary trends implied by their classification with the trends outlined by Mauseth in this chapter.

3. **Tissue Culture**

Surface sterilized fern spores may be grown in tissue culture using the same techniques employed for bryophytes. As an added variation, place plates of spores under different light spectra, red, blue and white. In many species of ferns, orientation of cell division is under photomorphogenetic control so students can induce filamentous gametophytes, as well as typical prothalli. Simple, colored cellophane filters are adequate if kept under low intensity fluorescent illumination.

An alternative procedure to obtain gametophytes for use in the laboratory is to soak "Jiffy Pellets" in water, then sprinkle spores from a frond containing ripe sori on the top surface of a pellet. Place the inoculated pellet in a petri dish bottom, filled with water, and invert a beaker over the top. Maintain these cultures under diffuse light and add water to the dish as necessary to keep the pellet from drying out. Once a carpet of gametophytes forms, you can induce them to form sporophytes by placing drops of water on the surface of the thalli. Young sporophytes should be apparent within a few weeks.

LABORATORY - MATERIALS, EQUIPMENT, AND PREPARATION

1. **Comparative Morphology and Anatomy**
    -dissecting microscope or hand lens
    -compound microscope, slides and cover glasses
    -toluidine blue and phloroglucinol (see ch. 5)
    -as many of the following specimens as possible: Psilotum (grows like a weed in greenhouses once it is established), *Lycopodium* (temperate species are difficult to maintain in culture),*Selaginella (S. kraussiana* and *S. emmelliana* grow well in terraria),

Isoetes (is locally abundant in some areas but difficult to grow), ferns (a variety, see list for Ch. 18, many grow well in greenhouses)

2. Fern Evolution
-same as above

3. Tissue Culture
-fresh fern fronds with ripe (brown or black) sori - Hyponex® media and equipment for sterile isolation of spores as for moss culture in previous chapter. Spores will germinate and grow without sterilization but, again, there may be a problem with overgrowth of fungi. To sterilize spores, place a fresh frond on a sheet of paper overnight, covered with a piece of plastic. The next day, transfer the released spores from the sheet of paper to a centrifuge tube, add bleach/surfactant and shake for one minute. Spin at high speed in a clinical centrifuge for one minute to pellet the spores, decant the bleach solution and replace with sterile water. Resuspend the pellet, then recentrifuge and decant off the liquid. Repeat with another change of sterile water.
-red and blue colored cellophane

REFERENCES
General:

Frankel, E. 1981. *Ferns: A Natural History*. The Stephen Green Press, Brattleboro, Vt. 264 p. A plant lover's guide to fern biology including interesting folklore and growing tips for selected species.

Grillos, S.J. 1974. *Ferns and Fern Allies of California*. University of California Press, Berkeley, CA. 104 p. A pocket guide to the 80+ species found in California.

Wherry, E.T. 1961 *The Fern Guide: Northeastern and midland United States and Adjacent Canada*. Doubleday and Co., Garden City, NY. 318 p. A good laymans guide to the common fern flora of this region.

Resources:

Bierhorst, D.W. 1971. *Morphology of Vascular Plants*. Macmillan, New York. 560 p. An encyclopedic work containing numerous original observations. Includes both fossil and extant groups.

Gifford, E.M. and A.S. Foster. 1989. *Morphology and Evolution of Vascular Plants, 3rd ed.*. W.H. Freeman and Co., New York. 626 p. The classic textbook of comparative morphology and evolution of vascular plants.

Taylor, T.N. 1981. *Paleobotany: An Introduction to Fossil Plant Biology*. McGraw-Hill Book Co., New York. 589 p. A technical textbook of paleobotany.

CHAPTER 24: GYMNOSPERMS

Many students will recognize the term gymnosperm and immediately think of conifers, pines in particular. In one sense this is fortunate, because you will have a familiar reference point for your discussions. On the other hand, this apparent familiarity masks the diversity of plant forms contained within the broad group commonly known by this term. Furthermore, the pines, although the most familiar taxon, are the least representative group in terms of general morphology. Like the seedless vascular plants, the fossil record contains many gymnosperms and the evolutionary history of the group is becoming much clearer. Unlike the cryptogams, however, many of the evolutionary trends characteristic of the group are not visible in extant members, so it is more difficult to provide convincing evidence in class. For instance, extant members do not show much evidence of the branch-like nature of the ovuliferous scale in conifer cones.

OBJECTIVES

1. Describe the structure and functional significance of the seed.
2. Explain the characteristic differences between the extant groups of gymnosperms.
3. Describe the typical life cycle of a gymnosperm.
4. Describe the ecological and economic importance of the gymnosperms.

CONCEPTS

Students will have a general functional understanding of what a seed is, but the concept that it consists of an embryo within a gametophyte, itself enclosed within a protective layer of parental sporophytic tissue will be entirely foreign. Similarly, students will know that the pollen grain is a reproductive structure somehow associated with the sperm. Some students even think that pollen and sperm are synonymous. Few students, however, will understand that a pollen grain is part of the gametophyte generation of the plant. Although the general reduction of the sporophyte from one group of land plants to the next is a general trend that should be obvious, you will specifically have to point out this pattern to students before they will see it.

The most visible difference between gymnosperms and the vascular cryptogams is the production of considerable amounts of wood in the sporophyte body. Although some of the fossil cryptogams produced "wood", their vascular cambium was limited in its capability. Mauseth described the function of typical vascular cambium in Chapter 8. However, he did not emphasize that the ring of vascular cambium increases in circumference as the diameter of the tree grows because the cells have the can divide in the radial plane, as well as tangentially. This capability apparently was not present in the woody lycopods or horsetails of the past.

We usually use pine to illustrate the typical life cycle of gymnosperms. Although these are the most familiar representatives to most students, it is in some ways an unfortunate choice. In particular, the vegetative morphology of pines is atypical in that two types of shoots are formed, long shoot and cryptic short shoots. The latter are the fascicles terminate in one to five needle leaves. The papery, white scales at the base of the fascicle are actually reduced leaves along a short branch shoot. Close examination will reveal a leaf or leaf scar subtending each axillary fascicle branch.

One of the first things students will think of when you mention the term gymnosperm is cones, specifically pine cones. The reproductive structures of several of the gymnosperm taxa consist of simple cones where specialized sporophylls attach along a cone axis. This is also true of the microsporangiate (male) cones of conifers. The megasporangiate (female) cones, with which students are most familiar) are a complex arrangement whose true nature is not obvious. Bracts attach along a primary cone axis, but except in a few genera, most notably *Pseudotsuga*, they are inconspicuous. The woody cone scales are actually fused branch systems in the axils of the much-reduced bracts. These compound cones are, thus, similar to a branching inflorescence where bracts along the primary axis subtend the branches that actually bear flowers.

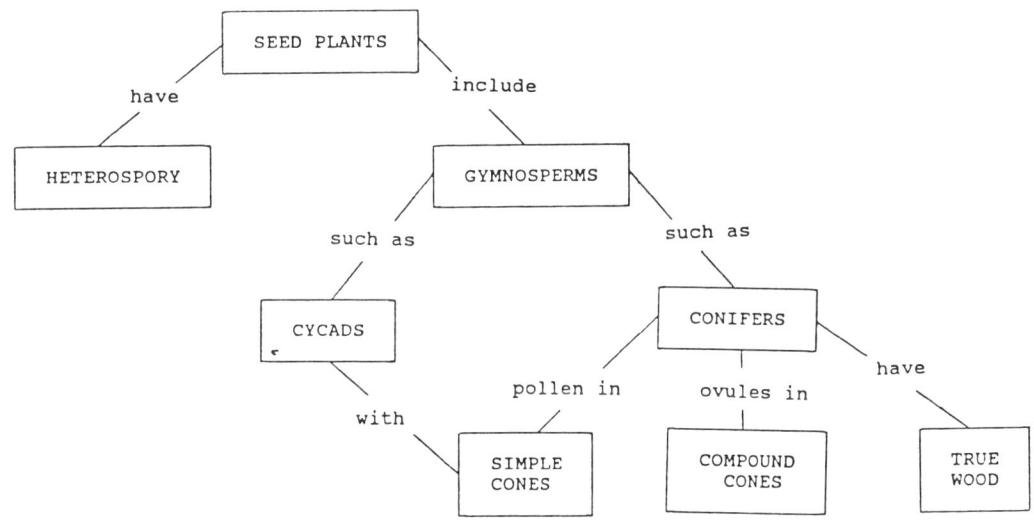

INSTRUCTIONAL AIDS

Lecture - The seed characterizes the remaining plant taxa, but the evolutionary nature of this structure will be obscure to your students. Begin by diagramming the structure of a typical mature seed and reviewing the function of each part. In the center will be an embryo, the young sporophyte of the new generation. Surrounding the embryo is a reserve of food (the reason so many seeds are important for human nutrition). The food in gymnosperm seeds is the remains of the female gametophyte. Finally, a protective layer surrounds the entire structure. This seed coat inhibits desiccation, and frequently provides considerable mechanical protection. Now ask

Chapter 24

your students to predict what the origin of these structures must be, given what they know about reproduction in the cryptogams and the fact that evolution works by modifying structures and processes that already exist.

This question will stymie most students. Go back to the bryophytes and ask them to review for you how the zygote and sporophyte formed. What was the relationship between this sporophyte and the gametophyte on which it was produced? When spores are formed by meiosis, they are released and each is capable of growing into an independent dominant gametophyte plant. Now move on to a fern and ask the same questions. Ferns still produce spores by meiosis and these spores grow into independent gametophytes. Nevertheless, they are small and inconspicuous. Point out that, in general, there is a reduction in the size and importance of the gametophyte from being dominant and independent (actually having the sporophyte dependent on it!) in the bryophytes to being independent but much reduced in the ferns and fern allies. Meanwhile, the sporophyte has become independent of the gametophyte and considerably more specialized.

Now extrapolate this tendency to the gymnosperms. The sporophyte becomes even larger and more specialized and the gametophyte is even more reduced - to a few hundred cells in the ovule and as few as four cells in the pollen grain! Meiosis still occurs in a sporangium, but they develop into small gametophytes still retained within the sporangium. The microgametophyte becomes specialized for wind pollination. The megagametophyte builds up stores of food reserves, produced by the parent sporophyte. The megasporangium wall changes to provide protection as well as a mechanism to trap the microgametophyte and draw it near to the enclosed megagametophyte. This wall completes development following fertilization to become the seed coat, surrounding nutritious female gametophyte tissue, itself enclosing the developing embryo.

You can now place development of the seed into the perspective of the life cycle. The seed germinates to produce a typical woody sporophyte with roots, monopodial stem and leaves. We usually show the pine life cycle in part because this is the genus with which most students are likely to be familiar. One unfortunate result of this choice is that shoot structure, having both long and short shoots, is very atypical of the group as a whole. Students will recognize pine needles as being leaves, but it is difficult for them to visualize the entire fascicle, with one to five needle leaves extending from the tip, as a short shoot. If they look closely, however, they will see that fascicles are always in the axil of a solitary needle leaf or leaf scar. The whitish scales encircling the base of the fascicle are rudimentary scale leaves.

Simple cones, with sporophylls arranged around a central axis, are typical of several of the taxa of extant gymnosperms. The compound female cones of conifers are unique. The woody scales of ovulate pine cones appear to be homologous to the ephemeral microsporophylls of the pollen cones. Instead, these scales represent compressed, secondarily fused branching systems in the axils of bracts. The latter are frequently inconspicuous. The most notable exception is in *Pseudotsuga* where the trilobed subtending bract extends beyond the woody ovuliferous scale. These compound cones are similar to complex inflorescences where flower-bearing branches are borne along the primary inflorescence axis.

Pre-lecture Review - All gymnosperms are woody with a typical vascular cambium capable of expanding in circumference as the diameter of the tree grows. Although some of the fossil vascular cryptogams had a vascular cambium, it had limited activity and stem diameter was apparently determinate. We studied vascular cambium and wood in Chapter 8, but it will be good to review their structure and development as an introduction to the primary characteristics of the group.

Post-lecture Review - As in the previous chapter, a good way to summarize the key characteristics of several related taxa is to create a table of character states. Although inclusion of fossil forms provides a clearer picture of evolutionary trends, it is best to concentrate on extant forms, especially when material for student examination is available.

LABORATORY - EXERCISES

1. **Gymnosperm Morphology**

The most enjoyable way to leaf morphology is to go out to the field and examine specimens *in situ*. Gymnosperms are common enough as landscape plants that you should be able to find several different genera within a short walk of the classroom. Make use of a key to local plants (see Appendix I) to familiarize students with the morphology associated with key terminology. You also can use these excursions to teach some natural history of the plants. For instance, by counting the number of branch whorls along the stem you can estimate the age of a young pine tree. One whorl is produced per year and the length of internode between whorls is also a good indicator of growth conditions that year. Similarly, cycads produce a single whorl of leaves per year, so by counting the number of leaves present and dividing this number into the number of leaf scars, you can estimate the tree's age.

2. **Anatomical Adaptations**

We frequently refer to gymnosperms as evergreens, although this term is neither technically correct nor exclusive to this group. Nevertheless, the term does imply that the plant does not drop its leaves annually in response to stress. We therefore expect to find xeromorphic adaptations in the leaves. A favorite leaf to examine microscopically is that of *Pinus*. Most lab manuals have labelled illustrations of this material to guide students' study. In fact, we can also find most of the adaptations characteristic of *Pinus* in other gymnosperm taxa. Have students identify: thick cuticle, sunken stomata, hypodermis, endodermis, transfusion tissue, and resin canals in fresh hand sections of *Podocarpus, Araucaria, Picea, Abies*, etc.

Chapter 24

3. **Conifer Life cycle**

The typical laboratory exercise for gymnosperms is to examine fresh and pickled material and microscope slides illustrating various stages of a pine's life cycle. Any laboratory manual will provide a guide to this approach.

LABORATORY - MATERIALS, EQUIPMENT, AND PREPARATION

1. **Gymnosperm Morphology**
    -hand lens
    -pocket knife and/or pruning shears
    -key to local trees (including horticultural varieties if possible)

2. **Anatomical Adaptations**
    -compound microscopes
    -razor blades
    -slides and cover glasses
    -dropper bottles of toluidine blue and phloroglucinol
    -dropper bottle of water
    -shoots of various greenhouse and/or locally grown gymnosperms

3. **Life Cycle**
    -dissecting and compound microscopes
    -living and/or preserved specimens of shoots and cones
    -prepared slides of vegetative and reproductive organs

REFERENCES

General:
Chamberlain, C.J. 1965. *The Living Cycads*. Hafner Publishing Co., New York. 172 p. This facsimile of the 1919 edition contains a wealth of information about the habitat, appearance, and life history of the various cycads.

Resources:
Chamberlain, C.J. 1966. *Gymnosperms: Structure and Evolution*. Dover Publications, Inc. New York. 484 p. This facsimile reprint of the 1935 edition contains a wealth of illustrations, line drawings, reconstructions, and photographs of habit and habitat. This is especially useful with the more obscure groups.

Sporne, K.R. 1965. *The Morphology of Gymnosperms*. Hutchinson and Company, London. 216 p. A brief textbook of gymnosperm morphology.

Steward, W.N. 1983. *Paleobotany and the Evolution of Plants*. Cambridge University Press, Cambridge. 405 p. Another popular interpretation of fossil plants.

CHAPTER 25: ANGIOSPERMS

Angiosperms, or flowering plants, consist of a single division, alternately known as the Division Anthophyta or Division Magnoliophyta. The various names used to characterize the group identify the most definitive characters of its members. "Angiosperm" literally means vessel-seed. This refers to the ovules, and resulting seeds, contained within a vessel, the carpel. "Anthos" (flower) is also the root of anthers, microsporangiate structures unique to this group. "Flowering plants" refers to the entire reproductive shoot, the flower, whose significant reproductive parts are the anthers and carpels.

Although angiosperms are the dominant and most ubiquitous vascular plants on earth, the evolutionary history of the flower is not well known. In large part, this is because the flower parts, sepals, petals, stamens, and carpels, are usually soft and fragile. As a result they are not abundant in the fossil record. Phylogenetic interpretation of the origin of flowering plants remains "the abominable mystery" which puzzled Charles Darwin.

The flower is the most morphologically stable organ of the plant. That is, even under stressful conditions when vegetative parts are stunted, flowers remain close to their normal size and condition. Flowers are also the most evolutionarily stable part of the plant; therefore, floral structure is an important clue to evolutionary relationships. Of course, the flower is also the most aesthetically pleasing aspects of plants. A greater understanding of their structure and variation will lead to a greater appreciation of their beauty.

OBJECTIVES

1. Describe the probable evolution of floral parts from structures found in gymnosperm taxa.
2. Explain the evidence supporting a monophyletic origin of flowering plants.
3. Differentiate between the two classes of flowering plants

CONCEPTS

The concept that floral parts are serially homologous to each other and to vegetative leaves dates to Goethe, the German philosopher and poet. It is not at all difficult for students to conceive of sepals and even petals as having a leaf-like character. Stamens and carpels, however, are not so intuitively obvious. In large part, this is because of the specialization of sexual parts in many flowers, especially horticultural varieties, with which students are most familiar. Simply introducing students to some of the extreme diversity of flowers will help to clarify this concept.

A basic philosophical assumption of evolutionary theory is that related groups always have a common ancestor; a natural classification system will be monophyletic. The diversity found in angiosperms has led some botanists to suggest that the group is really polyphyletic and artificial. Mauseth presents some arguments, for shared characters, favoring the monophyletic view. This is essentially a cladistic approach that is very powerful because it lends itself to graphic visualization. Nevertheless, it may still be difficult for students to grasp this concept.

Because the angiosperms are such a large and diverse group, it is very difficult to separate them into discrete groups. The two major classes, Liliopsida and Magnoliopsida, are relatively straightforward with a number of distinctive characteristics. Only occasionally are there problems differentiating between one or the other. Beyond this level is probably more than you will want to require from non-majors.

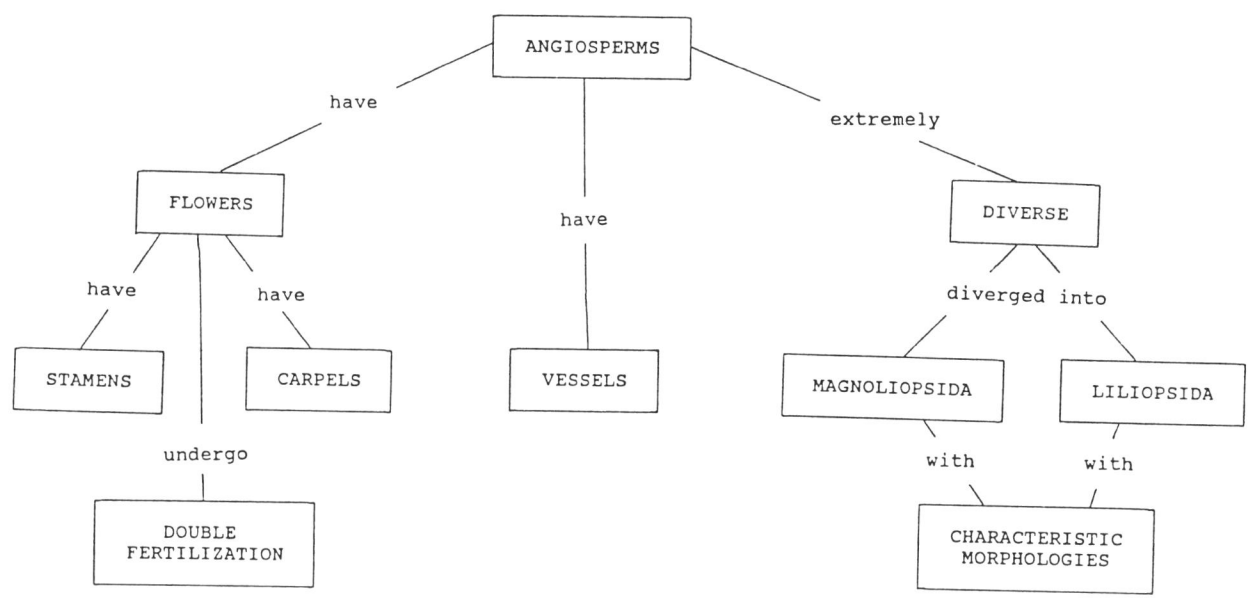

INSTRUCTIONAL AIDS

Lecture - Serial homology is a useful concept for interpreting the evolution of floral parts. It will not be difficult to convince students that sepal and petals have a leaf-like nature, stamens and carpels are another matter. If possible, have some demonstrations to show, or at least appropriate illustrations. A classic example of the homology of petals and stamens is the white water lily. As you move from the outside toward the inner flower parts there is a gradual transition from obvious petals to petals with rudimentary anthers near their tips, to petalloid stamens, to more-or-less typical stamens. Of the genera which typically have petalloid stamens, *use Southern Magnolia* as an example. Primitive carpels, suggesting their leaf-like nature, are less readily available. Again *Magnolia*, with its elongate semi-fused stigma/style is the best readily available example. The classic example, illustrated in most morphology texts, is *Drimys piperita*. The presence of flowers, and accompanying double fertilization (Chapter 9) are some of the strongest arguments for a monophyletic origin of flowering plants. You should also point

Chapter 25

out the role of pollinators as a selection force in the evolution of flowers. It is not a mere coincidence that the plant and animal groups that currently dominate the earth arose at the same time and provide the classic examples of co-evolution.

The structures of xylem and phloem are also distinctive in angiosperms. These tissues were studied previously (Chapter 5), but it is worth pointing out now that the features examined are characteristic of angiosperms. You should also point out that evolutionary trends in vascular tissues provide an independent index of evolution in the group, that generally parallels the patterns observed based on floral structure. In addition, evolutionary trends in vascular tissue extends into other vascular plants. For instance, vessel elements are derived from tracheids and within this continuum there is a gradual shortening, widening, increase in perforation of adjacent end walls, decrease in angle of end walls, and changes in pattern of pitting in sidewalls. The almost universal occurrence of sieve tube elements and the presence of vessels in all but the most relictual families of flowering plants argues strongly for the monophyletic origin of the division.

We have already discussed most of the characteristic features of dicots and monocots in appropriate chapters: two vs. one cotyledon (Chapter 9), reticulate vs. parallel leaf venation (Chapter 6), stem vascular bundles in a ring vs. scattered (Chapter 5), secondary growth vs. lack of true secondary growth (Chapter 8), and flower parts in 4s and 5s vs. parts in 3s (Chapter 9). Now is the time to synthesize this information as the definitive characteristics of the two classes of flowering plants.

Pre-lecture Review - Flowers are the distinguishing characteristic of the Division Magnoliophyta. Review the structure and function of a generalized flower to provide the basis for a discussion of the evolutionary origin of the flower and its parts.

Post-lecture Review - The best way to summarize the trends observed in flowering plants is to construct a concept map relating the classes (and subclasses) of angiosperms to a hypothetical ancestral angiosperm. Connecting lines should indicate major shared features while vertical progression through the map should indicate evolutionary trends.

LABORATORY - EXERCISES

1. Trends in Floral Evolution

Provide a variety of flowers for students to examine and attempt to arrange in a systematic fashion using the cladistic technique described in Chapter 18). A simpler approach to the same ends is to utilize the table of relictual and derived floral features (Table 25.1) in the text to rank the degree of specialization of a variety of flowers, then place them, in order of increasing specialization, within the classification framework provided for monocots (Fig. 25.12) or dicots (Fig. 25.23). To do this, students will have to first identify each of their flowers to the appropriate subclass using the following key.

KEY TO THE SUBCLASSES OF ANGIOSPERMS (after Cronquist)

A. Plants with net-veined leaves, flower parts in 4s or 5s ....Magnoliopsida

    B. Flowers polypetalous with numerous stamens; pollen uniaperturate .... Magnoliidae

    BB. Plants different from above, pollen with more than one aperture C

        C. Flowers reduced with inconspicuous perianth, oftenunisexualincatkins...Hamamelidae

        CC. Flowers not as above ....D

    D. Flowers sympetalous, stamens fewer than petal lobes, or alternating with petals if the samenumber...Asteridae

    DD. Flowers apopetalous, or if sympetalous with more stamens than petal lobes ...... E

        E. Flowers usually polypetalous and with axile placentation ....Rosidae

        EE. Flowers often sympetalous and, if axile placentation, with many ovules per carpel ....F

    F. Pollen trinucleate, plants usually herbaceous ....Caryophyllidae

    FF. Pollen binucleate, plants often woody ....Dilleniidae

AA. Plants with parallel-veined leaves, flower parts in 3s ....Liliatae

    G. Flowers usually apocarpous, plants herbaceous, often aquatic ....Alismataceae

    GG. Flowers mostly symcarpous, plants terrestrial ....H

        H. Flowers with well-differentiated sepals and petals ....Commelinidae

        HH. Petals and sepals similar or flowers reduced and clustered to form a spadix ensheathed with a spathe I

    I. Flowers numerous and small, often aggregated into spathe and spadix, leaves typically broad...Arecidae

Chapter 25

II. Flowers not aggregated, leaves narrow .... Liliidae

Assign each group of students one or two flowers to examine closely. The first thing they must do is key the flower out to the correct subclass and identify the subclass and species on a data card for that plant. You provide species names, or at least common names when you distribute the specimens. Next students examine the flower for each of the character states listed in Table 25.1 and record the character state on a card for that species. Assign a value of 0 for each relictual trait, 1 for each specialized trait, and 0.5 for an intermediate condition. Once each group has completed its data card, have groups rotate around the room to view each of the other specimens. Students should summarizing the data for each flower in their notebook. After examining all the specimens, students should arrange the species in the appropriate order on either Fig. 25.12 or 25.23 to reflect evolutionary specialization. List species with lower numerical totals for character states at the bottom of the diagram. Place species with higher totals nearer the top.

2. **Trends in Xylem Evolution**

Several xylem features provide useful information concerning phylogenetic relationships. The easiest to observe and quantify concerns vessel elements. Provide students with macerated wood of several different species. You can use either prepared slides or freshly macerated tissue. For each species, students should measure ten vessels, then determine: avg. vessel length; avg. vessel width; avg. angle of the perforation plate, from 10° to 90° (not including the tail which may extend on one side); average number of bars in the perforation plate, from many to none, and the type of lateral wall pitting, scalariform to alternate circular pits. Compare the relative primitiveness or specialization of wood of the various species by converting numerical values to % of the maximum observed. Plot the values for each character on a common axis.

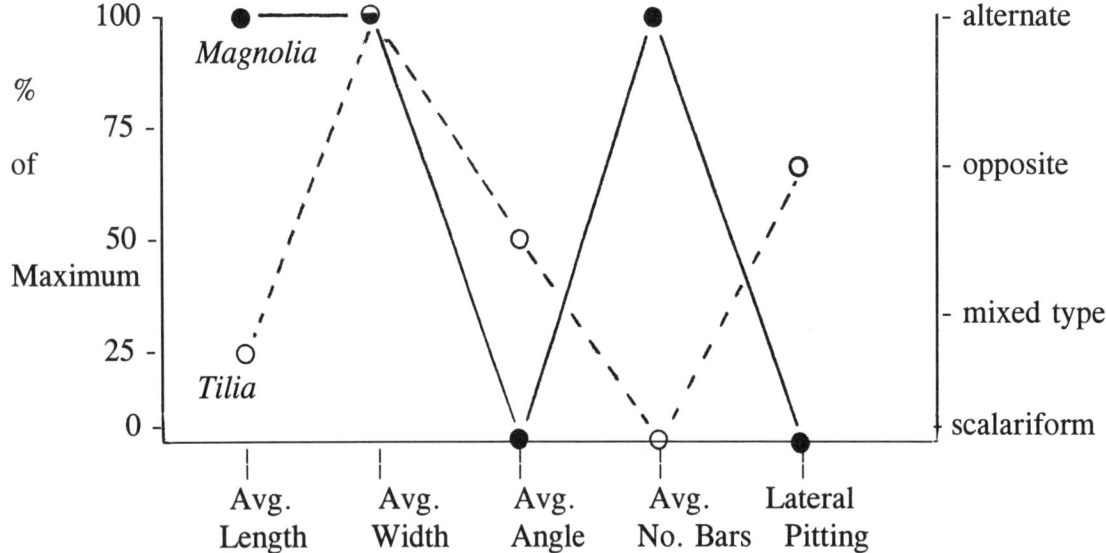

Plotted above are examples from two species with very different woods.

3. Flowering Plant Life Cycle: Microsporogenesis, Megasporogenesis, and Embryogenesis

Laboratory manuals typically include directions for observing appropriate slides of Lily anther development, Lily carpel development, and *Capsella* embryo development to illustrate the life cycle of flowering plants.

LABORATORY - MATERIALS, EQUIPMENT, AND PREPARATION

1. **Trends in Floral Evolution**
   -variety of flowers from the wild or greenhouse, preferably include some vegetative material as well to help identify subclasses.
   -dissecting microscope
   -razor blades and dissecting needles
   -compound microscope
   -slides and cover glasses
   -dropper of water
   -note cards
   -outline copies of Figs. 25.12 and 25.23

2. **Trends in Xylem Evolution**
   -prepared slides of a variety of macerated woods (alternatively you may make your own macerated woods; students apply a drop to a slide and add a cover glass. Collect small samples of appropriate woods,

   -compound microscopes

3. **Angiosperm Life Cycles.**
   -prepared slides of *Lilium* anthers and carpels at various stages of development
   -prepared slides of *Capsella* embryo development at various stages
   -compound microscope

REFERENCES

General:

Heywood, V.H. 1985. *Flowering Plants of the World*. Prentice-Hall, Englewood Cliffs, NJ. 335 p. This is a profusely illustrated atlas of the families of flowering plants, concentrating on key characteristics.

Thomas, Barry. 1981. *The Evolution of Plants and Flowers*. St. Martin's Press, New York. An introduction to the evolution of flowering plants.

Chapter 25

Resources:

Bailey, I.W. 1954. *Contributions to Plant Anatomy*. Chronica Botanica Co., Waltham, MA. 259 p. A collection of seminal papers from the botanist who contributed more than anyone else to our understanding of the xylem evolution and the relictual structure and evolution of primitive flowers.

Eames, A. 1961. *Morphology of Angiosperms*. McGraw-Hill, New York. 518 p. The standard reference to flowering plant structure, with emphasis on floral parts.

Leppik, E.E. 1977. *Floral Evolution in Relation to Pollination Ecology*. Today & Tomorrow's Printers and Publishers, New Delhi, India. 164 p. A short volume dedicated to floral evolution specifically as directed by insect selection pressure.

Missouri Botanical Garden. 1975. *The Bases of Angiosperm Phylogeny*. Ann. Mo. Bot. Gard. 62(3): 515-834. This number of the volume is dedicated to all aspects of angiosperm evolution.

CHAPTER 26: POPULATIONS AND ECOSYSTEMS

In many ways the unit on ecology will be the most easily understood by your students. Ecology studies the interactions between plants, other organisms and the environment. In many instances, the results of such interactions are readily observable, therefore concrete and easy to understand. Furthermore, ecology is frequently in the news so students have more familiarity with this topic than with many other areas of botany. Yet, there are also several difficult concepts for students to understand, such as niche and growth rate. There is sometimes a tendency for students to become overconfident in their ability because of familiarity and loose concentration.

Ecology is also an extremely broad and diverse subject. This makes it a difficult topic to cover in a limited amount of time. Confounding this is logistics. Course syllabi often save ecology until last to "tie everything together." Unfortunately, too little time is left in the semester to cover the subject adequately. If you find yourself in this situation, best to choose just one or two concepts. Try to concentrate on some relevant local problems that deal directly with these concepts, and concentrate on presenting the issues and complications involved with these. If you find yourself with sufficient time, you can "tie everything together."

OBJECTIVES

1. Describe the effects of abiotic and biotic factors on growth of plant populations.
2. Explain the concept of niche and the role of competition in populations and communities.
3. Explain the differences between r and k selection strategies.
4. Describe the factors which limit the distribution of individuals in populations and communities.
5. Describe the relationship of one trophic level with another in terms of energy, nutrient cycling, and biomass.

CONCEPTS

The role of abiotic factors in limiting plant growth seems obvious, yet some underlying concepts and principles prove to be very elusive for students. This is partly because, as pointed out in Chapter 1, growth itself is a complex concept. For populations, however, it is relatively straightforward. Population growth is an increase in the number of individuals in the population. The problem is in describing growth. We commonly use the term "rate" in the colloquial sense to describe the change in overall number of organisms. Thus, if there are two individuals to start with, four in the next generation and eight in the next, the impression of most students is that the rate of population growth is increasing. In fact the rate of growth is constant in this example. The number of individuals doubles each generation. The difference is that although the rate remains constant, 2, the number of individuals reproducing each generation increases by a factor of two. Thus the total population also doubles from one generation to the next.

Chapter 26

Tied to this problem is that several factors influence the rate of population growth. For instance, Mauseth mentions generation time and biotic potential. The latter may be further broken down into the number of viable offspring per generative cycle and the number of generative cycles.

Also associated with the concept of population growth rate is the concept of limiting factors. In general, students do not have a problem with the idea that the abiotic and biotic environment limits an individual's growth or that of a population. It is more difficult to visualize that under a particular set of conditions, a single factor is probably limiting. Although it is sometimes easy to identify what the limiting factor might be, there are other cases where it can be quite difficult. Understanding the limits of growth makes it much easier to understand the concept of carrying capacity and the results of exceeding that capacity.

Probably the single most difficult ecological concept to teach is that of niche. Again the problem is due to a misleading colloquial meaning of the same word. In common usage, a niche is a recess in a wall or a suitable place or position. Thus, students tend to associate niche with habitat, a physical place in their environment. In an ecological sense, niche is a much broader concept that includes not only this physical aspect but also all other interactions with the biotic and abiotic community.

Students do not have much difficulty with the idea of a hierarchy of levels within ecosystems, e.g., populations, communities, and ecosystems. Similarly the general idea of different trophic levels is straightforward, e.g., plants photosynthesize to produce their own food, animals eat plants or other animals that ate plants. The relationships between energy, biomass, and nutrient recycling becomes much more difficult, however, when tied to trophic levels in even a relatively simple food web.

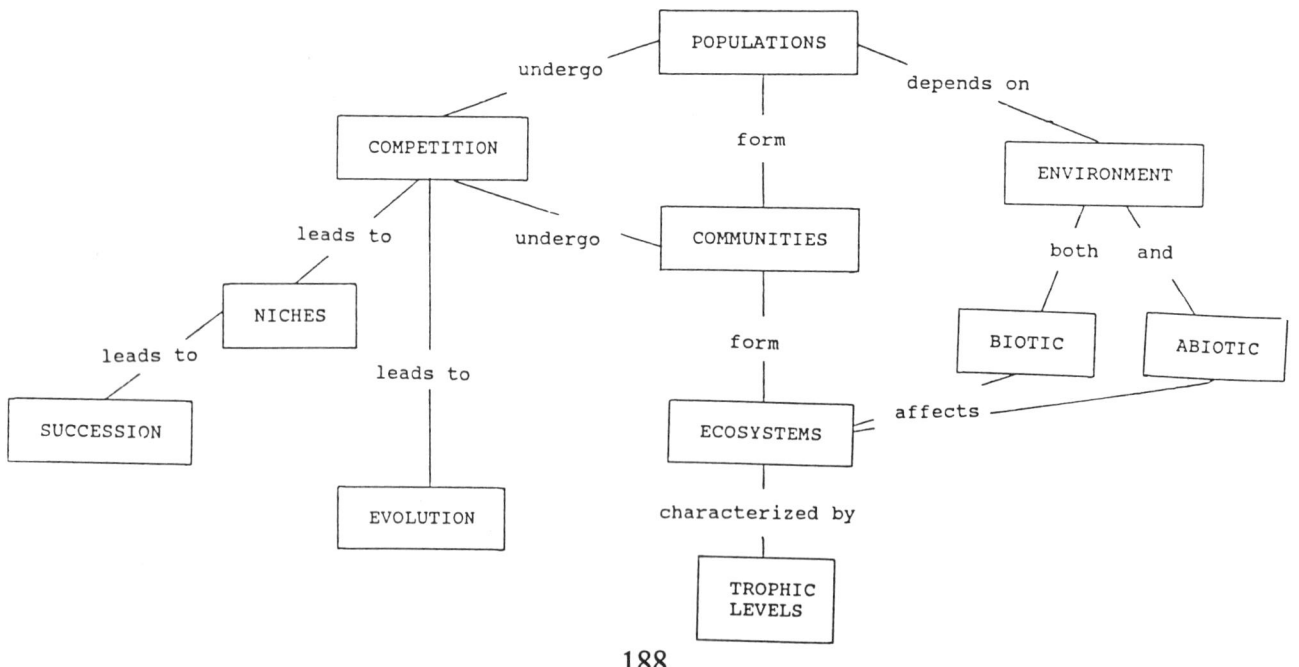

INSTRUCTIONAL AIDS

Lecture - The single most important abiotic factor affecting plants is the sun. In supplying energy, trapped directly by plants in photosynthesis, or providing heat to warm the earth, it also directly drives climate and indirectly is responsible for formation of the various soil provinces. The relationship between the earth and the sun seems obvious, until you ask your students to explain the differences in season. For instance, why is it that in summer the days are 2 hours longer in St. Paul, Minnesota than they are in Miami Beach, while in winter they are 2 hours shorter? The abiotic influence of the sun has a major effect on the distribution of biomes, the topic of the next chapter. The sun also directly affects lower levels of the hierarchy, all the way down to the level of the individual plant.

Given the conditions affecting growth, you can now begin to describe growth more precisely. If we sample a population periodically and estimate the total number of individuals, we can plot population increase on a graph. This increase, however, is not the growth rate. Make an analogy with compound interest earned in a bank account. You place a certain sum of money in the bank at a particular rate of interest. Money accumulates in exactly the same way a population increases. For instance, if you start with one dollar at 10% interest, after the first time it is compounded you will have added a dime and have $1.10 in the bank. The next time interest is compounded it will still be at 10%, but your principle is now $1.10 so $0.11 is added and you now have $1.21. The increment added at the second compounding was larger than at the first, and will continue to increase at each successive compounding. **The rate remains constant, 10%!** In the same way, even with a constant rate of reproduction, the increment of additional individuals may increase from one generation to the next in exponential fashion.

How rapidly a population increases, and what maximum level it eventually attains, will be due in part to inherent factors characteristic of that species, but also to the environment. In many situations, there will be a single environmental factor limiting population growth. If we remove that limitation, for instance irrigating a field during a drought, then growth will increase until another factor becomes limiting. In this example available nitrogen now may become limiting. Removing this limitation, by fertilizing, permits a further increase in growth, until something else becomes limiting. For instance, competition from weeds may become a factor, so you cultivate or use herbicides, etc.

In considering the plant's environment, average conditions tend to be less important than the extremes to which a plant may be exposed. These extreme conditions define the limits to which the plant must be well adapted.

These limits also help to define niche, although niche is more complex than this. In general terms, niche we define as the role of an organism in the ecosystem." Certainly, this includes the physical space occupied by an organism, as well as its requirements, but it also

Chapter 26

includes its contribution to the environment and its effect on other species. Niche therefore includes any waste products produced or secondary metabolites released to the environment that affect other organisms in the community.

Each of the above-mentioned factors lead to a similar result - competition. As a population increases in size, individuals begin to compete with other individuals of the same species as well as with each other. The theory of competitive exclusion suggests a less competitive species are eliminated from a particular niche by another, better adapted species. Such competition, given a range of diversity and the potential for overproduction of offspring, is also the basis for evolution by means of natural selection.

At this point you can introduce two very different strategies employed by plants in order to compete in their environment. We associate each strategy with a suite of characters related to energy. On the one hand energy is expended to reproduce as rapidly as possible. Alternatively, that energy is accumulated in biomass for extended periods of time. Table 26.2 lists some of the characteristics of r-selected and K-selected species. These names come from the growth formulae used to describe typical growth of the population. If a species expends energy in rapid reproduction, the rate of increase, r, is critical. We call this r-selection. The higher this value, the more rapidly a species can take advantage of an opening in the environment. Growth is typically exponential with the increase over time = rN (rate of increase times number of organisms present).

Species that channel their energy more into biomass are called K-selected. The term "K" refers to the carrying capacity reached by a population exhibiting logistic growth. During early stages of growth, increase in population size over time is essentially exponential. Demonstrate this by filling in some hypothetical numbers into the equation for logistic growth, = rN [(K-N)/K]. If the population, N, is small (K-N) approximates K and K/K = 1, therefore growth is approximated by rN as before. As the population increases in size, however, N begins to approximate K. For instance, if the carrying capacity, K, is 1,000,000 and the current population is 999,999, then [(K-N)/K] = 1/1,000,000 and the additional population increase will only be one millionth of rN. Growth levels off as population size approaches carrying capacity.

An interesting exercise is to collect a sample of green pond water, teeming with life. Ask students to sample it, using a series of water mounts, and sketch an example of all of the organisms they see. Each sample may have one or two worms and some insect larvae or small crustacea, and perhaps a good population of large protozoans. Each of these "beasts" will make it into your student's notebooks, but it is likely that not a single alga will be drawn, even though they may be the most numerous of all organisms, by several orders of magnitude! This is the first problem you will encounter with trophic levels and food webs. Plants, especially small ones, are invisible to most students. Once you point this out to students it will be obvious, especially when they recall the process of photosynthesis.

Ecology

The names attached to different trophic levels tend to cause confusion. The primary level of any food web consists of primary producers, or autotrophs, like plants, which can produce their own food. The second level consists of herbivores, that are primary consumers. Third are the carnivores, also known as secondary consumers. The problem students have relates to their tendency to see the terms primary, secondary, tertiary, etc., as nouns that should be memorized, rather than as adjectives used to modify a noun such as consumer or trophic level. Once this is clear, students will not have a problem understanding how a primary consumer can be at a secondary trophic level in the food web. The other problem, and this is more difficult, is that the same organism can be at different levels in the food web. Use human behavior as an example. If you're eating in a vegetarian cafe in San Francisco, you're feeding as a primary consumer. When you're dining on corn-fed beef at an Omaha steak house, you're feeding as a secondary consumer. Dine on alligator etouffee' in New Orleans, however, and you're feeding as a tertiary (or quaternary or higher, depending on what the 'gator ate) consumer!

Pre-lecture Review - To appreciate the essential role of plants in the environment, it is critical to understand the process of photosynthesis. Details are not important. However, remind students that not only is this how energy from an external source, the sun, is fixed into the chemical energy of carbon compounds, but it is also the source of the molecular oxygen aerobic organisms need for respiration. It is also useful to recall that competition between individuals is essential for natural selection to occur (Ch. 17). In this chapter students will look more closely at the nature of competition.

Post-lecture Review - Concept mapping is a device perfectly adapted to a broad, interrelated topic such as ecology. For every way of organizing five concepts, five different organizations will be equally justifiable. Even a cursory examination of ecology concept maps drawn by your students will point out to you concepts that are weak and where you did a good job of making connections.

LABORATORY - EXERCISES

1. **Population Growth**.

When grown under optimal conditions, the bacterium *Escherichia coli* may divide by binary fission even more rapidly than every 20 minutes. Luckily, optimum conditions can be maintained for only a short period of time As an interesting exercise have students calculate how many bacteria could theoretically be produced in 24 hr. and 1 week. As bacterial populations grow in liquid medium, the turbidity of the culture increases as a function of cell growth. Thus, we can use light absorbance as an index of cell number.

You can use the procedure outlined in Chapter 17 used to demonstrate early exponential, and longer term logistic, growth.

Chapter 26

2. Predator/Prey Relationships

Several different factors determine the effectiveness of herbivores feeding on plants. Use a simple simulation, setting up a blindfolded student as predator, to demonstrate some of these.

Blindfold one student in a group, then have the rest of the group scatter 50 beads, either randomly, clumped, or in a uniform distribution, on the surface of the pegboard. The blindfolded "predator" then taps with her/his finger around the board in search of prey. Each time the predator taps a bead, one of the students in the group removes that bead from the board. After one minute, call time and tally the number of captured prey. Repeat this at least three times for each of the different distribution patterns: random, clumped, evenly distributed. Set the pattern after the predator is blindfolded. Students will quickly see that some predators are more effective (aggressive?) than others and an analysis of the data will demonstrate that some patterns are more effective than others in protecting prey. You can make several interesting variations on this model. (see Waddell, 1981 for suggestions.)

3. Sampling a Terrestrial Community

It is usually impossible to examine every member of a community to determine such characteristics of the community as density, the number of individuals of a species in an area. For this reason, ecologists take small samples of the community and extrapolate information about the community as a whole. The goal of sampling is to obtain a representative cross-section of the community. One of the simplest and most widely used techniques for sampling is plot sampling.

In plot sampling, mark an area of known size and examine all individuals within this plot. The characteristics of the community determine the size of the plot. There are special techniques to determine the optimal plot size and number of samples that must be taken.

In wooded areas, a sample size of 1/100 hectare is usually appropriate. The easiest way to mark an area of this size is to cut a piece of cord or rope 5.64 meters long. If one person acts as a pivot holding one end of the rope and another person walks a circle holding the other end of the rope with the rope outstretched, the area inscribed will be 1/100 hectares.

The pivot person for each student group must be randomly positioned within the sample area.. The walker marks the sample area and, the remainder of the group identifies and counts each of the tree species found within the area. A recorder enters this data in a notebook. Take a minimum of four samples in each area. If you divide the class into four groups, each group need take only one sample.

Pool the data from each group and determine density and frequency of each species. Density is defined above. Frequency is the number of samples in which a species occurs expressed as a proportion of the total number of samples taken. For instance, if 12 of 100 trees sampled were white oaks and white oak appeared in 2 of 4 plots sampled, the frequency is 2/4 = 0.5 and the density is 12/0.04 ha = 300 white oak trees/ha. (remember, 4 samples of 1/100 ha each were sampled).

If students measure size of individual trees, they can also calculate an index of dominance. Ecologists define "importance value" as the sum of relative frequency, relative density, and relative dominance. Use it to provide an overall measure of the influence of a plant species in a community. (see Brower and Zar, 1977, for details)

4. Microecosystems

Microecosystems are set up in advance by collecting water samples from a local river or lake at weekly intervals for a month and every other day for the week before class. Sample these systems to determine, as best as possible, the food webs represented and changes in community structure associated with succession.

A different microecosystem should be available for each group in the laboratory. Their first task is to sample the ecosystem and record the kinds and relative numbers of each type of organism present. No identifications are necessary, simply place organisms into the following categories: phytoplankton, zooplankton, worms, arthropods, and "other large beasts." Label different species in a category simply $p1$, $p2$, $z1$, etc. Students should make a sketch to illustrate each of their organisms. This will permit students from different groups to compare their data.

It is important in sampling to randomize your location in the study site. Swirl the jar gently before sampling. Each student in the group should prepare at least one sample by placing a drop of sample on a clean microscope slide, adding a coverslip, and systematically scanning the preparation in search of organisms. Each student should make at least three replicate samples before the group pools its data. As organisms are "identified", estimate their relative numbers. In this way, by knowing the size of the organism, relative numbers, and whether or not it is photosynthetic, the group should be able to construct a probable food web.

By examining group data, collected at different times (therefore including cultures of different ages), students can see some indication of changes that occur during succession. Similar species will generally appear in two or three groups' data that were collected on successive collection dates. Some of the species that are very abundant in the youngest, most recently collected, samples will disappear from older samples, while "new" organisms, especially some of the larger and more active beasts, will suddenly appear in older cultures when there was no evidence of them previously. In general, the youngest cultures will be dominated by algae, the mid-aged cultures will probably have the greatest diversity, and the oldest cultures will be declining in species richness and numbers of individuals.

Chapter 26

LABORATORY - MATERIALS, EQUIPMENT, AND PREPARATION

1. **Population Growth**
   see Chapter 17

2. **Predator/Prey Relationships**
   per student group
   -2ft x 4ft sheet of pegboard
   -50 beads or seeds, large enough not to jam in pegboard holes
   -blindfold
   -watch or timer
   optional: a series of different-sized "walls" made out of furring strips and fitted with dowel pegs to hold them in position on the pegboard. These barriers are used to simulate varying degrees of habitat complexity.

3. **Sampling a Terrestrial Community**
   per student group
   -measuring tape or rope cord 5.64 m long
   -key to local tree species

4. **Microecosystems**
   -quart jars containing water samples from freshly collected to about 1 month old. At least one different-aged sample per student group.
   -droppers
   -compound microscopes
   -slides and cover glasses

REFERENCES

General:

Carson, R. 1962. *Silent Spring* Houghton Mifflin, New York. 304 p. The gloom is almost overbearing, but it shocked the country into thinking about cleaning up the environment. Highlights the role of biological magnification up trophic levels in food webs.

Ehrlich, P.R. 1968. *The Population Bomb*. Ballantine, New York. 223 p. The book is old but the arguments helped create the first Earth Day - and are essentially still valid.

Leopold, A. 1949. *A Sand County Almanac*. Oxford University Press, Oxford. 295 p. This is the book that began scientists thinking about a land ethic.

Resources:

Brower, J.E. and J.H. Zar. 1977. *Field and Laboratory Methods for General Ecology.* Wm. C. Brown Co., Dubuque, IA. 193 p. A clearly written manual of many commonly used ecological techniques, including worked out examples for all calculations.

Smith, R.L. 1980. *Ecology and Field Biology, 3rd ed.* Harper and Row, New York. 835 p. A standard textbook of ecology for background information.

Waddell, J. 1981. Predator-Prey Simulation Exercises for the Classroom, in: J.V. Glase, ed., *Tested Studies for Laboratory Teaching, Vol. 2.* Kendall/Hunt, Dubuque, IA. pp 169-180. Describes a number of variations on the laboratory exercise suggested here.

Wagner, C.K. 1984. The Use of Aquatic Research Microecosystems in the Biology Teaching Laboratory in: Harris, C.L., ed. *Tested Studies for Laboratory Teaching, Vol 4.* Kendall/Hunt, Dubuque, IA, pp 59-70. Includes optional exercises for the microecosystem laboratory.

CHAPTER 27: BIOMES

The chapter on biomes can be one of the most interesting chapters for students. Many of your students will have travelled to different parts of the country, perhaps camping and hiking in national parks, so they will be able to relate to the information you present. They might also have interesting contributions to make. Some students will have never been out-of-state. They will be interested in seeing and hearing about other parts of the country. We are lucky because all of the major biomes, except Tropical Grasslands and Savanna, are well represented in one or another of the states and territories. If you don't have a good selection of slides or videotapes to illustrate various biomes, check with the geographers on campus. They will probably have exactly the kind of materials you need.

OBJECTIVES

1. Describe the environmental factors that differentiate one biome from another.
2. Understand how the sun, earth's rotation, continental geography and wind and ocean current circulation patterns combine to produce major climatic patterns.
3. Describe how the overall characteristics of major biomes change as one travels farther away from the equator or as one ascends from sea level up a mountain.
4. Recognize the 12 major terrestrial biomes by the characteristic plant cover.

CONCEPTS

There are several conceptual difficulties at the beginning of this chapter related to the physical factors responsible for climatic patterns. The first is the relationship between the tilt of the earth's axis of rotation vs orientation to the sun during different seasons. Most students realize that the sun is higher overhead, and the days are longer, in the summer than they are during the winter. They also realize that the earth travels around the sun. Two common misconceptions are that either the earth is closer to the sun in summer or that the angle of rotation itself shifts from season to season. Both the angle of the earth's rotation and its trajectory around the sun stay the same, but because of the earth's angle, the relative position of the sun shifts from season to season.

Directly related to the position of the sun is the effect of heating on the atmosphere. Most students will know that warm air rises, but many of them will not realize that as it rises it also cools. Fewer students will realize that warm air holds more moisture than cold air until you ask them why moisture forms on the outside of a glass of ice water in summer or on a cold glass window in winter. Putting these concepts together, with the effects of the resulting convection currents, is difficult for many students.

The relationship between air temperature, moisture holding capacity, and elevation are also critical concepts to understanding rain shadow effects. The classic example of rain shadow occurs on the east side of the Cascade and Sierra Nevada Ranges of the western U.S., resulting in the cold and hot deserts of the west.

The final concept that causes some students difficulty is continental drift. When first proposed, early in this century, scientists were also skeptical of the idea of continents drifting over the surface of the earth. Today the theory of plate tectonics is widely accepted by scientists but still beyond comprehension by many students. Mauseth presents some of the good evidence that is used to determine previous location of continents, but more impressive to students is that we can literally measure that movement today.

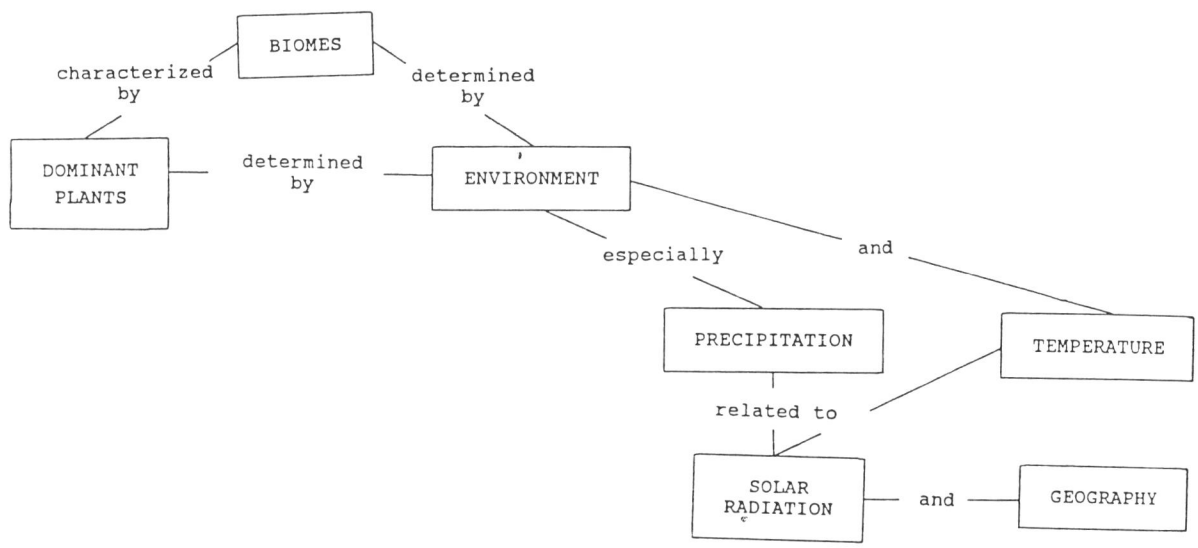

INSTRUCTIONAL AIDS

Lecture - Begin by asking your students to explain the relationship between the earth and the sun, first on a daily basis, then on a yearly basis. If the earth's rate of rotation around its axis is constant from day to day, why is it that in the U.S. days are longer in the summer than they are in the winter? If students suggest that the earth is closer to the sun in summer, ask why it can then be winter in the Southern hemisphere? If the earth is closer to the sun, shouldn't it also be warmer there? The best way to demonstrate the relationship between the earth and the sun is to bring a globe into class and walk it in a circle around the podium, keeping the angle of rotation constant. Position "observers" with a carpenter's level at key locations, summer and winter solstice and spring and fall equinox, to determine at which latitude the sun would be directly overhead and direct rays of light would be striking earth.

Now ask what happens to light energy when it strikes the earth? The absorption of light, then reradiation of infrared is the basis of warming of the air surrounding the earth. This is also the principle behind the greenhouse effect. What happens to the air as it is heated? Draw an analogy to a hot air balloon and why balloonists periodically fire the burners during a flight. In a building, warm air also rises such that it is warmer near the ceiling of a room than near the floor. Does the same thing happen in the atmosphere? Ask if any students have driven through the Rocky Mountains in summer. Ask one of your "volunteers" what the temperature was like

Chapter 27

at the top of one of the passes compared to the valley floor several thousand feet lower in elevation. Of course it is considerably cooler at higher elevations; as the air moves further and further from the source of radiant heat, it gradually cools, condenses and will begin to descend.

Related to the temperature of the air is its moisture holding capability. Students in most northern states will realize that the air is drier in the cold of winter while students in the southeastern states fully understand the moisture-holding capability of warm gulf air. In any area, the phenomenon of moisture condensation on a cold surface will be familiar at some time of year. Your challenge will be to relate the change in temperature with elevation to the moisture holding capability of that air - both as it rises and cools, and descends and rewarms. Take the process one step at a time, as Mauseth has done, beginning in the tropics with warming and expanding air picking up moisture and rising, then cooling, loosing moisture, compressing and falling in a giant cyclical current.

Once you establish the general relationships between temperature, elevation, and moisture- holding capacity, you are ready to consider the special case involving movement of air masses driven by prevailing winds over mountains. Especially near seacoasts, such patterns frequently result in rain forests on the windward slope and deserts on the leeward side - only miles apart. The usual example occurs in the Pacific Northwest with temperate rain forests on the west slope of the coast range and desert (or at least dry grassland) on the east side of the Cascade Range. The Hawaiian Islands provide an even better example of the effect of mountains on local rainfall. The windward side of Kawaikini Peak on the island of Kauai receives over 485 inches of rain per year (it reached 624 inches in 1948!), yet on the leeward side of each island it is dry. Similarly, tropical rainforest occurs on the windward side of the big island, while the Kau Desert on the leeward side of Mauna Kea and Mauna Loa is one of the most arid regions in the U.S.!

Extreme climatic fluctuations characterize the central regions of large land masses, especially if further isolated from the moderating effects of the oceans by mountain ranges. If you live in the west-central states, your students will be very familiar with the extreme conditions characteristic of a continental climate. Students from other parts of the country may not believe that in summer it can reach over 110° F in Fargo, North Dakota, while in winter it can reach below -30°F! A student from Boston, Miami, San Diego, or San Francisco has probably not experienced either extreme. The moderating effect of large bodies of water, as well as the topography of continental masses, have major impact on the climate of an area.

Whether or not a continental climate develops is, in part, a function of the position of the continent. The land mass of Brazil, for instance, is large enough to expect a more variable climate in the interior. Yet most of the country lies within the tropics and rain forest extends the entire length of the Amazon from the Atlantic coast to the eastern slope of the Andes. The location of continents is thus an important factor in determining their climate; and a factor that changes through time. Mauseth mentions some of the data we use to establish past continental positions. More impressive to students is that this continental movement continues today. Obvious examples include horizontal displacement along earthquake faults. At times, shifts of more than 40 feet have been recorded along fault lines and spreading can be observed along volcanic fissures of the Atlantic Ridge. With satellite and laser technology, we can measure the annual increase in distance between Europe and North America to within a fraction of an inch. Similarly, increases in the elevation of the Himalayan Mountains and islands of the Caribbean can be accurately measured with these techniques.

Given the characteristic climates of the world, it is a fairly simple process to predict the which biome will associate with that region. A simple relationship between average temperature and average precipitation has a high level of predictability in determining biome type. The best aid to talking about the various biomes is to have a collection of illustrations to demonstrate what you're talking about. Several commercial companies have sets of biome slides available, but perhaps the best resource is the plant geography slide set available through the Teaching Section of the Botanical Society of America. A slide-show field trip from the arctic tundra of Alaska, down through the great lakes states, around the Atlantic and Gulf coasts, across the prairies and plains, over the mountains, through the deserts, across the Sierra's to the chaparral, then up the coast to the Olympic Peninsula (with a quick final side trip to Hawaii and Puerto Rico) can easily provide two interesting days of biome identification and ecological diversions.

Pre-lecture Review - We briefly covered abiotic factors limiting plant distribution in the previous chapter. Call on students to recall these factors and list them on the board. Most of the factors will have at least an indirect relationship to energy from the sun. Use this relationship to prompt a more careful examination of the role of the sun in driving major climatic factors.

Post-lecture Review - You can use ecology, like evolution, to bridge a variety of subdisciplines and tie things together. Structural and functional adaptations, as well as the process of evolution itself, depend on the local environment in which a plant lives. At this point, students can summarize the field of botany by drawing a generalized concept map relating each of the topics covered during the course. Their map should now illustrate an interconnected network of relationships, rather than the linear arrangement of topics, as outlined in the table of contents. Food webs were once described as having everything connected to everything else. Student's concept maps of this course should illustrate this statement. Every aspect of the study of plants is connected to every other aspect.

LABORATORY - EXERCISES

1. **Biomes**
MacArthur suggested that the interrelationship of just two factors, average annual temperature and average annual precipitation, should provide a reliable prediction of the biome which will predominate in an area. Make copies of his graph available to students, then supply

Chapter 27

climatic data for a number of reasonably well-known cities from around the U.S. and ask students to plot the data for each city and predict the biome in that area.

Obtain climatic data from the weather bureau, from Walter's Atlas or from monthly tables published in the 1941 U.S.D.A. Yearbook of Agriculture, *Climate and Man*. The latter is especially useful because of the numerous recording sites within each state. Perhaps you can pick out interesting examples within your own state. Some interesting national examples, less than 100 miles apart, include: Seattle (grassland/shrub land) and Quinault (temperate rain forest) Washington; and Flagstaff (temperate forest) and Phoenix (desert), Arizona. Some surprises include: Juneau, Alaska (temperate rain forest) and Los Angeles (desert).

REFERENCES
General:
Farb, P. 1963. *Face of North America*. Harper and Row, Publishers, New York. 316 p. A brief, popular introduction to the geology and natural history of the continent.

Gleason, H.A. and A. Cronquist. 1964. *The Natural Geography of Plants*. Columbia University Press, New York. 420 p. A good non-technical introduction to the factors influencing plant distributions.

Hunt, C.B. 1974. *Natural Regions of the United States and Canada*. W.H. Freeman and Co., San Francisco. 725 p. Similar to the above, but slightly more technical.

Shelford, V.E. 1978. *The Ecology of North America*. University of Illinois Press, Urbana, IL. 610 p. A survey of the major biomes of North America, including characteristic animals associated with various plant formations.

Vankat, J.L. 1979. *The Natural Vegetation of North America: An Introduction*. John Wiley and Sons, New York. 261 p. Concentrates on the major vegetation types.

Resources:
MacArthur, R.H. 1972. *Geographical Ecology: Patterns in the Distribution of Species*. Harper & Row, New York. 269 p. This is a very technical treatment that considers the theoretical basis for basic patterns of distribution.

U.S.D.A. 1941. *Climate and Man*. United States Department of Agriculture, Washington, D.C. 1248 p. This yearbook of agriculture contains extensive state-by-state climatic tables and maps.

Walter, H. and H. Lieth. 1967. *Klimadiagramm-Weltatlas*. Gustav Fischer Verlag, Jena. This oversize volume is a collection of maps and climate diagrams from around the world. These diagrams have become the standard way to characterize the climate of a location.

APPENDIX A, KEYS TO SELECTED TREES OF THE EASTERN UNITED STATES

A KEY TO SELECTED TREES OF THE SOUTHEAST

A. Leaves usually needle-like and evergreen; plants usually conebearing; seeds not produced within a fruit..Gymnosperm

    B. Leaves clustered ....*Pinus* (Pines - see genus key)

    BB. Leaves opposite C

        C. Leaves feather-like, spreading in a flat plane on the sides of the branchlets, 6-8 cm long. A large tree with a buttressed (expanded) base and irregular root outgrowths (knees) ....*Taxodium distichum* (Bald Cypress)

        CC. Leaves scale-like or needle-like, in four rows along the stem, overlapping. Fruit a blue berry ....*Juniperus virginiana* (Red Cedar)

AA. Leaves usually flattened, deciduous or evergreen; plants bearing flowers with seeds in fruits ....Angiosperm

    D. Leaves opposite ....E

        E. Leaf margins entire (occasionally with 1 tooth per side), blade more than 9 cm wide, tubular flowers up to 5 cm long in dense clusters; fruit a pod 13-40 cm long with many winged seeds ....*Catalpa bignonioides* (Southern Catalpa)

        EE. Leaf margins toothed or lobed ....*Acer* (Maples - see genus key)

    DD. Leaves alternate ....F

        F. Leaves simple ....G

            G. Leaves lobed or undulate ....H

                H. Leaves 4 lobed, notched at tip ....*Liriodendron tulipifera* (Tulip Tree)

                HH. Leaves not as above I

                    I. Base of petiole enlarged, forming a "cap" over the lateral bud; leaves 3-5 lobed with sinate toothed margins, stipules leaf-like, fruit a spherical head about 2 cm in diameter ....*Platanus occidentalis* (Sycamore)

                    II. Petiole base not encapping bud ....J

                        J. Leaves "star-shaped" with 5-7 triangular lobes; fruit a spiny, woody ball .... *Liquidambar styraciflua* (Sweet Gum)

Appendix A

        JJ. Not as above ..... *Quercus* (Oaks - see genus key)

    GG. Leaves not lobed .... K

      K. Leaves entire, 12-25 cm long by 4-7 cm wide, evergreen, leathery, dark shiny green above. Flower large, 15-25 cm wide, white .... *Magnolia grandiflora* (Southern Magnolia)

      KK. Leaves toothed... L

        L. Leaves 4-10 cm long, evergreen; teeth, if present, tipped by a sharp spine. Fruits bright red, globose .... *Ilex opaca* (American Holly)

        LL. Leaves not spiny .... M

          M. Leaves with three main veins from the base of the blade, bark with corky warts; fruit orange-red, globose .... *Celtis laevigata* (Sugarberry)

          MM. Leaves with a single major vein, doubly toothed, with obliquely asymmetrical leaf bases. Fruits with an oval wing. .... *Ulmus americana* (American Elm)

  FF. Leaves Compound, fruit a nut-like drupe .... *Carya illinoiensis* (Pecan)

Genus Key

Acer

A. Leaf sinuses rounded at base, palmately 3-5 lobed, margin ± smooth. .... *Acer saccharum* (Sugar Maple)

AA. Leaf sinus sharply angular at base.. B

  B. Leaves deeply lobed (at least half the distance from the tip of the adjacent lobe to the insertion of the petiole), silvery white beneath .... *Acer saccharinum* (Silver Maple

  BB. Leaves not deeply lobed nor silvery beneath, margins toothed .. *Acer rubrum* (Red Maple)

Quercus

A. All leaves definitely lobed .... B

B. Lobes bristle tipped, leaves up to 25 cm long, 3-5 lobed, terminal lobe with smaller, bristle tipped teeth .... *Quercus falcata* (Southern Red Oak)

BB. Lobes rounded.. C

C. Leaves 15-25 cm long, 7-9 lobed, sinuses extending nearly to mid-rib. Cap of acorn enclosing 1/4 of length of nut .... *Quercus alba* (White Oak)

CC. Leaves 10-25 cm long with many shallow lobes. Acorn cap enclosing nearly half the length of nut .... *Quercus michauxii* (Cow Oak)

AA. Leaves not lobed or shallowly lobed at tip only .... D

D. Leaves evergreen, 5-15 cm long, oblong, dark shiny green above, margins entire .... *Quercus virginiana* (Live Oak)

DD. Leaves decidouos, not as above .... E

E. Leaves entire, narrow, 4-10 cm long by 1 cm wide, "willow-like" .... *Quercus phellos* (Willow Oak)

EE. Leaves entire or slightly lobed, broadest at tip and tapering toward base .... *Quercus nigra* (Black Oak)

Pinus

A. Needles in clusters of 2's only, 3-9 cm long; branchlets up to 3 cm diameter, smooth. Bark on trunk breaks into small, flaky plates ....*Pinus glabra* (Spruce Pine)

AA. Needles in clusters of 3's only, 15-30 cm long, cone sessile, often 2-3 in a cluster, bark breaks into large, deep plates .... *Pinus taeda* (Loblolly Pine)

AAA. Needles in clusters of 2's and 3's .... B

B. Leaves usually in 2's, 9-15 cm long, branchlets rough. Bark on trunk breaks into large flat plates .... *Pinus echinata* (Shortleaf Pine)

BB. Leaves at least 20 cm long C

C. Leaves usually in 3's. 20-50 cm long, in dense spreading tufts at the ends of branches. Bark irregular and scaly .... *Pinus palustris* (Longleaf Pine)

CC. Leaves 20-36 cm long, appressed to branches, not fanning out into tufts, cones stalked, usually solitary, bark breaks into large plates . *Pinus elliottii* (Slash Pine)

Appendix A

A KEY TO SELECTED TREES OF THE NORTHEAST

A. Leaves usually needle-like and evergreen; plants usually conebearing; seeds not produced within fruit .... Gymnosperms

    B. Leaves alternate, scattered, or clustered C

        C. Leaves all alternate or scattered .... D

            D. Trees .... E

                E. Leaves 4-sided, on prominent bases; cones pendant .... *Picea* (Spruce - see genus key)

                EE. Leaves flattened .... F

                    F. Leaves bright green on both sides; trifid bract scales projecting from the cones .... *Pseudotsuga taxifolia* (Douglas Fir)

                    FF. Leaves whitish beneath; bract scales included in cones G

                        G. Leaves sessile; cones large (5-13 cm), erect .... *Abies balsamea* (Balsam Fir)

                        GG. Leaves petioled; 2-ranked; cones small (<3 cm), pendant .... *Tsuga canadensis* (Hemlock)

            DD. Shrubs; leaves flat and sharp pointed .... *Taxus* (Yew-genus key)

        CC. At least some of the leaves clustered .... H

            H. Leaves deciduous, many per cluster .... *Larix laricina* (Larch)

            HH. Leaves evergreen, 2-5 per cluster .... *Pinus* (Pine - genus key)

    BB. Leaves opposite or in whorls of 3 .... J

        J. Leaves in whorls of 3, needle-like, jointed at the base .... *Juniperus* (Juniper - see genus key)

        JJ. Leaves opposite, scale-like or needle-like, not jointed .... K

            K. Leaves all scale-like; twigs flattened; pairs of leaves of two kinds alternating .... *Thuja canadensis* (Arbor vitae)

KK. Leaves scale-like or needle-like, all alike on the same twig, twigs not flattened .... *Juniperus* (Juniper - see genus key)

AA. Leaves usually flattened, deciduous or evergreen; plants bearing flowers with seeds in fruits ....Angiosperms

    L. Leaves opposite .... M

        M. Leaves simple .... *Acer* (Maples - see genus key)

        MM. Leaves compound .... N

            N. Petiole bases almost meeting around twig; leaflets often lobed; twigs often shiny green .... *Acer negundo* (Boxelder)

            NN. Petiole bases widely separated; leaflets not lobed; twigs gray .... *Fraxinus* (Ashes - see genus key)

    LL. Leaves alternate .... O

        O. Leaves simple .... P

            P. Leaves lobed or undulate .... *Quercus* (Oaks - see genus key)

            PP. Leaves not lobed .... Q

                Q. Petioles flattened near blade. *Populus* (Aspens - see genus key)

                QQ. Petioles not flattened .... R

                    R. Leaves broad, heart-shaped; petioles > 3 cm long .... *Tilia americana* (Basswood)

                    RR. Leaves longer than broad; petiole < 3 cm long .... S

                        S. Leaves with 3 major veins from base; bark with corky worts ...... *Celtis occidentalis* (Hackberry)

                        SS. Leaves with only 1 major vein .... T
                        T. Leaves singly toothed....................U

                            U. Brown hairs on under midrib; bark with large scales, petioles glandular .... *Prunus serotina* (Black Cherry)

Appendix A

        UU. Not as above, leaves at least 5 times as long as wide. .... *Salix* (Willow species)

    TT. Leaves doubly toothed V

        V. Leaf bases obliquely asymmetrical ...*Ulmus* (Elms - see genus key)

        VV. Leaf bases symmetrical .... W

        W. Bark papery *Betula*.(Birch - see genus key)

        WW. Bark not papery *Ostrya virginiana* (Hornbeam)

O. Leaves compound X

    X. 13-17 leaflets; chambered pith in twigs . *Juglans* (Walnut - see genus key)

    XX. 5-9 leaflets, pith not chambered, buds yellow .... *Carya cordiformis* (Yellowbud Hickory)

Acer

A. Lobes with rounded sinuses .... *Acer saccharum* (Sugar Maple)

AA. Lobes with sharp sinuses .... B

    B. Leaves deeply lobed, silvery beneath .... *Acer saccharinum* (Silver Maple)

    BB. Leaves not deeply lobed, not silvery .... *Acer rubrum* (Red Maple)

Betula

A. Young bark yellow, twigs with wintergreen flavor .... *Betula lutea* (Yellow Birch)

AA. Young bark white or red, twigs without wintergreen flavor .... B

    B. Leaves oval, young bark white .... Betula papyrifera (White Birch)

    BB. Leaves diamond-shaped, young bark reddish .... *Betula nigra* (River Birch)

Fraxinus

A. Twigs with four corky ridges, square in cross-section .... *Fraxinus quadrangulata* (Blue Ash)

AA. Twigs without corky ridges, round in cross-section .... B

    B. Leaflets sessile .... *Fraxinus nigra* (Black Ash)

    BB. Leaflets petiolate C

        C. Leaflets pale beneath; leaf scars crescent-shaped, notched at top
.... *Fraxinus americana* (White Ash)

        CC. Leaflets green on both surfaces; leaf scars semicircular, straight on top .... D

            D. Twigs and main rachis of leaves hairy . *Fraxinus pennsylvanica* (Red Ash)

            DD. Twigs and rachis smooth
.... *Fraxinus pennsylvanica* var. *lanceolata* (Green Ash)

Juglans

A. Leaflets with sticky hairs beneath, more or less straight-sided; nuts longer than broad, pith chocolate color .... *Juglans cinerea* (Butternut)

AA. Leaflets without sticky hairs, widest near the base, nuts nearly spherical, pith light brown
.... *Juglans nigra* (Black Walnut)

Juniperus

A. Prostrate shrubs .... B

    B. Leaves in whorls of 3, all needle-like ; very sharply pointed
.... *Juniperus communis* var. *depressa* (Prostrate Juniper)

    BB. Leaves mainly opposite, needle-like and scale-like
.... *Juniperus horizontalis* (Creeping Juniper)

AA. Upright trees and shrubs C

    C. Scale leaves acute; needle leaves without white stripes above; blue berries
.... *Juniperus virginiana* (Red Cedar)

Appendix A

    CC.   Scale leaves obtuse; needle leaves marked on top with 2 white stripes .... *Juniperus chinensis* (Chinese Juniper)

Picea

A.   Twigs spread horizontally or ascending; cones < 8 cm long .... B

    B.   Leaves to 2 cm long; projecting forward on the branches C

        C.   Leaves dark green; twigs pubsecent; found mainly in bogs .... *Picea mariana* (Black Spruce)

        CC.   Leaves pale green, sharply pointed; twigs glabrous .... *Picea glauca* (White Spruce)

    BB.   Leaves 2-3 cm long; projecting at right angles from the twigs .... *Picea pungens* (Blue Spruce)

AA.   Twigs pendulous; cones 10-20 cm long .... *Picea avies* (Norway Spruce)

Pinus

A.   Needles in clusters of 5's .... *Pinus strobus* (White Pine)

AA.   Needles in clusters of 2's, sometimes 3's .... B

    B.   Trees C

        C.   Needles 10-15 cm long .... D

            D.   Needles dull green, stiff; branchlets yellow-brown; cone scales with short spines .... *Pinus nigra* (Austrian Pine)

            DD.   Needles lustrous green, soft and flexible; branches red-brown; cone scales unarmed .... *Pinus resinosa* (Red Pine)

        CC.   Needles < 8 cm long .... E

            E.   Needles to 4 cm long; cones curved and pointed forward on branches; cone scales unarmed .... *Pinus banksiana* (Jack Pine)

            EE.   Needles 4-8 cm long; cones pointed backward on branches; cone scales with small prickles; branches bright orange-red .... *Pinus sylvestris* (Scotch Pine)

Populus

A. Leaf triangular in outline, base straight or slightly heart shaped . *Populus deltoides* (Cottonwood)

AA. Leaf more or less circular in outline, base rounded .... B

    B. Margin of leaf with large teeth (<12 per side)
        .... *Populus grandidentata* (Large Tooth Aspen)

    BB. Margin of leaf with small teets (>12 per side)
        .... *Populus tremuloides* (Quaking Aspen)

Quercus

A. Leaf lobes with bristles from points......................................B

    B. Sinuses narrow toward base, reach up to 3/4 of the way to mid-rib; acorns in saucer-shaped cups, buds smooth .... *Quercus rubra* (Red Oak)

    BB. Sinuses wide throughout, up to 9/10 of the way to midrib, acorns in deep cups, buds often hairy .... *Quercus ellipsoidalis* (Northern Pin Oak)

AA. Leaves without bristles from lobe points C

    C. Leaves with several deep sinuses; bark whitish and easily rubbed off
        .... *Quercus alba* (White Oak)

    CC. Leaves with only one deep sinus on each side near base, bark deeply fissured
        .... *Quercus macrocarpa* (Bur Oak)

Ulmus

A. Upper leaf surface very rough, twigs and buds hairy, bark layers all brown
    .... *Ulmus rubra* (Slippery Elm)

AA. Leaf only slightly rough, twigs sparcely hairy, bark layers alternating dark brown and white .... *Ulmus americana* (American Elm - now rare)

Appendix A

APPENDIX B. FIELD TRIPS

The ultimate way to study biomes is to visit them. Below are four suggested 3-week field trips that are appropriate for an interim or summer school session. For convenience, each begins at St. Louis, then visits a number of different biomes on a transect to a different region of the country. The minimum number of biomes visited is five; some trips pass through as many as seven. In order to gain a real appreciation for any particular biome, you plan to spend at least two, and preferably four days, in a particular location. This means that some days will be hard driving (literally). Plan your route and itinerary with care! The first day will be your best for traveling; your students will be fresh and excited and you can get an early start if you pack and load vehicles the day before you leave. Set an early departure time, the earlier the better. Have lunch (and if possible dinner) pre-prepared to minimize stopover time and set-up time in the evening. The first day or two out, anticipate that it will take at least an hour more than you think for the group to get up, eat, take down and pack. By the third day, groups will understand their responsibilities and begin to act as coordinated units.

Planning should begin at least 6 months in advance, earlier is better. If you plan to do any collecting (this is a great way to build up teaching collections of preserved and herbarium specimens), you will have to get permission from the appropriate agency well in advance. Take advantage of national forests and national parks. The former usually will let your group collect specimens, provided you notify them in advance what you plan to collect and report to them later what you actually collected. They also have great, although "primitive," campsites. National parks are great places to stay. You will not be able to collect specimens, but group camps are free to educational institutions (reserved in advance), and showers are available. On a three-week trip, regular showers and an occasional recreation stop go a long way toward improving morale. You can usually arrange for a park naturalist or scientist (several of the larger parks have their own and many parks have visiting scientists working on particular problems in the park) to speak to your group, either at your site or in a classroom building (the latter is an added luxury if your group is experiencing a late-season snowstorm in the Rockies). Certain states have very strict restrictions on collecting, notably Arizona. Be sure you contact the states you plan to visit to check on restrictions - they will also frequently supply you with additional information about areas of interest you may want to add to your itinerary.

At each of the major sites where you stay, have the class do a habitat analysis and at least one form of vegetative sampling. In some instances, you will want to try two different techniques to compare the effectiveness of each. At some sites, particularly in the deserts and shrub-lands of the west, it will be interesting to do an analysis of distribution patterns. Age-structure analysis can be interesting both in old-growth sites and disturbed areas. Estimate age by size, or, if you have increment borers you can count tree rings. Divide the class into groups and assign an ongoing project of making a field collection for your herbarium. You will want assigned groups anyway to assign rotating camp chores like cooking, cleaning up, loading and unloading vehicles, etc. specimens. Each group will need field presses and appropriate identification keys or manuals to identify and preserve its specimens.

Begin recruiting far in advance and obtain a non-refundable down-payment from students. This will give you a committed number to work with in logistical planning and in making reservations. Provide students with a list of everything they will need (from sunscreen and sunglasses to toothpaste, to how many changes of clothes). Limit them to one carry-one backpack, a sleeping bag, and one additional bag. Strongly discourage hair dryers, make-up, large radios, tape decks and the like. Be sure that everyone labels their clothing with permanent markers   You will have to do two or three group washes and labels will facilitate sorting afterwards.

A.   **The Pacific Northwest**

On this trip you will see temperate rain forests, montane and subalpine forests, alpine tundra, grasslands, temperate deciduous forest, shrublands, and cold desert.

With an early departure you can make it to the Konza Prairie, south of Manhattan, Kansas, by lunch. The Konza is one of the largest prairie preserves in the U.S. and has been termed "the most significant" tall grass prairie remnant. Traveling across what was originally prairie will probably "be boring" to your students, but in the right season, you may be able to get a feel for the "prairie as a garden" as expressed by some of the early sodbusters. You should be able to reach Colorado in time to set up camp. Plan to spend at least two days in Rocky Mountain National Park. In early summer there will still be snow at higher elevations so this may limit your itinerary. The Front Range is impressive, rising out of the plains. You will have numerous opportunities to examine the influence of microhabitat on plant distribution. This will also be your first opportunity to do a transect up a mountain to above tree line.

By one route or another, head toward Salt Lake City. A northern route along I-80 allows you to follow the historic Oregon Trail. A southerly route along U.S. 40 will provide a better "feel" for the Rocky Mountains, but it probably will cost you an extra day of travel time. In the Salt Lake area, you will want to stop at the lake itself and probably make a side-trip to the old Bingham Canyon Copper Mine. The latter is one of the largest man-made pits in the world, and the associated smelter was the largest point-source air polluter in the country.

From Salt Lake, follow I-84 to Pendleton, Oregon. This transect across the northern great basin and cold desert region is a good place to look closely at patterns of plant distribution and to discuss the effects of overgrazing. Sagebrush, *Artemesia*, is easily the most abundant and conspicuous component of the "natural" flora. There is also big agriculture - and big irrigation.

Just out of Pendelton, you will have to make a decision. You could stay on I-84 along the Columbia River valley through the gorge. This is an interesting geological and biogeographical feature in itself. Alternatively, you can cross the river and catch U.S. 12 near Pasco. The former route will permit you to stop at Mt. St. Helens, the latter will bring you to Mt. Ranier. Either place is worth at least a day's stop-over, but your real destination at this time is Olympic National Park, where you will want to have three or four days. The drive on U.S. 101 from Aberdeen is educational in the first-hand views of clear cutting at its "best." In the park itself, plan to spend least one full day in the rain forest and at least another full day on the coast. Be sure to check the tide schedules. Even then you may have to wade on some of the coastal hikes. The transect drive to Hurricane Ridge can take another full day as can a visit to the indian villages between the park and the coast. There is an interesting archeological excavation on the Ozette Indian Reservation at Cape Alava.

At this point you are half way out and ready to begin heading east. It is a good time to take off a day and allow your students a day to be tourists. You have two excellent choices. One alternative is to head directly east and drive (and/or ferry) to Seattle. The areas between Waterfront Park (and Seattle Aquarium) and Pioneer Square have lots of interesting things to see and do. My preference is to add a bit of an international flavor and take the ferry North to Victoria, B.C. for the day. If you take this

Appendix B

alternative, be forwarned that Canadian customs will want to inspect any plants being brought into the country. This includes pressed herbarium specimens. If you already have collections started, it will save much time to have fruit species and gymnosperms easy to pull out for inspection. Plant quarantine is looking primarily for rust fungi. High tea at the Empress Hotel is by reservation. Make reservations ahead of time if some of your students are be interested in partaking of this British custom.

If you ferry back from Victoria to Anacortes, try to schedule a tour of the Oceanographic Laboratory at Friday Harbor. Later, check out the Breazeale Interpretive Center and the Padilla Bay National Estuarine Sanctuary in Bay View, WA, when you disembark on the mainland. From Seattle, just head north. The transect across the Cascades will be more dramatic from west to east than it was from east to west, especially after having visited the Olympic Peninsula. Your students will now really understand what is meant by rain shadow. The Grand Cooley Dam is a good lunch stop and a chance to see how western rivers have been "harnessed" to provide electrical power and irrigation water. As soon as you hit Idaho, you are back in the Rockies and there will be plenty of places to stop. Your main objective at this point, though, is Yellowstone, where you will want to spend at least three or four days. You can't visit Yellowstone without going to the usual tourist sites - but add a bit of a twist. Request permission ahead of time to sample water temperatures in some of the pools, rivers, or hot springs to do a correlation of color vs temperature (if you collect microscopic samples you may even be able to correlate species to temperature). You will definitely want to do some hiking in unburned and burned areas to note the difference in cover and associated wildlife.

From Yellowstone make your way to Casper, then follow the old Oregon Trail along the Platte River back toward Missouri. Along the way are a few relics of virgin prairie, the best is on the summit of Scottsbluff. Gauge your pace on the way back and plan stops accordingly. If your students are like mine have been, after nearly three weeks on the road, they will just as soon keep driving and not stop until you reach home!

B. The Desert Southwest

On this trip you will see grasslands, montane and subalpine forest, alpine tundra, temperate deciduous forest, shrublands, three deserts (Great Basin, Sonoran, and Chihuahuan - if you loop the Grand Canyon to the west through Las Vegas you will also see Mojave Desert), and possibly southern evergreen forest (if you take the southerly route home)

This trip begins with the same first-day itinerary as the trip to the northwest. A shorter alternative to a stay at Rocky Mountain is to stay on I-70 through Denver. Spend a day or two in the Arapaho National Forest Near Loveland Pass. Your first major stop will be the cold desert region of southern Utah. Plan to spend at least a day at Arches and two at Zion. Arches looks like a desert, an image that is enhanced by the unusual rock formations. Make note of plant distributions and adaptations as you hike the trails. Zion is like an oasis in the wilderness; the valley is a stark contrast to the plateau. This will probably be the first place you can point out *Ephedra*, which grows in the "Watchman" campground. Schedule an afternoon of R & R swimming at the Emerald Pools.

Your next destination is the low desert, but you can't pass the Grand Canyon without stopping. If you leave Zion after breakfast, you can have lunch at the Glen Canyon Dam and be on the South Rim of the canyon for dinner. The road from Cameron into the park passes through some classic *Pinon-*

Juniper woodland; you will want to get out to sample and make observations. From the south rim through Flagstaff is a high plateau covered with coniferous forest.

South from Flagstaff, on the way to Phoenix, you have to make a difficult choice. To make the best time you should stay on the freeway, but Alt. U.S. 89 through Oak Creek Canyon is one of the most scenic drives in the country. In either case, you will soon start descending from the plateau into the great valley of southern Arizona. About half-way down, almost like columns of soldiers, Saguaro cacti will suddenly appear. With their appearance you know that you are now in the hot desert.

Phoenix is a typical big city that would generally not be of much interest on a trip like this. From an ecological perspective, there are a few features that you should point out. First is the Central Arizona Project Canal. You will drive over it on the north side of the city. Water from this, and other canals, makes possible the palm-lined boulevards, golf courses, lagoons and pools associated with larger buildings and housing projects, the citrus groves south and west of the city, and of course the water park, "Big Surf." Irrigation is not a twentieth-century invention in the valley of the sun, however. Just east of the airport, off I-10, is the Pueblo Grande Museum where there are visible remains of a once extensive irrigation system from 700-900 A.D. Only a mile or so further east is the exit to Papago Park and the Desert Botanical Garden. Although specializing in the desert southwest of North America, there are sections devoted to xerophytes from around the world. The Garden has an excellent scientific and education staff, so be sure to contact them ahead of time for a "behind the scenes" tour.

From Phoenix, there are two directions you can choose for an interesting three- or four-day stopover (or do both if you can spare the time). Literally on the Mexican border is Organ Pipe National Park. The park is named after the rare organ pipe cactus that, together with the saguaro and various species of agave and cholla cactus, represent the typical Arizona desert. In the Senita Basin region, you can also find the Senita cactus and elephant tree, two species characteristic of the Sonoran Desert of Mexico. On the west side of the park is Quitobaquito Springs, one of the few watering holes on the old Spanish trail from Sonoyta, Sonora (on the park's southern boundary) to Yuma, Arizona. This manmade desert oasis seems strangely out of place. Even in May the days are hot; plan most of your activities for the morning and evening and be sure to have plenty of water available on hikes.

From Organ Pipe drive through the Papago Indian Reservation to Tucson. On the west side of town is the Tucson Mountain Unit of Saguaro National Park and the Arizona-Sonora Desert Museum. The plant collections at the museum are restricted to native plants and are therefore not nearly as interesting as those of the Desert Botanical Garden. The main attraction of the Museum is its animal displays - a desert zoological garden. If you have time, you will want to stop, and while you're there, at least drive through this unit of the Park. The Rincon Mountain Unit of the Park, on the east side of the city, contains an older and more diverse (and perhaps senescing) saguaro community. Plan to spend at least half a day here.

The principle objective at Tucson, however, is Mt. Lemmon. As you travel between the two units of Saguaro National Park you will be driving on Speedway Blvd. Seven miles after passing under the interstate you will come to Wilmot road. Half a mile to the north, Tanque Verde Road branches right and three miles later, Catalina Highway splits to the left. If you follow this road for the next thirty-some miles, you will end up in a fir forest at above 9150 ft. elevation overlooking the desert and the city. Along the way you will make a transect through all but one of the biomes (alpine tundra) you have seen

Appendix B

so far on the trip. Be sure to contact the Department of Ecology and Evolutionary Biology at the University of Arizona before hand. They have an excellent guide to the various communities you will pass through on your ascent. Plan to spend three or four days here. There are excellent national forest campgrounds and you can choose the elevation with a temperature to suit your students.

Heading east from Tucson, you will begin to climb back out of the valley and leave the Sonoran Desert behind. This higher land is an arid grassland with interspersed cacti, agaves, and yucca. Chiricahua National Park, in the Coronado National Forest, is a good stopover point as you make the transition to the Chihuahuan Desert. Like Mt. Lemmon, the Chiricahua Mountains are an island of diversity in the desert. To save backtracking, you can take the dirt forest service road over the mountains and come out at Portal, just before the New Mexico border.

The short stretch across southern New Mexico crosses the northern reaches of the Chihuahuan Desert. If you haven't made some good collections of male and female *Ephedra* for your teaching collection, now is the time. In places along the freeway it grows "like a weed." Your goal is El Paso for some well-deserved time off and a cultural learning experience. In El Paso you can park on the street near the Mesa Street border crossing and walk across the Rio Grande to Ciudad Juarez. Juarez is the largest Mexican city along the border, over 1 million inhabitants, and provides a brief introduction to U.S. college kids of what life is like in developing countries. The central market, city square, cathedral, bullring, etc., are all within easy walking distance.

Your final major stop will be Big Bend National Park, an excellent example of the Chihuahuan Desert. On the way to Big Bend from El Paso, you may want to stop at the Chihuahuan Desert Research Institute in Alpine. Again, make reservations ahead of time and they will arrange a program for you. They have also produced some award-winning films about the desert that you may want to rent ahead of time for promoting the course (or as a very poor substitute if you can't actually make the trip). In Big Bend, you will probably want to stay in Rio Grande Village for access to the showers and laundry, if nothing else. At first glance Big Bend will seem barren compared to the Sonoran Desert of Arizona. There are none of the large cacti so typical of the latter. As the class gets out and begins to walk the trails, however, they will begin to discover the large number of smaller cactus species scattered on the ground. You will have to observe closely, some seem half buried in the desert floor. Plan to spend a day in the area between the Village and Boquillas Canyon and at least another day in The Basin area.

Your route home will depend on time. The more direct path will be to take I-20 through Dallas, then cut diagonally through the Ouachita and Ozark mountains. Once you cross the Pecos River in West Texas, you begin to loose the desert features as you move through high plains grasslands. Approaching Fort Worth, you begin to pick up scattered groves of oaks in the oak savanna region. Driving north and east from Dallas, you move into the southwestern most extension of deciduous forest - the oak-hickory formation of the Ozark region. If you stay on a more easterly course out of Dallas before heading north, you will pass through fine stands of Southern pine lands.

A more southerly route along I-10 takes you over the Edwards Plateau to the hill country between Austin and San Antonio. East from San Antonio you drop onto the humid coastal plain. Galveston Island State Park provides an opportunity for your group to relax and play for a day before starting for home. The Big Thicket of East Texas, north of Beaumont, is a unique area that deserves at least a day of study if you have time. Then on through the pine forests near Shreveport, the oak-hickory of Arkansas and the Ozarks, and home.

C. **The Southeastern Forests**

On this trip you will see grassland, deciduous forests (several different formations), southern evergreen forest, montane and subalpine forest, as well as balds and swamps.

One advantage of eastern trips is that distances are generally shorter than one encounters in the west. With an early start you can reach your first major destination by the end of the day. Plan to spend two or three days at Lilley Cornett Woods in southeastern Kentucky. This may be the only virgin mixed mesophytic forest left in the U.S. The mixed mesophytic formation was the most diverse association of plants in the eastern deciduous forest and covered extensive areas of Kentucky, Ohio, Tennessee, Virginia and West Virginia. There is more than 900 ft of relief from bottomland flood plain with sugar maple and tulip poplar to stands of hemlock on the ridge crests and peaks.

After leaving Lilley Cornett, it is a short drive to Roan Mountain, east of Johnson City, Tennessee. Spend the afternoon analyzing vegetation in the vicinity of the state park campground. At one time, chestnut dominated this forest, but they have succumbed to chestnut blight. Among the dominants now are maple, basswood, beach, and tulip poplar. Get an early start to go "over the top" the next day. At higher elevations beech-maple gives way to spruce-fir communities. On top of Roan Mountain, 6300 ft., is a huge natural rhododendron garden in balds between subalpine species. Grassy balds and alder balds are also common near the summit. Continue over Roan Mountain toward Asheville and the Smokies.

Plan at least four days in Great Smoky Mountains National Park, the finest extant stand of Southern Appalachian forest, nearly half of which is virgin timber. Plan to spend one day being tourists and visiting Cades Cove, Clingman's Dome and either Gatlinburg or Cherokee. The latter is on a smaller scale and offers some historical perspective at the Cherokee Indian Museum and Indian Village (the latter is interesting for the native uses of many plants). The North Carolina side of the park is generally less crowded than the Tennessee side, and the campgrounds on the north and east sides of the park are isolated enough that campsites should be available. Reservations are required at the more popular Cades Cove, Elkmont and Smokemont campgrounds. Elevation in the park ranges from less than 900 feet to more than 6,600 feet with corresponding changes in vegetation. The high country is wet (over 85 inches of rain per year - the highest in the contiguous states outside of the temperate rain forests of the northwest) and cool, up to 20°F cooler than at lower elevations. At lower elevations, chestnut once dominated, along with the tulip poplar, maples and oaks which remain co-dominants. Rhododendron is an important understory species in many areas. At mid-elevations basswood, buckeye, beech and cherry become more important while tulip poplar drops out. Red maple is replaced by sugar maple. At the highest levels of hardwood basswood, buckeye, sugar maple and birch are co-dominant with balsam fir. Spruce-fir dominates the highest peaks.

After leaving the Smokies, head east over the Piedmont towards the coast and Carolina Beach State Park. The Piedmont and coastal plain are entirely second growth. Oaks and pines predominate. Through Columbus and Brunswick Counties watch for bogs near the road. Not only will you find three species of pitcher plants and several sundews, but these counties hold the world's only wild populations

Appendix B

of venus flytrap. Green Swamp, north of U.S. 17 near Supply, North Carolina, possesses all of the species of insectivorous species that grow in the state.

Carolina Beach State Park is on an Atlantic barrier island. You will stop at a gulf coast barrier island later in the trip, so this will be your first chance to study this unique and dynamic ecosystem.

Your next major stop is Laura S. Walker State Park in Southern Georgia bordering the Okefenokee Swamp. Pine plantations of loblolly and slash pine abound in this area, including those tapped for "naval stores." Also abundant in these wet sites will be cypress, and tupelo. The most noticeable characteristic of these southern forests, however, will be the epiphytic Spanish moss. Count on at least a day to tour the swamp and another to work in the pine woods. This may be your students' first opportunity to see alligators in the wild.

Your next stop is Florida, but unless you have at least a week to spare, you will only see the panhandle. If you do have time to get to the southern tip (the vegetation changes noticeably south of Palm Beach-Ft Myers) spend at least two days in the everglades and a day at John Petticamp State Park at the top of the keys. John Petticamp has the northernmost coral reefs along the coast.

Most likely, however, you will have to start heading west along the gulf coast. The next major stop is the Dauphin Island Sea Lab at the mouth of Mobile Bay. Make reservations well ahead of time. They have comfortable dormitories and an excellent dining facility. Plan to spend at least two or three days. This is a large, old barrier island with well developed dune systems dominated by evergreen forests, and a freshwater lake, on the largest rise. On the east end, erosion is obvious and the island is extending west where the last few miles of dune is a huge seabird rookery. On the Mississippi Sound side of the island, and the corresponding mainland, are extensive salt marshes. The Lab has a well-equipped classroom and laboratory with an array of preserved plant and animal specimens. One of several vessels may be chartered to collect specimens from the Gulf.

Continue west from Dauphin Island to New Orleans for a "day off." Group camping is available at Fontainebleau State Park on the north side of Lake Ponchartrain near the Causeway Bridge. The French Quarter (Vieux Carre), with the French Market on one end and the Aquarium of the Americas on the other, is readily accessible from I-10 downtown.

Cross the Mississippi at New Orleans and take the Bayou Highway, U.S. 90, west. Except for the freeway between Baton Rouge and Lafayette, which has 20 miles of causeway, U.S. 90 is still the only way to drive east to west across the Atchafalaya Swamp. At Houma plan two hours to take a swamp tour into the bayous. You will cross the Atchafalaya River at Morgan City. This is probably the future mouth of the Mississippi and is the only place along the Louisiana Coast where sediment is being deposited faster than it is eroding. The Atchafalaya River begins where the Red River joins the Mississippi. At this point the Corps of Engineers has constructed a large dam, the Old River Control Structure, which forces the Mississippi to stay in its current channel. If the control structure breaches, as almost occurred in the 1973 flood, the Mississippi would change course and leave Morgan City underwater in the middle of the new channel. The Mississippi River and Atchafalaya Swamp provide an effective isolation mechanism for many species of plants. To identify plants in Louisiana, you need the *Flora of Texas* for the western half of the state and the *Flora of the Carolinas* for the east.

Follow U.S. 90 through sugar cane country to Lafayette, the capital of cajun country, then pick up I-49 north to Alexandria. At Alexandria cross the Red River and head north. When you cross the Red River you will begin driving out of the Mississippi delta region and back into pine country, an extension of the coastal pine forests that lined the freeway in Alabama, Mississippi and the Florida panhandle. Your last major stop is in the Quachita Mountains west of Little Rock, the western-most formation of the Eastern Deciduous Forest where you began.

D. **The Northeastern Forests.**

On this trip you will visit most of the major subdivisions of the eastern deciduous forest. You also will visit subalpine forest and alpine tundra, a rocky coast with well-developed intertidal communities, a portion of boreal forest and tall-grass prairie.

Your first day out has the same itinerary as the southeast tour, to get to Lilley Burnett Woods where you can spend several days studying mixed mesophytic forest. Then cross the mountains and pick up I-81 north through the Shenandoah Valley. Drive at least part of the Blue Ridge Highway in Shenandoah National Park, which is about half way to your next main stop. You can reserve group campsites in the park and you can hike sections of the Appalachian Trail. It parallels the highway along the ridge. The Blue Ridge region is part of the oak-chestnut formation, similar but less luxuriant than you would find in the Smokies. Tulip poplar and occasional black and white oak are found at lower elevations . This entire area has been heavily lumbered so much of what you see is secondary growth. Red oak, maple, and basswood, with abundant rhododendron and mountain laurel understory, dominate the higher slopes.

Plan an additional two or three days at Hickory Run State Park, Pennsylvania (Group cabins are available here as well as group tent sites - make reservations ahead of time). The forest is the northernmost extension of the Oak-Chestnut formation observed in the Blue Ridge region. There is considerably less diversity than was observed in Kentucky. This may not be immediately obvious to your students but will become so as soon as they begin sampling. Several different oaks dominate the lower elevations. At higher elevations you will find birch, hemlock and white pine.

Continue north along the interstate to reach the southern Adirondacks by evening. You will now begin to see the influence of the boreal forests of Canada. The dominant species over much of this region is red spruce. Swamps and bogs, dominated by black spruce, balsam fir, and birch, with occasional oaks, occupy low spots. On higher ground, hemlock and red maple appear. A bog walk, especially on a floating *Sphagnum* bog, is worth an extra day. If you have equipment to do water sampling, it would be worth an extra day to examine some of the lakes in this area for algae, plankton, and small invertebrates. You will also want to compare the pH with other lakes and streams you encounter on this trip. The lake region of the Adirondacks is among the hardest hit by acid deposition in the United States.

Your next major stop is the White Mountains, in the vicinity of Mt. Washington, New Hampshire. Mount Washington, at nearly 6300 ft., is the highest point in the northeast and is well known for its severe weather. There are well marked trails to the treeless summit where wind speeds in excess of 200 mph have been recorded! Severe decline of conifers is also evident, a decline associate with acid deposition. The peak is also accessible by road from Glen House. Lower elevations are a mixture of maple forest, the northern most extent of the great Eastern Deciduous Forest, and birch.

Appendix B

These species give way to red spruce at higher elevations that in turn are replaced by balsam fir. Near timberline is a dwarf fir forest, a good eastern example of "krumholz." A number of alpine tundra types, dwarf shrubs and herbs, predominate at these elevations. From the Dolly Copp Campground you can also hike several short sections of the Appalachian Trail. You also will want to arrange a tour of the Hubbard Brook Experimental Forest, near West Thornton. Much of what we know about energy relationships and ecosystem dynamics has been learned at this site. Check with the director of the Northeast Forest Experiment Station.

Plan at least three or four days at Acadia National Park. The rocky coastline is as rugged as anywhere else in the country and perfectly suited for a study of the intertidal communities. Boreal forest covers the higher ground; you can hike a transect over Cadillac Mountain. You will also want to arrange a tour of the Jackson Laboratory at Bar Harbor. This is normally the eastern most extent of the trip, however, depending on your time schedule, you may want to drive one day further east to Fundy National Park in New Brunswick. Neither the terrestrial vegetation nor the intertidal species will be different from what you observed at Acadia, but 40-foot tides are impressive. You will have to spend a full day observing a full ebb and maximum tide to really appreciate this phenomenon. The tides ebb and flow twice each day. At low tide you can survey tidal pools. As the tide comes in be sure you have a route back to camp away from the beach.

Plan a "day off" in Quebec City. You will, of course, be in another country. In much of Canada there does not seem to be much of a difference from the United States. In Quebec you will definitely know you've gone international. Both the lower and upper towns are worth exploring. Quebec City is just south and east of the transition zone between the mixed conifer-hardwood forest which nearly encircles the Great Lakes, and the boreal forest of Canada. To experience the latter, drive west to spend the night at La Mauricie National Park.

Your next main stop will be in the Alleghenies of western Pennsylvania, but plan to spend at least one day in the Finger Lakes region of New York. This is wine country, so if economic botany is an aspect you stress, you may want to stop at a winery along Seneca or Keuka Lakes. At Geneva, on the north shore of Seneca Lake, is the New York State Agricultural Extension Station specializing in grapes and apples. Arrange a tour of their labs and research plots.. Watkins Glen, at the south end of the lake has an impressive gorge, characteristic of this region.

The Alleghenies represent a return to a more southerly-influenced, mixed mesophytic community. You can reserve group sites at Chapman State Park which will serve as a base for a study of the mixed forests associated with different elevations. At lower levels are a variety of oaks and other hardwoods which blend into birches, then hemlock and pines at higher elevations. Some magnificent white pines dominate the landscape. Thirty miles south, at Cook Forest, you will see many of the same species observed on your first day out at Lilly Cornett Woods.

The Indiana Dunes area on the eastern shore of Lake Michigan is your last major stop. These dune areas are the site of classic early studies on succession by botanists from the University of Chicago. The dunes are constantly shifting, and constantly stress the plant species growing on them. Moving east over the dunes, some of which reach nearly 200 ft. high, you move into beech-maple forest, which was the dominant forest type over most of Ohio and half of Michigan and Indiana. Several oak species are also locally abundant.

Northeast

The last stretch toward home crosses the former tall-grass prairie peninsula that covered most of Illinois. Take I-55 from the Chicago area and stop at the Goose Lake Prairie Nature Preserve. At over 1500 acres, this is the largest expanse of prairie east of the Mississippi River. There is an interpretive center and a marked trail system.

MATERIALS, EQUIPMENT, AND PREPARATION

Group Equipment
- tents
- mallet
- assorted pans
- ecological equipment
- compasses
- soil test kit
- Brunton Compass
- tarps with grommets
- shovels
- kettles
- thermometers
- pH paper
- hatchet, axe
- gas stoves
- grill
- plant presses
- wind gauges
- measuring tapes

Personal Equipment
- sleeping bag
- duffle bag
- 5 pr socks
- warm jacket
- 2 pr shorts
- air mattress
- 2 pr jeans
- 1 pr boots
- sweatshirt
- sunglasses
- bag liner or sheet
- 4 shirts with pockets
- 1 pr sneakers
- rain suit or poncho
- 7 changes underwear

- hat
- lip balm
- pocket knife
- several pens/pencils
- swimming suit
- plastic bag (laundry)
- cup, utensils
- misc. toilet supplies
- bath towel
- gloves
- notebook
- hand lens
- canteen
- compass
- insect repellent
- brush or comb
- flashlight
- pocket notepad
- binoculars
- camera
- calculator
- suntan lotion

A. The Pacific Northwest

District Program Officer
Agriculture Canada
Plant Health and Plant Products Division
Food Production and Inspection Branch
118-816 Government Street
Victoric, B.C. V8W 1W9 CANADA

B. The Desert Southwest

Director
Natural Resources Conservation Division

Appendix B

Arizona State Land Department
1624 West Adams
Phoenix, Arizona 85007

Director
Chihuahuan Desert Research Institute
P.O. Box 1334
Alpine, Texas 79830

C.   The Southeastern Forests
Director of Natural Areas
Eastern Kentucky University
Richmond, Kentucky 40745

Director
Dauphin Island Sea Lab
P.O. Box 369-370
Dauphin Island, Alabana 36528

REFERENCES

General:
Farb, P. 1963. *Face of North America*. Harper and Row, Publishers, New York. 316 p. A brief, popular introduction to the geology and natural history of the continent.

Hunt, C.B. 1974. *Natural Regions of the United States and Canada*. W.H. Freeman and Co., San Francisco. 725 p. Similar to the above, but slightly more technical.

Watts, M.T. 1957. *Reading the Landscape: An Adventure in Ecology*. Macmillan, New York. 230 p. This is a must-read book before taking any of the trips outlined above. What can you tell about the ecological history of a region as you look out the window of a moving vehicle.

Resources:
Grasslands
Weaver, J.E. 1954. *North American Prairie*. Johnsen Publishing Co., Lincoln, NE. 348 p. This text discusses the tall grass prairies of the prairie peninsula of Illinois and Indiana through eastern Kansas and Nebraska.

Weaver, J.E. and F.W. Albertson. 1956. *Grasslands of the Great Plains*. Johnsen Publishing Co., Lincoln, NE. 395. Discusses the mid and short grass plains vegetation west from the prairies to the Rocky Mountains.

Deserts

Jaeger, E. C. 1957. *The North American Deserts*. Stanford University Press. Stanford, CA. 308 p. This popularly written account describes the most distinguishing characteristics of each of the desert formations of North America.

Larson, P. 1977. *A Sierra Club Naturalist's Guide: The Deserts of the Southwest*. Sierra Club Books, San Francisco. 286 p. Not only does this describe the characteristic plants and animals of the region, but provides essential information on how to survive comfortably in this inhospitable environment.

McGinnies, W.G. 1981. *Discovering the Desert: Legacy of the Carnegie Desert Botanical Laboratory*. University of Arizona Press, Tucson, AZ. 276 p. Most of the classic work on the Sonoran Desert, including the original survey of Mt. Lemmon, was done out of this laboratory. Here is a capsule summary of their research.

Eastern Forests

Braun, E.L. 1950. *Deciduous Forests of Eastern North America*. Macmillan, New York. 596 p. This is the definitive study of the forested regions of the Eastern U.S.

Lindsey, A.A. and L.K. Escobar. 1971. *Eastern Deciduous Forest, Volume 2: Beech-Maple Region: Inventory of Natural Areas and Sites Recommended as Potential Natural Landmarks*. National Park Service Publication, Washington, D.C. 238 p. An invaluable resource for the return portion of the northeastern trip.

Waggoner, G.S. 1975. *Eastern Deciduous Forest, Volume 1: Southeastern Evergreen and Oak-Pine Region: Inventory of Natural Areas and Sites Recommended as Potential Natural Landmarks*. National Park Service Publication NPS 135. U.S. Government Printing Office, Washington, D.C. 206 p. An invaluable reference, especially for the southeastern trip.

Seacoasts

Carefoot, T. 1977. *Pacific Seashores: A guide to Intertidal Ecology*. University of Washington Press, Seattle, WA. 208 p. This is a good, general introduction to the ecology of this specialized ecosystem.

Fotheringham, N. and S. Brunenmeister. 1989. *Beachcomber's Guide to Gulf Coast Marine Life: Florida, Alabama, Mississippi, Louisiana, & Texas, 2nd ed.* Gulf Publishing Co., Houston. 142 p. Similar to the above for the gulf coast.

Kozloff, E.N. 1983. *Seashore Life of the Northern Pacific Coast: An Illustrated Guide to Northern California, Oregon, Washington, and British Columbia*. University of Washington Press, Seattle. 370 p. Unlike the above volumes, this concentrates on the features of common organisms characteristic of different habitats.

CHAPTER 1: INTRODUCTION TO PLANTS AND BOTANY

MULTIPLE CHOICE QUESTIONS

d 1. A membrane is a thin sheet of fat and protein molecules that surrounds each cell and most organelles inside a cell. Our knowledge of the spatial arrangement of fat and protein molecules in a membrane is a result of
   a. religious methods
   b. metaphysics
   c. speculative philosophy
   d. scientific methods

a 2. Which of the following is not part of the scientific method?
   a. explaining phenomena controlled by unknowable forces
   b. testing of all proposed explanations
   c. controlled experiments
   d. explaining observable phenomena
   e. carefully documented observations

d 3. From the choices below, which would not be an absolute requirement of life?
   a. nonrandom organization
   b. a system of heredity and reproduction
   c. metabolism
   d. a vital force
   e. growth

c 4. Photosynthesis first arose in
   a. eukaryotic algae
   b. multicellular plants
   c. prokaryotic bacteria
   d. eukaryotic bacteria
   e. fungi

d 5. Most algae are classified in the Kingdom
   a. Plantae
   b. Animalia
   c. Monera
   d. Protista
   e. Mycetae

a 6. The statement "All plants are composed of cells" is a
   a. theory
   b. fact
   c. observation
   d. expectation
   e. assumption

b 7. Which of the following could not be studied scientifically?
   a. the contribution of carbon dioxide in the atmosphere to global warming
   b. the meaning of Shakespeare's Hamlet
   c. the meaning of spots on a flower
   d. a rainbow
   e. the existence of the Loch Ness monster

e 8. The probable sequence in which organisms evolved is
   a. prokaryotic bacteria ---> eukaryotic algae ---> cyanobacteria ---> land plants
   b. eukaryotic bacteria ---> cyanobacteria ---> eukaryotic algae ---> land plants
   c. cyanobacteria ---> prokaryotic bacteria ---> eukaryotic algae ---> land plants
   d. cyanobacteria ---> eukaryotic algae ---> prokaryotic bacteria ---> land plants
   e. prokaryotic bacteria ---> cyanobacteria ---> eukaryotic algae ---> land plants

a 9. Mycoplasmas are the smallest organisms composed of cells and are basically bacteria without a cell wall around the cell. Into which Kingdom would you put mycoplasmas?
   a. Monera
   b. Mycetae
   c. Protista
   d. Animalia
   e. Plantae

e 10. To fully understand how an organism functions, biologists must understand principles of
   a. chemistry
   b. mathematics
   c. physics
   d. a and c only
   e. a, b, and c

c 11. If plants are exposed to light from only one direction, many will bend toward the light. Experimentation has shown that this is due to an unequal distribution of a chemical in the plant. This explanation was derived using
   a. metaphysics
   b. religious methods
   c. the scientific method
   d. speculative philosophy
   e. all of the above

d 12. Which of the following is false about an hypothesis?
   a. it should be consistent with future observations
   b. it may become a theory
   c. it should be able to predict experimental results
   d. it is either totally right or totally wrong
   e. it is a product of observation and experimentation

a 13. The first organisms to evolve have been dated at about
   a. 3.5 billion years and were much like present-day bacteria
   b. 2.8 billion years and were photosynthetic like present-day cyanobacteria
   c. 3.5 billion years and had a nucleus
   d. 1.5 billion years and were terrestrial
   e. 3.5 billion years and possessed organelles which had specialized functions

d 14. All of the following are true concerning the effects that humans have on other organisms except
   a. Driving your car produces carbon dioxide that contributes to global warming and changing climate
   b. Using solar cookers in some areas could decrease desertification
   c. Increased farming would increase food production for people, but destroy plant and animal habitats
   d. The introduction of exotic species into areas by people rarely has an effect on native species
   e. Biotechnology can change the genetic characteristics of other species

e 15. The trees in a forest
   a. carry out a metabolism that is unique to plants and has no relationship to the general principles of chemistry
   b. carry out a metabolism that is random and uncontrolled
   c. are all genetically identical
   d. grow in order to take in the maximum amount of sunlight
   e. are adapted to grow and reproduce in forest conditions

TRUE-FALSE QUESTIONS

F _____ 1. If a phenomenon cannot be explained by science at present, it can never be explained using the scientific method.

F _____ 2. The scientific method can, ultimately, explain any concept.

T _____ 3. Organisms in one generation of a population are usually not identical, but show some variation.

F _____ 4. The idea that mitochondria and chloroplasts originated as independent, prokaryotic organisms is now a fact.

T _____ 5. Eukaryotic organisms evolved from prokaryotic organisms.

F _____ 6. A set of observations always leads to the same interpretation.

T _____ 7. If an individual possesses a characteristic that enables it to survive in its environment better than others, it will probably produce more offspring than the others.

F _____ 8. If a species is evolving, all features of that species must evolve at the same rate.

T _____ 9. If the leaf size in the first generation of a species is small and is large in the hundredth generation, leaf size is a derived feature.

F _____ 10. All biological knowledge applies to all organisms.

MATCHING

Match each statement from List A with one term from List B. Terms in List B may be used more than once.

	List A	List B
c	1. creation of species by God	a. scientific method
d	2. logical extension of an observation	b. metaphysics
a	3. controlled experiments	c. religious method
b	4. phenomena unexplained by natural processes	d. speculative philosophy
a	5. testable hypothesis	

ESSAY QUESTIONS

1. How would you use the scientific method to determine the water requirements of your houseplants?

2. From your own experience, list some differences between plants and animals. Do your characteristics apply to all plants? All animals?

3. If you were an aquanaut and discovered a new object deep in the ocean, what criteria would you use to determine whether the object was living or non-living?

4. Design a series of experiments to test the hypothesis that plants are alive.

5. Microtubules are long, tubular structures found in virtually all eukaryotic cells. They are composed of a protein called tubulin, which is remarkably similar in all of the eukaryotic cells studied so far, whether they are fungal, plant, animal, or algal cells. Is tubulin a derived or relictual feature? Why?

6. Which are more people likely to agree on, observations about a particular plant or explanations of the observations. Why?

7. Why might someone think that a sports car is alive? Would you agree with their opinion? Why or why not?

CHAPTER 2: INTRODUCTION TO THE PRINCIPLES OF CHEMISTRY

MULTIPLE CHOICE QUESTIONS

c 1. An atom of nitrogen has 14 protons in a nucleus which is surrounded by how many electrons?
   a. 6
   b. 7
   c. 14
   d. 18
   e. 28

a 2. An atom is composed of
   a. a nucleus of protons and neutrons surrounded by electrons
   b. a nucleus of protons surrounded by neutrons and electrons
   c. a nucleus of electrons and neutrons surrounded by protons
   d. a nucleus of protons and electrons surrounded by neutrons
   e. a nucleus of electrons surrounded by protons and neutrons

b 3. Each molecule of molecular nitrogen, the principal form of nitrogen in the atmosphere, is composed of two nitrogen atoms that share two electrons. The type of bond holding these two atoms together is a(an)
   a. polar bond
   b. covalent bond
   c. hydrogen bond
   d. ionic bond
   e. hydrophobic bond

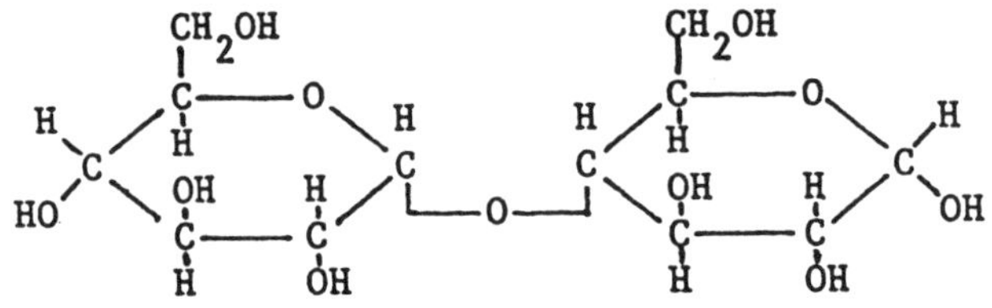

Questions 4 and 5 refer to the accompanying diagram:

a 4. What type of glycosidic bond is pictured in this molecule?
   a. alpha-1,4
   b. alpha-1,6
   c. beta-1,4
   d. beta-1,6
   e. none of the above

b 5. This type of bond would be found in
   a. amylose only
   b. amylose and amylopectin
   c. cellulose
   d. amylose, amylopectin, and cellulose
   e. amylose and cellulose

d 6. The reason that people can utilize the energy in starch but not in cellulose is because cellulose molecules
   a. are too large to hydrolyze
   b. contain no energy
   c. are branched
   d. contain a different bond than starch
   e. all of the above

e 7. Monosaccharides are used by organisms as
   a. energy storage molecules
   b. energy transport molecules from one part of an organism to another
   c. monomers
   d. a and b
   e. a, b, and c

e 8. If a protein exhibiting quaternary structure is hydrolyzed, which levels of structure are destroyed?
   a. primary
   b. primary and secondary
   c. primary and tertiary
   d. secondary, tertiary, and quaternary
   e. primary, secondary, tertiary, and quaternary

e 9. The tertiary structure of a protein is often maintained by
   a. disulfide bridges
   b. hydrogen bonds
   c. ionic bonds
   d. a and b
   e. a, b, and c

b 10. A biochemist was working with an enzyme known to be a single polypeptide chain. When the pH around the enzyme was lowered slightly, the enzyme's activity slowed. When the pH was returned to the original value, the enzyme resumed normal activity. Lowering the pH probably altered the enzyme's
   a. primary structure
   b. tertiary structure
   c. quaternary structure
   d. peptide bonds
   e. all of the above

c 11. Nucleotides are composed of
   a. a phosphate group, a 6-carbon sugar, and an organic N-containing base
   b. a phosphate group, a 5-carbon sugar, and an organic P-containing base
   c. a phosphate group, a 5-carbon sugar, and an organic N-containing base
   d. a nitrate group, a 5-carbon sugar, and an organic N-containing base
   e. a 5-carbon sugar and an organic N-containing base

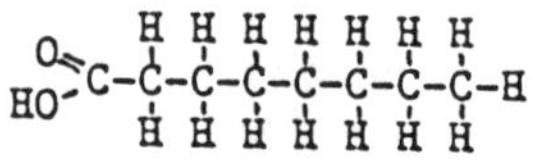

Questions 12 through 15 refer to the accompanying diagrams:

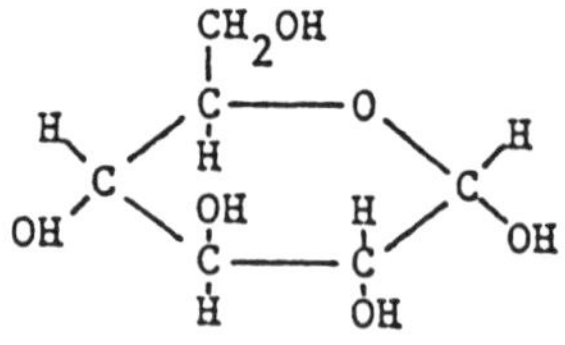

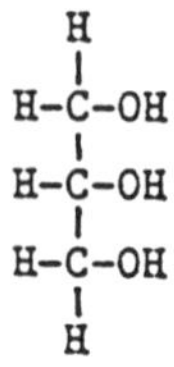

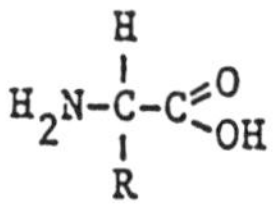

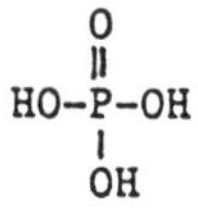

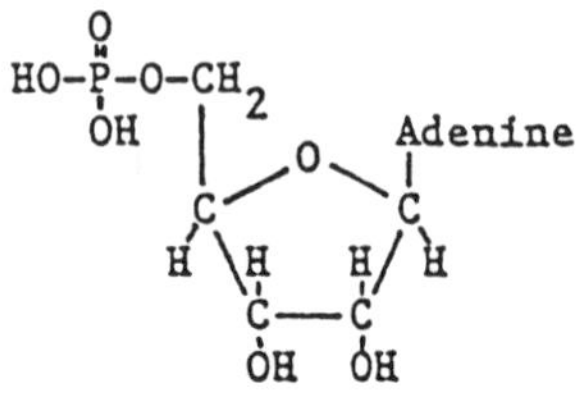

c 12. The building block(s) for proteins is(are)
- a. 1
- b. 2
- c. 3
- d. 4
- e. 2 and 6

d 13. The building block(s) for triglycerides is(are)
- a. 1
- b. 3
- c. 5
- d. 2 and 6
- e. 2, 4, and 6

a 14. The building block for amylose is
- a. 1
- b. 2
- c. 3
- d. 5
- e. 6

d 15. The building block(s) for RNA is(are)
- a. 1
- b. 2
- c. 3
- d. 5
- e. 1 and 4

d 16. The two major components of membranes are
- a. triglycerides and proteins
- b. proteins and monosaccharides
- c. fatty acids and proteins
- d. phospholipids and proteins
- e. glycerol and proteins

d 17. ATP
   a. is an energy carrier
   b. has high energy phosphate bonds
   c. can be transported from one cell to another
   d. a and b
   e. a, b, and c

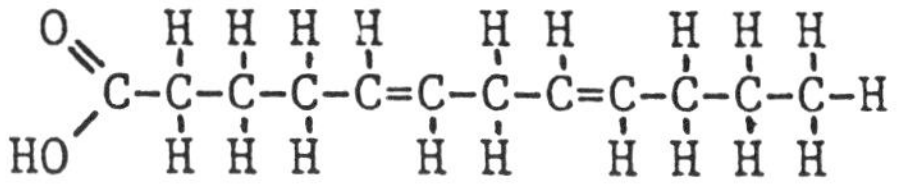

$$\underset{HO}{\overset{O}{\diagdown}}C-CH_2-CH_2-CH_2-CH=CH-CH_2-CH=CH-CH_2-CH_2-CH_2-H$$

a 18. What type of molecule is shown in the accompanying diagram?
   a. unsaturated fatty acid
   b. saturated fatty acid
   c. unsaturated triglyceride
   d. saturated triglyceride
   e. unsaturated phospholipid

e 19. Which of the following is not used to control enzyme activity in cells?
   a. altering production of an enzyme
   b. changing pH in a cell
   c. changing ion concentration in a cell
   d. production of an inhibitor
   e. secreting the enzyme from the cell

b 20. If an endergonic reaction occurs in a plant cell, the most probable direct source of energy for the reaction would be
   a. glucose
   b. ATP
   c. $NADH + H^+$
   d. $NADPH + H^+$
   e. $FADH + H^+$

d 21. Within a molecule
   a. all of its atoms are electrically neutral
   b. its atoms are always held together by ionic bonds
   c. electrons are always shared equally by atoms
   d. atoms may share two electrons to form a double bond
   e. all of the atoms are the same element

b 22. The process of photosynthesis may be studied by monitoring changes in oxygen concentrations. This study would be considered in vitro in which situation?
   a. submersing oxygen electrodes into a lake to monitor algal photosynthesis
   b. extracting chloroplasts from spinach leaves and measuring oxygen evolution under varying light levels
   c. monitoring the oxygen evolution in a test tube containing Chlorella, a unicellular alga
   d. measuring oxygen changes in a sealed glass box containing a tobacco plant
   e. all of the above

a 23. The body of a person who is a vegetarian has to manufacture which of the following essential compounds because plants cannot be relied on to provide it?
   a. cholesterol
   b. some amino acids
   c. vitamin C
   d. niacin
   e. some fatty acids

b 24. Acids
   a. contribute hydroxyl ions to a solution
   b. contribute protons to a solution
   c. contribute protons and hydroxyl ions to a solution
   d. decrease the concentration of protons in a solution
   e. do not affect the charge of other molecules in a solution

c 25. In the reaction A + B ----> C + D, each substance contains the following amount of energy:

   A: 5 kcal/mole   C: 7 kcal/mole
   B: 12 kcal/mole   D: 2 kcal/mole

   This reaction is
   a. endergonic because A + B contains more energy than C + D
   b. endergonic because A + B contains less energy than C + D
   c. exergonic because A + B contains more energy than C + D
   d. exergonic because A + B contains less energy than C + D
   e. definitely exothermic

TRUE-FALSE QUESTIONS

F _____ 1. In plant cells, all of the DNA is found in the nucleus.

T _____ 2. While there are several type of RNA, they all function some way in protein synthesis.

T _____ 3. The temperatures on Earth change slowly because over 3/4 of the surface of the planet is covered by water.

F _____ 4. A solution that contains a high concentration of hydroxyl ions has a low pH.

F _____ 5. If two different exergonic reactions released the same amount of energy, the activation energy required for each reaction to proceed would be identical.

F _____ 6. Glucose, fructose, and galactose all have the same chemical formula so organisms can use these molecules interchangably.

T _____ 7. Magnesium atoms, which have 12 electrons, tend to lose two electrons to become more stable.

T _____ 8. The higher the electronegativity of an atom, the higher the tendency to gain electrons.

F _____ 9. In molecules held together by covalent bonds, electrons are always shared equally between atoms.

T _____ 10. If all of the necessary chemicals and ribosomes are put into a test tube, protein synthesis will occur. This would be *in vitro* protein synthesis.

T _____ 11. Until the 1800's, most scientists believed that life could arise spontaneously from a mixture of nonliving chemicals.

F _____ 12. Non-polar substances dissolve easily in water.

MATCHING

Match each term in List A with a diagram in List B.

	List A		List B
f	1. methyl	a. -OH	
e	2. amino	b. -SH	
b	3. sulfhydryl	c. -H	h. -O-P(=O)(OH)-OH
a	4. hydroxyl	d. -C(=O)-	
g	5. aldehyde	e. -NH$_3$	i. -C(=O)OH
i	6. carboxyl	f. -CH$_3$	
		g. -C(=O)H	

Match each term in List A with one in List B. Terms in List B may be used more than once.

	List A		List B
a	1. protein	a.	polymer
b	2. deoxyribonucleotide	b.	monomer
a	3. amylose	c.	oligomer
a	4. cutin		
b	5. ribose		
b	6. amino acid		

ESSAY QUESTIONS

1. Explain why each different enzyme must have a unique primary structure to function correctly.

2. Name the important polymers in organisms and the advantage(s) to the organism that they are polymers.

3. Distinguish between the structure and function of monosaccharides, polysaccharides, and oligosaccharides.

4. Relate the chemical properties of a phospholipid to its function in a cell.

5. What is the advantage to a cell of having enzymes that show high substrate specificity?

CHAPTER 3: CELL STRUCTURE

MULTIPLE CHOICE QUESTIONS

c 1. Which of the following would be most likely to cross a cell membrane by diffusion?
   a. Na^+
   b. $PO_4^=$
   c. CO_2
   d. glucose
   e. amino acid

c 2. The concentration of iodine ions in cells of seaweeds or kelps can be up to 30,000 times higher than in seawater. This is due to
   a. diffusion
   b. facilitated diffusion
   c. active transport
   d. endocytosis
   e. all of the above

d 3. Seeds that store energy in fat molecules can convert that fat to the carbohydrates needed for growth in
   a. nuclei
   b. chloroplasts
   c. dictyosomes
   d. glyoxysomes
   e. peroxisomes

b 4. If you wanted to study membrane structure, you might isolate and study a plant cell's
   a. hyaloplasm
   b. tonoplast
   c. ribosomes
   d. mitotic spindle
   e. microfilaments

e 5. The function of microtubules include(s)
   a. the cytoskeleton
   b. mitotic spindle
   c. guiding vesicle movement
   d. a and c
   e. a, b, and c

d 6. A sclerenchyma cell must be very strong to function, so it has a
   a. primary wall only
   b. primary wall, plus a secondary wall that contains only cellulose
   c. primary wall, plus a secondary wall that contains cellulose and hemicelluloses
   d. primary wall, plus a secondary wall that contains cellulose, hemicelluloses, and lignin
   e. secondary wall only

b 7. You would expect mature cells of a white potato to contain abundant
   a. proplastids
   b. amyloplasts
   c. chromoplasts
   d. chloroplasts
   e. no plastids at all

d 8. A common unit of measure in microscopic work is the micrometer, which equals
   a. 10^3 m
   b. 10^{-2} m
   c. 10^{-3} m
   d. 10^{-6} m
   e. 10^{-9} m

a 9. The red color of beets is due to a water soluble pigment called anthocyanin, which is most likely found in a beet cell's
   a. vacuole
   b. chromoplasts
   c. cytoplasm
   d. nucleus
   e. chloroplasts

e 10. Indian pipes are completely colorless flowering plants that are parasitic on tree roots. Cells of these plants lack
   a. a cell wall
   b. mitochondria
   c. ribosomes
   d. plastids
   e. chloroplasts

d 11. All of the following are characteristics of membranes except
   a. growth
   b. the ability to fuse with other membranes
   c. lateral movement of molecules within the membrane
   d. they are static
   e. semi-permeability

c 12. Prokaryotic cells lack
   a. DNA
   b. ribosomes
   c. mitochondria
   d. a plasma membrane
   e. protoplasm

c 13. Cells on the surface of the leaves of the carnivorous plant sundew secrete enzymes to digest insects by the process of
   a. translocation
   b. endocytosis
   c. exocytosis
   d. active transport
   e. facilitated diffusion

e 14. The total genetic composition of a plant is determined by the
   a. nucleus
   b. nucleus and mitochondria
   c. nucleus and plastids
   d. nucleus, mitochondria, and microbodies
   e. nucleus, mitochondria, and plastids

a 15. If the mitochondria were removed from a plant cell, what process would immediately stop in the cell?
   a. respiration
   b. photosynthesis
   c. lipid synthesis
   d. protein synthesis
   e. starch synthesis

b 16. The pathway that an amino acid takes after entering a cell until found in a nectar droplet is
   a. smooth endoplasmic reticulum ---> ER vesicle --->dictyosome ---> secretion vesicle ---> droplet
   b. rough endoplasmic reticulum ---> ER vesicle ---> dictyosome ---> secretion vesicle ---> droplet
   c. ER vesicle ---> rough endoplasmic reticulum ---> smooth endoplasmic reticululm ---> dictyosome ---> droplet
   d. dictyosome ---> rough endoplasmic reticulum ---> ER vesicle ---> smooth endoplasmic reticulum ---> secretion vesicle ---> droplet
   e. rough endoplasmic reticulum ---> dictyosome ---> droplet

a 17. If you scrape some cells from the inside of your cheek and examine them microscopically, you would not expect to find
   a. cell walls, plastids, central vacuoles, and glyoxysomes
   b. cell walls, mitochondria, and plastids
   c. plastids, central vacuoles, and glyoxysomes
   d. plastids, dictyosomes, and central vacuoles
   e. cell walls, plastids, centrioles, and microtubules

e 18. If a cell is capable of protein synthesis, it must contain
   a. a nucleus
   b. mitochondria
   c. chloroplasts
   d. rough endoplasmic reticulum
   e. ribosomes

c 19. Cells that have a nucleus, are colorless, and are surrounded by a chitin-containing cell wall are
   a. root cells
   b. interior stem cells
   c. fungal cells
   d. fruit cells
   e. flower cells

b 20. If isolated intact, which of the following would appear green in color?
   a. plasma membranes
   b. thylakoids
   c. cristae
   d. tonoplasts
   e. nuclear envelopes

e 21. The cells of many plants contain crystals that
   a. are composed of calcium oxalate
   b. may function to protect the cells
   c. may regulate calcium concentration in the cells
   d. a and b
   e. a, b and c

d 22. Regarding cell walls
- a. all contain cellulose
- b. all contain chitin
- c. some contain cellulose + chitin
- d. some contain cellulose, but not chitin
- e. cell walls contain neither cellulose nor chitin, which are nutritional polysaccharides

b 23. The yellow color of daffodil petals is due to
- a. yellow chlorophyll in chloroplasts
- b. lipid pigments in chromoplasts
- c. lipid pigments in leucoplasts
- d. water-soluble pigments in the central vacuole
- e. water-soluble pigments in amyloplasts

a 24. Seeds of legumes, such as beans and peas, contain large amounts of protein that are stored in the cells. The cells of these seeds contain
- a. numerous ribosomes and abundant rough endoplasmic reticulum in which the protein is stored
- b. numerous ribosomes and abundant smooth endoplasmic reticulum in which the protein is stored
- c. numerous ribosomes and cytoplasmic granules of stored protein
- d. few ribosomes and numerous microbodies in which the protein is stored
- e. few ribosomes and numerous dictyosomes in which the protein is stored

c 25. Membranes are very effective at compartmentalization. Which of the following is not compartmentalized by one or more membranes?
- a. photosynthesis
- b. cellular respiration
- c. the cytoskeleton
- d. storage of waste products
- e. detoxification of hydrogen peroxide

TRUE-FALSE QUESTIONS

F _____ 1. DNA is found only in the nucleus of a plant cell.

F _____ 2. While a membrane may be quite different with regard to the proteins found on each side, both sides are always the same in terms of lipid content.

T _____ 3. When information to build a plant cell is retrieved from DNA, the copies are in the form of messenger RNA.

F _____ 4. The only process involved in growth of a plant cell is vacuole enlargement.

T _____ 5. Once an ion from the soil crosses a plasma membrane into a root cell, that ion could move to the top of the plant without crossing another plasma membrane.

F _____ 6. The only metabolic process that occurs in plastids is photosynthesis.

F _____ 7. Once a plant cell has formed a cell wall, it can alter that wall by hydrolyzing the cellulose in the wall.

F _____ 8. Because of their cell walls, plant cells cannot communicate with one another.

T _____ 9. You would expect to find the same basic types of cells in flowers and roots.

T _____ 10. If the arrangement of microtubules in a flagellum is called "9 + 2", the arrangement in a centriole would be called "9 + 0".

F _____ 11. All plant cells contain chloroplasts.

MATCHING

Match each structure in List A with a chemical in List B. Chemicals in List B may be used more than once.

List A

c _____ 1. microtubule
e _____ 2. chloroplast
d _____ 3. ribosome
f _____ 4. amyloplast
h _____ 5. chromosome
b _____ 6. cell wall
i _____ 7. microfilament
c _____ 8. flagellum
a _____ 9. microbody
c _____ 10. centriole
g _____ 11. middle lamella

List B

a. catalase
b. cellulose
c. tubulin
d. rRNA
e. chlorophyll
f. starch
g. pectic compounds
h. histones
i. actin

Match each structure in List A with a process in List B.

List A

j _____ 1. chromosome
l _____ 2. glyoxysome
b _____ 3. middle lamella
i _____ 4. plasmodesma
c _____ 5. chloroplast
a _____ 6. mitochondrion
e _____ 7. ribosome
h _____ 8. smooth ER
f _____ 9. microfilament
d _____ 10. cilium
k _____ 11. amyloplast

List B

a. respiration
b. "glues" cells together
c. store light energy in sugar
d. cell movement
e. protein synthesis
f. cytoskeleton
g. material packaging
h. lipid synthsis
i. cell communication
j. major site of DNA
k. starch storage
l. convert fat to carbohydrate
m. storage of cell waste

ESSAY QUESTIONS

1. If pigments from spinach leaves are separated using a technique called chromatography, five pigments are isolated: 2 green chloroplylls, 2 yellow xanthophylls, and an orange carotene. Why do spinach leaves look green?

2. Devise an experiment to demonstrate that photosynthesis has occurred in a leaf.

3. Describe the process by which each of the following would be moved across a membrane: $PO_4^=$, O_2, glucose.

4. Why is a redwood tree a collection of cells rather than one cell?

5. Animal muscle cells are much more active than fat tissue cells. Would you expect muscle cells or fat cells to contain more mitochondria per cell? Cristae per mitochondrion? Why?

6. What organelles would be present in a carnivorous plant cell actively making and secreting digestive enzymes? Why?

7. Mature plant leaf cells of a particular plant species contain approximately 50 chloroplasts and 100 mitochondria. Immature cells, actively dividing at the growing points, are smaller and do not contain as many of these organelles as the mature cells. How does the number of chloroplasts and mitochondria increase as the cells mature?

CHAPTER 4: GROWTH AND DIVISION OF THE CELL

MULTIPLE CHOICE QUESTIONS

c 1. If a zygote contains 12 chromosomes, how many chromosomes did the original egg cell contain?
   a. 3
   b. 4
   c. 6
   d. 12
   e. 24

a 2. The growth phase of a cell cycle includes
   a. G_1, S, and G_2
   b. G_1, S, G_2, and prophase
   c. telophase, G_1, and S
   d. telophase, G_1, S, G_2, and prophase
   e. G_1 and S

b 3. If a chromosome is hydrolyzed, the products are
   a. nucleotides and fatty acids
   b. nucleotides and amino acids
   c. nucleotides, amino acids, and fatty acids
   d. monosaccharides and amino acids
   e. fatty acids, glycerol, and phosphates

a 4. During meiosis, DNA replication occurs during
   a. G_1
   b. S
   c. G_2
   d. interkinesis
   e. S and interkinesis

e 5. High concentrations of colchicine cause microtubules to depolymerize preventing
   a. interphase
   b. prophase
   c. metaphase
   d. anaphase
   e. c and d

a 6. The pool of free nucleotides necessary for DNA replication is mainly synthesized during
   a. G_1 of interphase
   b. G_1 of prophase
   c. S of interphase
   d. S of prophase
   e. G_2 of interphase

d 7. If you examine a root tip, you would find most of the cells in
   a. anaphase
   b. telophase
   c. metaphase
   d. interphase
   e. prophase

d 8. In onion cells, the diploid number of chromosomes is 16. How many chromosomes are present in a cell right after anaphase?
    a. 4
    b. 8
    c. 16
    d. 32
    e. 64

e 9. Meiosis and mitosis differ in that
    a. DNA replicates during prophase in mitosis
    b. meiosis occurs only in animals
    c. meiosis produces only gametes
    d. chromosomes condense only in mitosis
    e. mitosis produces body cells

b 10. Crossing over occurs during
    a. prophase
    b. prophase I
    c. prophase II
    d. interkinesis
    e. interphase

b 11. One cell divides mitotically, then each daughter cell divides mitotically. At the end of this process, how many genetically different types of cell are there?
    a. 0
    b. 1
    c. 2
    d. 3
    e. 4

a 12. Which of the following does not occur during prophase?
    a. DNA replication
    b. disappearance of the nucleolus
    c. chromosome coiling
    d. nuclear envelope breakdown
    e. spindle formation

c 13. If a normal body cell of a plant contains 5 picograms of DNA, then that cell at the end of prophase contains
    a. 2.5 picograms
    b. 5 picograms
    c. 10 picograms
    d. 15 picograms
    e. 20 picograms

e 14. The order of events in prophase I is
    a. leptotene ---> pachytene ---> diplotene ---> zygotene ---> diakinesis
    b. diakinesis ---> diplotene ---> leptotene ---> pachytene ---> zygotene
    c. diplotene ---> pachytene ---> leptotene ---> zygotene ---> diakinesis
    d. leptotene ---> pachytene ---> zygotene ---> diplotene ---> diakinesis
    e. leptotene ---> zygotene ---> pachytene ---> diplotene ---> diakinesis

d 15. When a cell divides, new cell membranes must form where the two daughter cells meet. The source of this new membrane material is
   a. the nuclear envelope
   b. smooth endoplasmic reticulum
   c. mitochondria
   d. dictyosomes
   e. no organelle, just chemical precursors in the hyaloplasm

b 16. During anaphase of mitosis, chromatids
   a. become entwined
   b. separate
   c. break down
   d. remain together
   e. a and d

c 17. The order of events in one cell cycle is
   a. interphase ---> metaphase ---> anaphase ---> prophase ---> telophase
   b. prophase ---> metaphase ---> anaphase ---> telophase
   c. interphase ---> prophase ---> metaphase ---> anaphase ---> telophase
   d. prophase ---> metaphase ---> anaphase ---> interphase ---> telophase
   e. interphase ---> prophase ---> anaphase ---> telophase

b 18. Cell division in eukaryotic and prokaryotic cells is different in that
   a. prokaryotic cytokinesis is an outward extension of a cell plate
   b. mitosis occurs only in eukaryotes
   c. meiosis occurs only in prokaryotes
   d. microtubules form only in prokaryotes
   e. a, c, and d

d 19. As many plant cells enlarge, the number of mitochondria and chloroplasts in each cell increases. The origin of these organelles in the cell is from
   a. small, microscopic, chemical precursors
   b. the cell's endomembrane system
   c. division of microbodies
   d. division of pre-existing mitochondria and chloroplasts
   e. unknown sources

a 20. The last part of a chromosome to replicate or form during cell division is(are) the
   a. centromere
   b. histones
   c. linking pieces of DNA
   d. kinetochores
   e. genes

d 21. The growth pattern of a plant is
   a. arithmetic only its entire life
   b. geometric only its entire life
   c. mainly arithmetic when an embryo, then geometric whe older
   d. mainly geometric when an embryo, then arithmetic when older
   e. an equal mixture of arithmetic and geometric its entire life

a 22. Interphase is the stage in the cell cycle
   a. where a cell synthesizes nucleotides for DNA replication
   b. when a cell carries on little metabolism
   c. which is the shortest
   d. a and b
   e. a, b and c

c 23. During the cell cycle
   a. DNA replication occurs during early prophase
   b. at the end of G_1 phase a chromosome is composed of two chromatids
   c. at the end of anaphase a chromosome is composed of one chromatid
   d. only the DNA increases to form new chromatids
   e. none of the above

b 24. Meiosis II
   a. most closely resembles interphase
   b. most closely resembles prophase through telophase of mitosis
   c. most closely resembles interphase through telophase of mitosis
   d. is identical to meiosis I
   e. is unique and does not resemble any other process

a 25. Which of the following is <u>incorrect</u> about the cell plate in a dividing plant cell?
   a. it grows from the region of the parent cell wall inward, dividing the cytoplasm
   b. its structure includes the phragmoplast, a system of microtubules parallel to the spindle
   c. it traps dictyosome vesicles that are added to it
   d. the component vesicle membranes become part of each daughter cell's plasma membrane
   e. all of the above are correct

TRUE-FALSE QUESTIONS

F _____ 1. All of the DNA in a chromosome replicates during S phase.

F _____ 2. Cytokinesis occurs in the same direction in plant and bacterial cells.

T _____ 3. Metabolically, a cell is most active during interphase.

F _____ 4. A cell cannot divide unless centrioles are present in the cell.

T _____ 5. If karyokinesis occurs without cytokinesis, multinucleate cells result.

F _____ 6. In plant cells, centrioles migrate to the spindle poles during prophase.

F _____ 7. All cells of a plant continue to divide during the lifetime of that plant.

T _____ 8. At the end of anaphase I, a dividing cell contains the diploid number of chromosomes.

T _____ 9. Mitosis produces genetically identical cells and meiosis produces genetically different cells.

F _____ 10. You cannot see chromosomes with a light microscope in a non-dividing cell because they are absent and only form just prior to division.

F _____ 11. The replication of nuclear DNA and plastid DNA are closely coordinated.

F _____ 12. Cancer is equally life threatening to plants and animals because it disrupts normal organ structure in both.

MATCHING

Match each term from List A with an event from List B.

	List A	List B
e _____	1. telophase	a. chromatids become chromosomes
d _____	2. metaphase	b. DNA replication
b _____	3. interphase	c. nucleolus disappears
c _____	4. prophase	d. chromosomes align at cell's equator
a _____	5. anaphase	e. spindle disappears

Match each event from List A with a term from List B. Terms in List B may be used more than once.

	List A	List B
a _____	1. chiasmata form	a. prophase I
a _____	2. synapsis	b. metaphase II
b _____	3. centromere division	c. telophase II
e _____	4. separation of homologs	d. prophase II
c _____	5. reappearance of nuclear envelope	e. anaphase I
g _____	6. separation of chromatids	f. metaphase I
c _____	7. cytokinesis completed	g. anaphase II

ESSAY QUESTIONS

1. Which is more like mitosis - meiosis I or II? Why? In what way(s) is it different?

2. Distingiush beween: endoreduplication and gene amplification; meiosis I and II; anaphase and anaphase I; interphase and interkinesis; eukaryotic and prokaryotic cell division; karyokinesis and cytokinesis.

3. Body cells of *Machaeranthera gracilis* have four chromosomes. Draw a cell of this plant in each of the following stages: interphase, anaphase, prophase I, anaphase I, anaphase II.

4. Does mitosis or meiosis or both produce genetically different daughter cells? Include your reasoning.

5. Could a prokaryotic cell divide by meiosis? Why or why not?

6. The amount of DNA per cell at the beginning of G_1 phase is 0.40 picograms/cell. Draw a graph plotting amount of DNA/cell vs. phases of the cell cycle.

7. Why are genes packaged in chromosomes rather than existing as individual genes?

CHAPTER 5: TISSUES AND THE PRIMARY GROWTH OF STEMS

MULTIPLE CHOICE QUESTIONS

d 1. The three basic types of plant cells are
   a. parenchyma, collenchyma, and tracheary element
   b. parenchyma, sieve element, and epidermis
   c. collenchyma, sclerenchyma, and sieve element
   d. parenchyma, collenchyma, and sclerenchyma
   e. sclerenchyma, tracheary element, and sieve element

c 2. If a plant does not wilt when not watered, the principal type of supporting tissue present in the plant is
   a. parenchyma
   b. collenchyma
   c. sclerenchyma
   d. phloem
   e. epidermis

b 3. A peach pit consists of the inner layer of fruit surrounding a seed, giving the seed protection. The fruit inner layer is most likely composed of
   a. parenchyma cells with thick primary walls
   b. sclerids
   c. fibers
   d. collenchyma
   e. tracheids

d 4. If each stem node has two leaves and each leaf pair is at right angle to the pair below it, the phyllotaxy is said to be
   a. alternate
   b. alternate and distichous
   c. opposite and distichous
   d. opposite and decussate
   e. spiral

a 5. At maturity a sieve tube member
   a. has no nucleus
   b. has a secondary wall
   c. is spherical in shape
   d. is dead
   e. is isolated from adjacent cells

e 6. The main characteristic of mature collenchyma cells is that they
   a. are alive
   b. are dead
   c. have a secondary wall
   d. have a thick primary wall that exhibits plasticity
   e. have a primary wall that is thick in some places, thin in others

d 7. On a per unit basis, which of the following would be the most expensive to build energetically?
   a. a moss plant that is entirely parenchyma
   b. a herbaceous plant that consists mainly of parenchyma, collenchyma, and vascular tissue
   c. a cactus that consists mainly of parenchyma, but also contains vascular tissue and sclerenchymous spines
   d. a woody plant composed mainly of xylem, plus some parenchyma and phloem
   e. an aquatic plant that contains mainly aerenchyma, in addition to vascular tissue

b 8. The vessel elements of an aquatic plant would most likely have what type of secondary wall?
   a. sclariform thickenings
   b. annular thickenings
   c. solid with circular bordered pits
   d. reticulate thickenings
   e. all of the above are equally likely

c 9. For a plant to survive, its stems must contain
   a. parenchyma and fibers
   b. sclerids and phloem
   c. phloem and xylem
   d. collenchyma and xylem
   e. parenchyma and collenchyma

e 10. Which of the following mature types of cells would be most active metabolically?
   a. vessel element
   b. sclerid
   c. collenchyma
   d. sieve tube member
   e. parenchyma

c 11. Onion meristematic cells contain 16 chromosomes. How many chromosomes would you expect to find in a mature stem parenchyma cell?
   a. 4
   b. 8
   c. 16
   d. 32
   e. none, these cells have no nuclei at maturity

c 12. If you were the first person to observe plants on another planet similar to Earth, what plant parts would you expect those plants to have, based upon your knowledge of Earth plants?
   a. roots and stems only
   b. roots and leaves only
   c. roots, stems, and leaves
   d. stems and leaves only
   e. stems only

a 13. Trichomes are advantageous to a plant because they may do all of the following except
   a. absorb CO_2
   b. trap water molecules
   c. deter insects from chewing on a plant part
   d. act as a sunscreen
   e. secrete irritating compounds

d 14. The only living cells in a flowering plant that lack a nucleus at maturity are
   a. vessel elements
   b. companion cells
   c. fibers
   d. sieve tube members
   e. collenchyma cells

b 15. A plant's cuticle would be thickest in a(n)
   a. aquatic plant
   b. desert plant
   c. tropical rainforest plant
   d. temperate forest plant
   e. plant that loses its leaves when environmental conditions are unfavorable

d 16. In what structure would you most likely expect to find sclerenchyma as the main supporting tissue?
   a. root tips
   b. mature parts of roots
   c. a young tendril
   d. mature stems
   e. young petioles

c 17. The major functions of aerial stems would not include
   a. transport of water and sugars
   b. production of leaves
   c. absorption of water
   d. support
   e. food storage

a 18. In a pine tree, the type of cell used to move water and minerals from one part to another is a
   a. tracheid
   b. vessel element
   c. fiber
   d. sieve cell
   e. sieve tube member

c 19. A bulb is a
   a. vertical, fleshy stem with thin leaves
   b. thin, horizontal stem with long internodes
   c. short stem with thick, fleshy leaves
   d. fleshy, horizontal stem
   e. long, thin, vertical stem with thick fleshy leaves

c 20. The vascular system of a plant differs from the vascular system of an animal. One way they differ is that in plants
   a. all transport tubes are composed of living cells
   b. cells actually move
   c. cells do not move
   d. flow is circular
   e. there is only one type of transport tissue

a 21. Which of the following is not a function of parenchyma?
   a. long distance transport of water
   b. short distance transport of sugars
   c. photosynthesis
   d. production and secretion of fragrances
   e. the opening of a stamen to release pollen

c 22. In humans, cartilage gives firm but flexible support. The comparable tissue in a plant would be
   a. parenchyma
   b. sclerenchyma
   c. collenchyma
   d. phloem
   e. xylem

e 23. If you pinch off the terminal apical meristem of a stem, branches often develop, making a plant bushier. The origin of those branches is
   a. leaf tissue of the node
   b. leaf scars
   c. stem tissue at mid-internode
   d. stem tissue of the leaf axil
   e. a and d

c 24. Mature tracheary elements are unique in a plant because they
   a. have an elongated shape
   b. are alive and transport mainly sugars
   c. are dead and transport mainly water and minerals
   d. have a primary wall only and are dead
   e. are alive and lack a nucleus

a 25. Herbaceous stems are important economically and supply us with bast fibers, which are
   a. groups of soft fibers, mainly from phloem
   b. individual cells, mainly from phloem, that are used to produce cotton clothing
   c. groups of hard fibers, mainly from xylem, that are used to make rope
   d. groups of collenchyma cells that are used to make linen
   e. groups of parenchyma cells with a thickened primary wall that are used to make writing paper

TRUE-FALSE QUESTIONS

T _____ 1. All specialized cells in a plant are initially parenchyma cells.

F _____ 2. Pits are areas in a cell wall where there is no primary wall.

F _____ 3. All plants have substantial root and aerial stem systems.

F _____ 4. In most plants, stomates are open at night, when it is coolest, so less water is lost by evaporation.

T _____ 5. All vascular bundles contain both xylem and phloem and are said to be collateral.

F _____ 6. Cells in an apical meristem divide by meiosis to produce either stem or root cells.

F _____ 7. If a plant is to grow larger, the only process required is cell division.

T _____ 8. Protoxylem cells have annular secondary walls that enable them to stretch as cells around them grow.

F _____ 9. The only economic uses of plants are their food value, lumber, and use as ornamentals.

T _____ 10. If trichomes start to appear on leaves of the third node, are fully mature on leaves of the sixth node, and the plastochron of the plant is 3 days, then it takes 9 days for the trichomes to develop.

F _____ 11. All roots are underground, therefore they are incapable of photosynthesis.

T _____ 12. Although many plants appear to be rootless, stemless, or leafless, they really are not.

MATCHING

Match each term in List A with a compound in List B.

	List A	List B
b	1. strength	a. pectins
a	2. plasticity	b. lignin
e	3. waterproofing	c. cutin
c	4. protection against microorganisms	d. suberin
		e. suberin, cutin, and lignin

Match each process in List A with the term in List B that is the <u>best</u> match.

	List A	List B
d	1. water transport	a. chlorenchyma
c	2. strong support	b. collenchyma
a	3. photosynthesis	c. fibers
b	4. flexible support	d. xylem
e	5. sugar transport	e. phloem

ESSAY QUESTIONS

1. Discuss the relationship between structure and function of: parenchyma tissue; collenchyma; xylem; phloem.

2. What type of secondary wall would you expect to find in the xylem cells of the trunk of a redwood tree? Why?

3. How could you make a shrub bushier?

4. Explain why terrestrial plants need: transporting tissues; supporting tissues; parenchyma.

5. Starting with the outer surface of a plant and moving inward, compare the sequence of tissues you would encounter in a mature dicot stem and its subapical meristem.

6. Distinguish between: node and internode; leaf axil and axillary bud; lysigeny and schizogeny; plasticity and elasticity; fiber and sclerid; tracheid and vessel element; sieve pore and plasmodesma.

7. Compare the structure of phloem in conifers to that in a flowering plant.

8. Both tracheids and vessel elements conduct water. Compare the structure of these two types of cells and explain the advantage to a plant of using vessel elements rather than tracheids.

CHAPTER 6: LEAVES

MULTIPLE CHOICE QUESTIONS

b 1. In a foliage leaf, the majority of photosynthesis occurs in the
   a. upper epidermis
   b. palisade parenchyma
   c. bundle sheath parenchyma
   d. spongy mesophyll
   e. lower epidermis

d 2. The only type of leaf used for support is a
   a. spine
   b. bud scale
   c. trap leaf
   d. tendril
   e. sclerophyllous leaf

b 3. If a plant produces leaves that are oriented horizontally, which of the following might you expect to find in greater abundance on the upper side relative to the lower side?
   a. stomates
   b. cuticle
   c. spongy mesophyll
   d. epidermal cells
   e. veins

d 4. Spines and tendrils are structurally similar because they both
   a. contain abundant parenchyma
   b. have a prominent petiole
   c. lack vascular tissue
   d. lack a lamina
   e. are photosynthetic

c 5. If you discovered a new plant that had sharp projections on its stems, what type of cells would you expect to find in abundance in those projections?
   a. parenchyma
   b. sieve-tube members
   c. fibers
   d. vessel elements
   e. collenchyma

a 6. A mature grass leaf differs from a mature dicot leaf in that it
   a. can continue to grow at its base
   b. can continue to grow at its tip
   c. can continue to grow along its margins
   d. a and c
   e. a, b, and c

d 7. In most cases, bud scales
   a. have a long petiole
   b. have a very thin cuticle
   c. are compound
   d. often have a thin layer of corky bark
   e. have extensive chlorenchyma

b 8. Stipules are outgrowths of the
   a. stem
   b. petiole
   c. stem and petiole
   d. blade
   e. blade and petiole

e 9. Which of the following leaf adaptations would you not expect to find in desert plants?
   a. hairy leaves
   b. stomates confined to the lower epidermis
   c. thick cuticle
   d. stomates in epidermal cavities
   e. thin palisade layer

a 10. The order of tissues encountered in the midrib section of a dicot leaf, from top to bottom would be
   a. upper epidermis ---> palisade parenchyma ---> xylem ---> phloem ---> spongy mesophyll ---> lower epidermis
   b. upper epidermis ---> spongy mesophyll ---> xylem ---> phloem ---> palisade parenchyma ---> lower epidermis
   c. upper epidermis ---> palisade parenchyma ---> phloem --->xylem ---> spongy mesophyll ---> lower epidermis
   d. upper epidermis ---> spongy mesophyll ---> phloem ---> xylem ---> palisade parenchyma ---> lower epidermis
   e. lower epidermis ---> palisade parenchyma ---> xylem --->phloem ---> spongy mesophyll ---> upper epidermis

c 11. Veins with a bundle sheath in dicot leaves are mainly involved in
   a. conduction
   b. support
   c. conduction and support
   d. conduction, support, and water release
   e. conduction, support, and sugar loading

a 12. Where would you expect to find stomates in the floating leaves of water lilies?
   a. upper epidermis only
   b. mainly in the upper epidermis with a few in the lower epidermis
   c. equally distributed in the upper and lower epidermis
   d. mainly in the lower epidermis with a few in the upper epidermis
   e. lower epidermis only

d 13. Sessile leaves are usually
   a. large and broad
   b. small
   c. long and narrow
   d. b or c
   e. a or c

e 14. A petiole is advantageous to a leaf because it can
   a. increase the amount of light available to a leaf
   b. keep a leaf cool
   c. retard insect damage
   d. increase available CO_2
   e. all of the above

c 15. Which of the following types of leaves would most likely minimize leaf damage due to insects?
   a. simple and petiolate
   b. simple and sessile
   c. compound and petiolate
   d. compound and sessile
   e. all leaves are equally vulnerable to insect damage

e 16. Which of the following is not a function of some types of leaves?
   a. protection
   b. support
   c. storage
   d. nitrogen procurement
   e. wood production

d 17. Most foliage leaves are flat and thin for maximal
   a. absorption of light
   b. absorption of CO_2
   c. absorption of water
   d. a and b
   e. a, b and c

a 18. The major type of tissue found in foliage leaves is
   a. parenchyma
   b. xylem
   c. phloem
   d. epidermis
   e. collenchyma

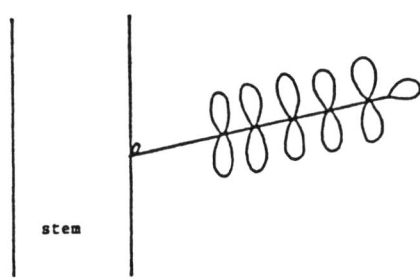

e 19. The leaf in the accompanying diagram is
   a. simple
   b. palmately simple
   c. pinnately simple
   d. palmately compound
   e. pinnately compound

a 20. One characteristic that distinguishes spongy mesophyll from palisade parenchyma is
   a. the loose arrangement of cells in the spongy mesophyll
   b. the tight packing of cells in the spongy mesophyll
   c. the presence of cutin on the surface of palisade parenchyma cells
   d. that spongy mesophyll cells are alive, while palisade parenchyma cells are dead at maturity
   e. spongy mesophyll is the main photosynthetic part of a leaf

d 21. Which of the following is not a leaf adaptation to protect the plant?
   a. the presence of a cuticle
   b. stomates sunken into epidermal cavities
   c. a translucent epidermis
   d. a greater number of stomates in the upper epidermis than in the lower epidermis
   e. the presence of trichomes

c 22. Leaves are quite vulnerable to damage by fungi and chewing insects. A leaf adaptation that would not be a deterent to those pests would be
   a. a cuticle on the surface of the epidermis
   b. a bundle sheath around leaf vascular bundles
   c. stomates in the epidermis
   d. trichomes on the surface
   e. a blade divided into leaflets

a 23. When trying to identify a vegetative plant, it would be most useful to examine its
   a. leaves because their shape is characteristic of a species
   b. stems because their structure varies greatly from one species to another
   c. roots because root structure is species specific
   d. a and c
   e. b and c

e 24. It is easy to mistake a compound leaf for a stem with simple leaves. To decide whether or not a structure is a compound leaf or a shoot, look for
   a. the placement of axillary buds
   b. the presence of a terminal bud
   c. phyllotaxy
   d. a and b
   e. a, b and c

b 25. Due to the chemicals they produce, many leaves are a source of herbs and spices that we use to flavor foods. The major advantage of these chemicals to a plant is that they
   a. attract seed dispersers
   b. are an antiherbivore defense
   c. reduce competition between plants for light, water and minerals
   d. act as pigments, giving flowers their color
   e. function as food storage molecules

TRUE-FALSE QUESTIONS

F _____ 1. All leaves have the same structure and function.

T _____ 2. The main reason all large leaves are compound is to minimize wind damage.

F _____ 3. Leaf shape is completely determined genetically, never by environment.

F _____ 4. Stem shape is routinely used to identify plants.

F _____ 5. The arrangement of tissues is uniform throughout a petiole.

T _____ 6. Stipules may contribute to the net photosynthesis of a plant.

T _____ 7. Initially, all leaf cells are meristematic.

F _____ 8. The leaves of insect-trapping plants are not capable of photosynthesis.

T _____ 9. Depending on the type of plant, a bundle sheath around a leaf vein is composed of either parenchyma or sclerenchyma.

T _____ 10. Most leaves are thin and flat to maximize exposure to light and surface area for CO_2 absorption.

F _____ 11. The mesophyll of all leaves is divided into palisade and spongy regions.

T _____ 12. Some plants produce leaves of more than one shape.

T _____ 13. Leaf shape varies among plant species more than stem or root shape.

MATCHING

Match each statement from List A with one term from List B. Terms in List B may be used more than once.

	List A	List B
a	1. Water distribution	a. vein
e	2. Leaflet support	b. petiole
b	3. Blade attachment to stem	c. blade
a	4. Sugar collection	d. abscission zone
c	5. Photosynthesis	e. rachis
d	6. Leaf detachment	

Match each term in List A with a function in List B. Functions in List B maybe used more than once.

	List A	List B
c	1. Foliage leaf	a. protection
a	2. Bud scale	b. C_4 photosynthesis
e	3. Trap leaves	c. "normal" photosynthesis
d	4. Tendril	d. support
b	5. Kranz anatomy	e. nitrogen procurement
a	6. Spine	

ESSAY QUESTIONS

1. Discuss the relationship between structure and function in leaves vs. stems.

2. What are the advantages of compound leaves?

3. Explain why cells are packed differently in the palisade parenchyma vs. spongy mesophyll of a leaf.

4. How could you distinguish between a compound leaf and a twig with several simple leaves?

5. Why can trees seem to develop leaves overnight in the spring?

6. Distinguish between active and passive trap leaves and explain why the adaptation of trapping insects is beneficial to some plants.

7. Many plants produce chemicals that are flavorful or toxic to people. Of what advantage is production of these compounds to the plants?

8. Compare the structure of simple and compound leaves. What are the advantages of compound leaves? Why are most large leaves compound?

9. What type of leaf adaptations are found in plants living in deserts?

10. Why is it advantageous for a woody stem to produce secondary wood and bark? Is it also advantageous for leaves? Why or why not?

11. Explain how an insect walking on the surface of a Venus' flytrap leaf can get caught. What happens to the insect once it is caught?

CHAPTER 7: ROOTS

MULTIPLE CHOICE QUESTIONS

d   1. Root hairs
    a. increase surface area
    b. are long-lived
    c. help release ions from the soil
    d. a and c
    e. a, b and c

d   2. Where would you look in a root to study cell division?
    a. apical meristem only
    b. apical meristem and root cap
    c. root cap only
    d. apical meristem, root cap, and zone of elongation
    e. zone of elongation only

b   3. Pericycle cells are
    a. parenchyma cells that control mineral absorption
    b. parenchyma cells that divide to form lateral roots
    c. fibers that give support
    d. endodermal cells that control mineral absorption
    e. a and b

c   4. The main function of a root cap is
    a. absorption of water and minerals from the soil
    b. root anchorage
    c. protection of the root apical meristem
    d. hormone production
    e. plant vegetative reproduction

e   5. A typical monocot plant has a root system consisting of
    a. a radicle, a tap root, and lateral roots
    b. a radicle, a tap root, and adventitious roots
    c. a taproot and lateral roots
    d. a radicle, lateral roots and adventitious roots
    e. adventitious roots and lateral roots

b   6. If a leaf is cut off of an African violet plant and placed in water, new roots form at the base of the petiole. The first root formed this way is
    a. a taproot
    b. adventitious
    c. a radicle
    d. a lateral root
    e. none of the above

c   7. You would expect roots to do all of the following <u>except</u>
    a. absorb water
    b. absorb minerals
    c. photosynthesis
    d. produce and export hormones to the shoot
    e. hold a plant in one place

c 8. Lateral roots originate in the
   a. phloem
   b. epidermis
   c. pericycle
   d. cortex
   e. apical meristem

d 9. If a water molecule entered a mature root at the root surface, it would cross what tissues, in their correct order, on its way to the center?
   a. epidermis, endodermis, cortex, phloem, xylem
   b. epidermis, cortex, pericycle, endodermis, phloem, xylem
   c. cortex, endodermis, phloem, pericycle, xylem
   d. epidermis, cortex, endodermis, pericycle, phloem, xylem
   e. endodermis, epidermis, cortex, pericycle, phloem, xylem

a 10. Perennial plants are likely to store carbohydrates in storage roots during the winter because
   a. temperature and humidity are more stable in the soil than above ground
   b. nutrient supply is more abundant in the soil
   c. leaves often are shed from the plant during the winter
   d. stems are too woody for carbohydrate storage
   e. stems cannot be winterproofed

d 11. Palm trees, which are monocots, can increase in diameter every year due to the production of additional
   a. wood
   b. lateral roots
   c. taproots
   d. adventitious roots
   e. monocots cannot increase in diameter

b 12. The rapid diffusion of oxygen to submerged portions of roots of some marsh plants occurs through
   a. chlorenchyma
   b. aerenchyma
   c. parenchyma
   d. xylem
   e. phloem

e 13. Roots that can anchor or brace a plant more firmly than normal roots are
   a. storage and mycorrhizal roots
   b. nodular and prop roots
   c. contractile and storage roots
   d. mycorrhizal and nodular roots
   e. prop and contractile roots

c 14. If the gladiolus bulbs in your garden remain at the same soil depth from year to year, this is most likely due to
   a. prop roots attached to the bulbs
   b. the fact that the stem is a bulb
   c. contractile roots attached to the bulbs
   d. root nodules that provide nitrogen for the stem to grow to the correct depth
   e. your gardening ability

a 15. If the apical meristem of a root is damaged, a new apical meristem forms from cells from
   a. the quiescent center
   b. the root cap
   c. the zone of elongation
   d. a lateral root
   e. a new apical meristem cannot form and the root stops growing

c 16. In dicot root xylem, metaxylem cells would be the
   a. innermost, narrow cells in vascular bundles
   b. innermost, wide cells in vascular bundles
   c. innermost, wide cells in a solid mass of xylem
   d. outermost, wide cells in a solid mass of xylem
   e. outermost, narrow cells in vascular bundles

a 17. The process of converting atmospheric nitrogen into a usable form of nitrogen for organisms to use is
   a. nitrogen fixation performed by some prokaryotes
   b. nitrogen fixation performed by some eukaryotes
   c. nitrification performed by some prokaryotes
   d. nitrification performed by some eukaryotes
   e. denitrification performed by some prokaryotes

c 18. A root is not uniform along its length, but has distinct zones. Starting at the tip, these zones are
   a. apical meristem ---> root cap ---> zone of elongation ---> zone of maturation/root hairs
   b. root cap ---> zone of elongation ---> apical meristem --->zone of maturation/root hairs
   c. root cap ---> apical meristem ---> zone of elongation --->zone of maturation/root hairs
   d. root cap ---> apical meristem ---> zone of maturation/root hairs ---> zone of elongation
   e. none of the above because roots are too variable in structure

b 19. Roots which have a structure most different from normal absorptive roots are
   a. storage roots
   b. haustorial roots
   c. prop roots
   d. nodular roots
   e. contractile roots

c 20. The major direct advantage to a plant of a mycorrhizal relationship is increased
   a. sugar production
   b. water absorption
   c. phosphorus absorption
   d. nitrogen fixation
   e. anchorage

b 21. The main functions of roots include
   a. anchorage and storage
   b. anchorage, absorption and hormone production
   c. absorption, hormone production, and food production
   d. absorption and storage
   e. hormone production and anchorage

e 22. Parasitic plants have roots highly modified into
   a. fibrous root systems
   b. lateral roots
   c. nodular roots
   d. taproots
   e. haustorial roots

c 23. As foods, roots are richest in
   a. protein
   b. fats
   c. carbohydrates
   d. vitamins
   e. minerals

a 24. Root nodules form primarily in which tissue?
   a. cortex
   b. xylem
   c. phloem
   d. pericycle
   e. endodermis

d 25. A mycorrhizal association develops between the roots of plants and
   a. bacteria
   b. parasitic plants
   c. algae
   d. fungi
   e. protists

b 26. Root caps secrete
   a. water to lubricate root growth through the soil
   b. mucigel to lubricate root growth through the soil
   c. mucigel as an antibacterial agent
   d. oxygen as a byproduct of photosynthesis
   e. minerals into the soil

TRUE-FALSE QUESTIONS

F _____ 1. All root cap cells are identical.

F _____ 2. Casparian strips are found in roots, but not in stems or leaves.

T _____ 3. Root cap cells function by being expendable to a plant.

F _____ 4. The most absorptive part of a root is right at the tip.

F _____ 5. Like stem trichomes, root hairs may be multicellular.

F _____ 6. If a plant is not growing in soil, it will not have roots.

T _____ 7. The formation of mature root nodules in plants such as legumes is due to a combination of bacterial and plant genetics.

T _____ 8. Not only do plants make use of food stored in roots, people do as well.

F _____ 9. Spines are always modified leaves.

T _____ 10. Many plants form roots with a specialized function in addition to more "typical" roots.

F _____ 11. The most common type of stored food in taproots is protein, which has a high nutritive value.

T _____ 12. Plasmodesmata connect cells of the root.

F _____ 13. Applying nitrogen fertilzer is a waste of money by farmers because all crop plants can form root nodules containing nitrogen fixing bacteria.

MATCHING

Match each root structure in List A with its major function in List B.

List A

g _____ 1. Root cap
d _____ 2. Root hair
b _____ 3. Apical meristem
f _____ 4. Parenchyma cells
e _____ 5. Endodermis
a _____ 6. Pericycle
c _____ 7. Xylem

List B

a. lateral root origin
b. cell production
c. mineral transport
d. increased surface area
e. control of mineral absorption
f. food storage
g. protection
h. support

Match each structure in List A with its major function in List B.

List A

g _____ 1. Mycorrhizal root
a _____ 2. Prop root
d _____ 3. Storage root
b _____ 4. Contractile root
f _____ 5. Haustorial root
c _____ 6. Root nodule

List B

a. stabilization and support
b. pull stem downward
c. nitrogen fixation
d. accumulation of carbohydrate
e. increase surface area
f. absorption from a host
g. increased phosphorus absorption
h. vegetative reproduction

ESSAY QUESTIONS

1. Discuss the possible routes a mineral ion might take from the soil to the xylem in the zone of maturation.

2. Distinguish between radicle, tap root, lateral root, and adventitious root. Would you expect the internal distribution of tissues to be the same or different in these roots? Why?

3. Would grass or carrots be most effective for control of erosion? Why?

4. Could a plant survive for long without roots? Why or why not?

5. Relate the overall shape of a root to its function and environment.

6. Why do farmers often rotate legumes and non-legumes from year to year in a field?

7. Relate the development and structure of the endodermis to its function.

8. For each of the following specialized roots, describe how their metabolism and structure differ from that of absorptive roots: storage roots, prop roots, contractile roots, haustorial roots, nodular roots. For each of these specialized roots, what is the selective advantage of their modifications?

9. Diagram the zones in the growing region of a root. Label each with its name and major characteristic.

10. List and discuss the functions of the root cap.

CHAPTER 8: STRUCTURE OF WOODY PLANTS

MULTIPLE CHOICE QUESTIONS

d   1. One way herbaceous plants can increase their conduction capacity is to produce
    a. additional leaves
    b. storage roots
    c. secondary tissues
    d. adventitious roots
    e. contractile roots

e   2. True secondary growth occurs in
    a. all ferns, gymnosperms, dicots, and monocots
    b. some ferns, gymnosperms, dicots, and monocots
    c. all gymnosperms, dicots, and monocots
    d. all dicots and some monocots
    e. all gymnosperms and some dicots

d   3. The phloem cells of a woody stem get their required oxygen through
    a. plasmodesmata
    b. normal cork
    c. phelloderm
    d. lenticels
    e. aerenchyma

e   4. Annual rings from the base of a tree trunk can be used to determine
    a. tree age
    b. climatic conditions
    c. dates of ancient settlements
    d. a and c
    e. a, b and c

c   5. The cells in primary tissues were produced by cell divisions in a(n)
    a. vascular cambium
    b. cork cambium
    c. apical meristem
    d. a and b
    e. a, b and c

c   6. As present-day woody plant stems and roots increase in length and width, the vascular cambia
    a. stay the same size
    b. must increase in size by enlargement of existing cambial cells
    c. must increase in size by cambial cell division and production of new cambial cells from non-cambial cells
    d. must increase in size by cambial cell division only
    e. must increase in size by enlargement of existing cambial cells, cambial cell division, and production of cambial cells from non-vascular cambial cells

a 7. Many people buy furniture made of oak. The wood of this tree is particularly good for furniture construction because it contains large amounts of
   a. fibers
   b. parenchyma
   c. sclerids
   d. tracheids
   e. vessel elements

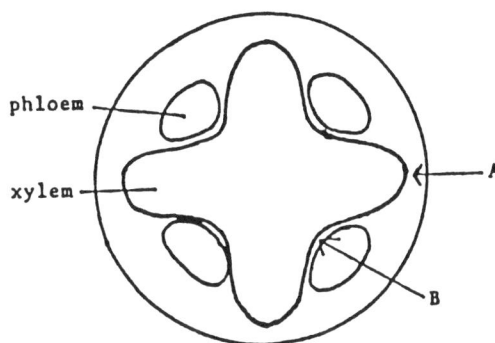

d 8. If a vascular cambium develops in the accompanying root, the initial rate of production of xylem and phloem cells in areas A and B will
   a. be the same
   b. differ; more xylem will be produced in area A than in area B
   c. differ; more phloem will be produced in area A than in area B
   d. differ; more xylem will be produced in area B than in area A
   e. differ; more phloem will be produced in area B than in area A

a 9. The cells which transport sugars from the leaves to the roots of a woody flowering plant are
   a. sieve tube members produced by fusiform initials of the vascular cambium
   b. sieve tube members produced by fusiform initials of the cork cambium
   c. tracheids produced by fusiform initials of the vascular cambium
   d. companion cells produced by ray initials of the vascular cambium
   e. fibers produced by fusiform initials of the cork cambium

b 10. Anticlinal divisions of fusiform initials may produce
   a. vessel elements
   b. new fusiform initials
   c. ray initials
   d. sieve tube members
   e. storage parenchyma cells

a 11. Periclinal division of ray initials produces
   a. storage parenchyma cells
   b. vessel elements
   c. companion cells
   d. sieve tube members
   e. none of the above

d 12. The maple sap tapped in the spring to make maple syrup comes from the xylem of sugar maple trees. The sugar in this sap was stored in
   a. root cortex cells
   b. phloem parenchyma cells
   c. xylem axial parenchyma cells
   d. xylem ray upright cells
   e. xylem ray procumbent cells

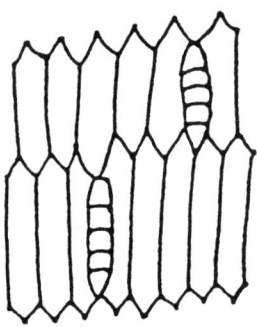

a 13. The accompanying diagram shows a
   a. storied cambium with uniseriate rays
   b. storied cambium with biseriate rays
   c. storied cambium with multiseriate rays
   d. nonstoried cambium with uniseriate rays
   e. nonstoried cambium with biseriate rays

d 14. Summer wood differs from spring wood in that
   a. there are more fibers present
   b. cell diameter is narrower
   c. cell diameter is wider
   d. a and b
   e. a and c

c 15. In a tree whose sapwood functions for nine years, how many annual rings have been converted to heartwood in a trunk section that is thirty five years old?
   a. 1
   b. 9
   c. 24
   d. 34
   e. 35

e 16. The origin of a stem cork cambium may be the
   a. cortex
   b. secondary phloem
   c. epidermis
   d. primary phloem
   e. all of the above are possible

d 17. The vascular cambium in a dicot root arises from
   a. primary phloem cells
   b. primary phloem and parenchyma cells
   c. parenchyma and primary xylem cells
   d. parenchyma and pericycle cells
   e. primary phloem, parenchyma, and pericycle cells

b 18. The origin of a root cork cambium is the
   a. epidermis
   b. pericycle
   c. phloem
   d. cortex
   e. all of the above are possible

c 19. Carrots and sweet potatoes are roots that have secondary growth. We find these roots easy to eat because most of that secondary tissue consists of
   a. cork
   b. phelloderm
   c. parenchyma
   d. phloem
   e. pericycle

e 20. As sapwood changes to heartwood, the changes that occur include
   a. cell death
   b. the formation of tyloses
   c. the deposition of insecticidal and fungicidal compounds
   d. b and c
   e. a, b and c

b 21. In monocots
   a. a vascular cambium forms like that of gymnosperms
   b. a cambium that produces only parenchyma arises outside the outermost vascular bundles
   c. a cork cambium only arises
   d. wood forms but is very fibrous
   e. secondary growth does not occur

e 22. The origin of the vascular cambium in dicot roots is from the
   a. cells between the pith and xylem
   b. pericycle
   c. endodermis
   d. cells between the xylem and phloem
   e. b and d

e 23. Secondary growth results in an increase in
   a. photosynthetic capacity
   b. seed production
   c. defensive chemicals
   d. a and b
   e. a, b, and c

a 24. Fasicular cambium is found
   a. between the metaxylem and metaphloem
   b. between the metaxylem and pith
   c. within the cortex
   d. within the phloem
   e. within the hypodermis

d 25. Xylem parenchyma
   a. serves as a reservoir of CO_2
   b. is the site of vascular cambium formation
   c. is the site of cork cambium formation
   d. serves as a reservoir of water
   e. provides strength and flexibility to wood

TRUE-FALSE QUESTIONS

T _____ 1. A woody plant is initially composed of primary tissues only.

F _____ 2. A mature woody plant is composed of secondary tissues, such as secondary xylem and phloem, but no primary tissues.

T _____ 3. It is possible for fusiform initials to become ray intitals and vice versa.

T _____ 4. If a woody plant is growing, the vascular cambium must enlarge in circumference each year.

T _____ 5. Tracheids may occur in the axial system of both gymnosperms and dicots, but in the radial system only of gymnosperms.

F _____ 6. All woody trees produce distinguishable annual rings.

F _____ 7. Once a tracheary element is produced in a woody plant, it will function in water transport for the lifetime of that plant.

F _____ 8. In plants that produce annual rings in secondary xylem, annual rings will also be seen in secondary phloem.

T _____ 9. Phloem and xylem rays are produced by the same ray initials.

F _____ 10. You can distinguish a cross section of a woody root from a cross section of a woody stem because the secondary xylem of the root is star-shaped.

F _____ 11. Palm trees are an exception to the idea that monocots cannot produce secondary tissues and wood.

F _____ 12. In a cross section of a tree, the oldest xylem is found near the center and the oldest phloem is found next to the xylem.

MATCHING

Match each structure in List A with a term from List B.

	List A	List B
h	1. Annual ring	a. sapwood
b	2. Non-conducting xylem	b. heartwood
a	3. Conducting xylem	c. softwood
f	4. Xylem on upper side of branch	d. hardwood
		e. ring porous wood
c	5. Xylem composed mainly of tracheids	f. tension wood
		g. compression wood
d	6. Xylem containing a large number of fibers	h. spring wood + summer wood
g	7. Xylem on underside of branch	

Match each structure in List A with the cambial cell from List B that produces it. Terms in List B may be used more than once.

	List A	List B
b	1. Tracheid	a. cork cambium cell
a	2. Cork	b. vascular cambium fusiform initial
c	3. Upright cell	
b	4. Fiber	c. vascular cambium ray initial
b	5. Axial xylem parenchyma	
c	6. Phloem ray	d. cork cambium fusiform initial
b	7. Companion cell	
c	8. Procumbent cell	e. cork cambium ray initial
a	9. Phelloderm	
b	10. Sieve tube element	

ESSAY QUESTIONS

1. Why are woody plants more likely to survive adverse environmental conditions than non-woody plants?

2. What is(are) the advantage(s) of a plant being a perennial rather than an annual? Disadvantage(s)?

3. What climatic conditions produce annual rings and which do not? Why?

4. Often in storms with high winds, trees fall over, revealing a hollow center. Why could the tree function normally before it was blown over?

5. What are the advantages of cork on the surface of a woody stem or root?

6. Compare cork and vascular cambia with respect to: origin, structure, function, longevity, relation to increased trunk circumference.

7. What are the protruding, horizontal lines in the bark of cherry trees? What is their function?

8. Compare <u>one</u> type of anomalous secondary growth with normal secondary growth.

9. Could a woody plant survive if the vascular cambium had no fusiform initials? Ray initials? Why?

10. Describe the kind of secondary growth exhibited by monocots such as joshua and dragon trees. Do you think that the "wood" produced would be useful for construction compared with the wood of pine or oak?

11. Would you expect the length of the sieve tube members to be related to the length of the fibers and vessel elements of the same tree? Why?

12. One way to determine the age of a living tree is to drill a core into the trunk. List the tissues that you would encounter as you drilled from the outside to the very center of the tree.

CHAPTER 9: FLOWERS AND REPRODUCTION

MULTIPLE CHOICE QUESTIONS

c 1. In the life cycle of a flowering plant, which of the following would be diploid?
   a. microspores
   b. egg
   c. megasporocyte
   d. gametophyte
   e. endosperm

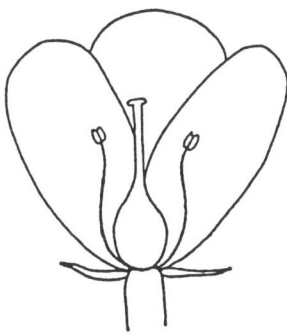

d 2. The accompanying diagram is a(n)
   a. epigynous flower
   b. apogynous flower
   c. perigynous flower
   d. hypogynous flower
   e. hypergynous flower

d 3. A dioecious plant produces flowers that are
   a. perfect and complete
   b. perfect and incomplete
   c. imperfect and complete
   d. imperfect and incomplete
   e. it depends on the plant

a 4. Coconut milk and meat are the result of one sperm fusing with
   a. polar nuclei
   b. an antipodal cell
   c. a synergid cell
   d. a syntipodal cell
   e. an egg

e 5. A flower that is more likely to be pollinated by a specific pollinator would have
   a. radial symmetry
   b. bilateral sylmmetry
   c. a specific fragrance
   d. no petals
   e. b and c

c 6. If you remove a leaf from an African violet plant and place it in water, a new plant will develop at the base of the leaf. This type of reproduction is
   a. asexual by the production of stolons
   b. asexual through fragmentation
   c. asexual through the production of adventitious shoots
   d. sexual through the production of bulbils
   e. sexual through the production of plantlets

b 7. In a typical plant life cycle, gametes are produced by a
   a. haploid gametophyte by meiosis
   b. haploid gametophyte by mitosis
   c. haploid sporophyte by meiosis
   d. diploid sporophyte by mitosis
   e. diploid gametophyte by mitosis

d 8. An orange is a(n)
   a. achene
   b. pome
   c. berry
   d. hesperidium
   e. legume

c 9. Which of the following plants would most probably produce seeds having the least genetic variability?
   a. a plant where the carpels mature several days after the stamens in each flower
   b. a plant where the stamens are much shorter in height than the carpels in each flower
   c. a monoecious plant where all flowers mature at the same time
   d. a dioecious plant where flowers mature at the same time
   e. a plant where the stamens produce pollen that matches an incompatibility gene present in the plant

e 10. Wind pollinated flowers usually are or have
   a. large in size
   b. zygomorphic
   c. complete flowers
   d. bright and colorful
   e. stigmas with a large surface area

e 11. The individuals of most species of plants in tropical rain forests are widely scattered from one another. These plants must rely on what for pollination?
   a. insects
   b. birds
   c. bats
   d. wind
   e. a, b and c

b 12. If a bird eats a fruit and its enclosed seeds then deposits the seeds at a site removed from the parent plant, the seeds have been dispersed by
   a. autochory
   b. ornithochory
   c. chiropterochory
   d. anemochory
   e. hydrochory

a 13. The two halves of a lima bean or pea are the
   a. cotyledons
   b. epicotyl
   c. embryo
   d. radicle
   e. hypocotyl

b 14. The order in which flower parts are attached to the receptacle, from lowest to highest, is
   a. petals ---> sepals ---> stamens ---> carpels
   b. sepals ---> petals ---> stamens ---> carpels
   c. sepals ---> stamens ---> petals ---> carpels
   d. sepals ---> petals ---> carpels ---> stamens
   e. these parts may occur in any order depending on the flower

a 15. Many organisms are oogamous and produce
   a. small, motile sperm cells and large, non-motile egg cells
   b. large, motile sperm cells and small, non-motile egg cells
   c. sperm and egg cells that are the same size and are both motile
   d. small, motile sperm cells and large, motile egg cells
   e. small, non-motile sperm cells and large, non-motile egg cells

c 16. A typical plant life cycle differs from an animal life cycle in that
   a. mitosis is part of the plant life cycle, but not the animal
   b. meiosis is part of the animal life cycle, but not the plant
   c. meiosis produces gametes in the animal life cycle, but not in the plant
   d. the animal life cycle produces two types of animals, one diploid, one haploid, but the plant life cycle produces only one type of plant
   e. b and c

e 17. For a structure to be considered a flower, it must have
   a. sepals, petals, stamens, and carpels
   b. sepals and petals
   c. petals
   d. stamens and carpels
   e. stamens or carpels

a 18. Which of the following plants would be most likely to produce a multiple fruit?
   a. one that produces an inflorescence of closely spaced flowers, where all floral parts become fleshy during fruit development
   b. one that produces an inflorescence of widely spaced flowers
   c. one that produces flowers with multiple carpels
   d. one that produces single flowers with fused ovaries
   e. one that produces single flowers with a single ovary

b 19. The major reason showy flowers have evolved is to
   a. give people pleasure
   b. attract the correct pollinators
   c. attract and please animals
   d. protect the seeds
   e. all of the above

c 20. A pollen grain, as it leaves the anther, is a(n)
   a. immature megagametophyte
   b. mature megagametophyte
   c. immature microgametophyte
   d. mature microgametophyte
   e. microspore

c 21. The layers of a fruit wall are, in order from the outside to the inside
   a. mesocarp, endocarp, and exocarp
   b. endocarp, mesocarp, and exocarp
   c. exocarp, mesocarp, and endocarp
   d. exocarp, endocarp, and mesocarp
   e. mesocarp, exocarp, and endocarp

e 22. Which of the following correctly matches a fruit type to an example of that type?
   a. capsule - green beans
   b. samara - peanut
   c. pome - orange
   d. achene - poppy
   e. drupe - cherry

b 23. In the plant life cycle, meiosis results in the production of
   a. gametes
   b. spores
   c. zygotes
   d. sporophytes
   e. carpels

d 24. Pollen grains are caught and germinate on the
   a. anther
   b. ovary
   c. style
   d. stigma
   e. filament

d 25. Which of the following statements is <u>false</u> about the pollen tube?
   a. the pollen tube grows through the style
   b. the pollen tube enters one synergid cell
   c. the pollen tube contains three nuclei
   d. the nuclei in the pollen tube are diploid
   e. the pollen tube contains two sperm

b 26. Wind pollinated flowers, in contrast to insect pollinated flowers, are usually characterized by
   a. a greatly reduced stigma and style
   b. reduced or no petals and sepals
   c. producing few pollen grains
   d. brightly colored sepals or stamens
   e. zygomorphy

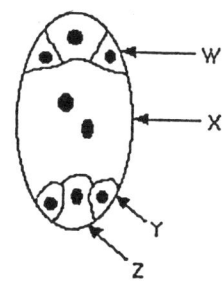

b 27. In the accompanying diagram, the endosperm will form, after fertilization, from the structure labeled
   a. W
   b. X
   c. Y
   d. Z
   e. all of the above

c 28. The "nib" of a lima bean or pea is
   a. the cotyledon
   b. part of the seed coat
   c. the epicotyl and radicle
   d. the endosperm
   e. the suspensor

TRUE-FALSE QUESTIONS

F _____ 1. In a population of flowering plants, all of the seeds produced will be genetically identical and equally well adapted to the environment.

T _____ 2. For most sexually reproducing plants, the potential advantages of sexual reproduction outweigh the potential disadvantages.

F _____ 3. Cell division by meiosis always produces sperm and egg cells.

F _____ 4. A new plant produced by a runner will be more fit for its environment than its parent.

F _____ 5. A sunflower is a single, large flower that has evolved to attract a specific pollinator.

T _____ 6. Dry fruits are much more likely to be wind-distributed than fleshy fruits.

F _____ 7. A peach pit is a seed.

T _____ 8. A major advance in the development of human civilization and culture was due to the beginning of agriculture.

T _____ 9. If flower color is controlled by a gene in the plastids of a plant, then the offspring of two plants of different color will be the same color as the maternal parent.

T _____ 10. In both monocots and dicots, nutrients in the endosperm are absorbed into the embryo by the cotyledons.

F _____ 11. A fruit is an enlarged, mature flower.

T _____ 12. The three components of most mature seeds are a seed coat, an embryo, and stored food.

T _____ 13. Wind pollinated flowers are often incomplete flowers.

T _____ 14. Pollen tubes can be considered parasitic on the carpels.

F _____ 15. Many flowers last a long time because they produce secondary tissues and become woody.

MATCHING

Match each term in List A with the type of reproduction in List B. Terms in List B may be used more than once.

	List A	List B
a	1. Zygote	a. sexual reproduction
b	2. Runner	b. asexual reproduction
b	3. Adventitious shoot	
a	4. Gamete	
b	5. Leaf margin plantlets	
b	6. Bulbil	
a	7. Syngamy	

Match each function in List A with the appropriate structure in List B.

	List A	List B
a	1. Usually attracts pollinators	a. corolla
c	2. Produces pollen	b. calyx
f	3. Flower parts attached	c. androecium
b	4. Protects the flower bud	d. gynoecium
d	5. Contains the ovules	e. pedicel
e	6. Supports the flower	f. receptacle

ESSAY QUESTIONS

1. Diagram a typical plant life cycle and compare it to a typical animal life cycle. Be sure to include where mitosis and meiosis occur.

2. What are the advantages and disadvantages of a strawberry plant reproducing by stolons? By seeds?

3. Why are immature fleshy fruits unpalatable?

4. Discuss the advantages and disadvantages to a plant of producing large flowers; small flowers.

5. Draw and label a megagametophyte of a flowering plant and compare it to the microgametophyte.

6. Why are sepals and petals sometimes referred to as nonessential flower parts?

7. Distinguish between hypogynous, perigynous, and epigynous flowers.

CHAPTER 10: ENERGY METABOLISM: PHOTOSYNTHESIS

MULTIPLE CHOICE QUESTIONS

e   1. The most unstable way to store energy in a cell is in the form of
   a. glucose
   b. starch
   c. fat
   d. sucrose
   e. ATP

b   2. To promote plant growth, lights called "grow lights" should produce light enriched in what wavelengths?
   a. red and green
   b. red and blue
   c. yellow and blue
   d. blue and violet
   e. orange and violet

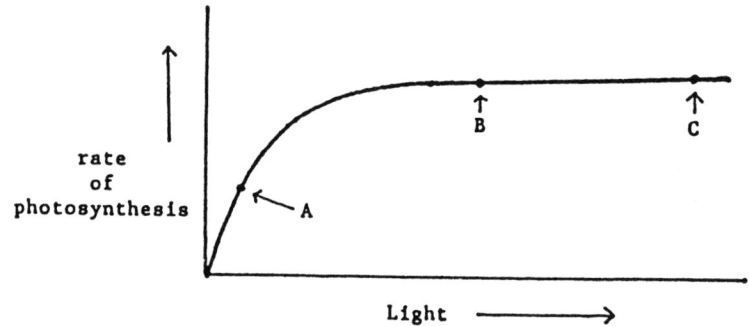

c   3. Plants were grown in growth chambers containing 330 ppm CO_2 and varying amounts of light. The accompanying graph indicates that the rate of photosynthesis was limited by
   a. light at points A, B and C
   b. light at points A and B and by CO_2 availability at point C
   c. light at point A and by CO_2 availability at point C
   d. CO_2 availability at points A, B and C
   e. you cannot tell what limits the rate of photosynthesis by this graph

b   4. During a twenty four hour period where the day is bright and sunny, there will be how many light compensation points?
   a. 1
   b. 2
   c. 3
   d. 4
   e. there is no way to tell how many

c   5. Which of the following is not a protective adaptation against intense sunlight?
   a. vertical leaves
   b. accessory pigments
   c. abundant chlorophyll
   d. a heavy coating of wax on the leaves
   e. hairy leaves

276

a   6. The main advantage of Kranz anatomy is that
      a. RuBP carboxylase is found only in bundle sheath chloroplasts
      b. RuBP carboxylase is found only in palisade parenchyma chloroplasts
      c. RuBP carboxylase is found only in spongy mesophyll chloroplasts
      d. RuBP carboxylase is found only in mesophyll chloroplasts
      e. PEP carboxylase is found only in bundle sheath chloroplasts

b   7. Photosystem I and photosystem II must be connected for the formation of
      a. ATP
      b. NADPH
      c. NADH
      d. H_2O
      e. reduced ferredoxin

c   8. ATP is synthesized from ADP and phosphate
      a. in the hyaloplasm using the oxidation by oxygen of organic molecules as the energy source
      b. in mitochondria using light energy as the energy source
      c. in chloroplasts using light energy as the energy source
      d. in chloroplasts using the oxidation of organic molecules as the energy source
      e. c and d

d   9. Which of the following is not a photoautotroph?
      a. a white pine tree
      b. a geranium
      c. *Anabaena*, a cyanobacterium
      d. *Agaricus*, the common field mushroom
      e. *Spirogyra*, a green alga

c   10. The link(s) between the light dependent reactions and the stroma reactions is(are)
      a. CO_2 and H_2O
      b. ATP
      c. ATP and NADPH
      d. chlorophyll a and NADPH
      e. RuBP and ATP

b   11. The ultimate source of electrons for photosynthesis is
      a. sunlight
      b. H_2O
      c. ATP
      d. NADPH
      e. chlorophyll a

b   12. The two most important long term storage forms of energy and carbon in plants are
      a. polysaccharides and proteins
      b. polysaccharides and fats
      c. disaccharides and fats
      d. fats and ATP
      e. ATP and NADPH

c 13. The strongest evidence supporting the hypothesis that chlorophyll is the essential pigment for photosynthesis is that
   a. all plants are green
   b. it is present in all plants
   c. its absorption spectrum matches the action spectrum of photosynthesis
   d. its absorption spectrum does not match the action spectrum of photosynthesis
   e. chlorophyll is green in color

e 14. The major function of carotenoids is to
   a. absorb light wavelengths different than chlorophyll and pass most of that energy to chlorophyll a
   b. absorb the same light wavelengths as chlorophyll a and pass most of that energy to chlorophyll a
   c. pass absorbed light energy to chlorophyll by fluorescence
   d. reflect excessive light, protecting chlorophyll molecules from intense light
   e. absorb excessive light, protecting chlorophyll molecules from intense light

e 15. A compound with which redox potential would most likely donate electrons to another compound?
   a. 0
   b. -0.35
   c. +0.35
   d. -0.50
   e. -0.70

e 16. The yellow pigment molecules in daffodil petals absorb what wavelenth(s) of visible light?
   a. red
   b. orange
   c. yellow
   d. blue
   e. all wavelengths except yellow

b 17. If a light source is placed between your eye and a flask of chlorophyll, the chlorophyll appears red instead of green. This is due to
   a. absorption of green light
   b. fluorescence
   c. passing an excited electron to "X"
   d. reflection of red light
   e. none of the above

e 18. Crab grass, a C_4 plant, grows very efficiently and can usually outcompete the more desirable Kentucky blue grass, a C_3 plant, in yards. Crab grass grows so well because
   a. it contains PEP carboxylase, which has a high affinity for CO_2
   b. CO_2 concentrations inside the leaves are low, so CO_2 diffuses into the leaves rapidly
   c. photorespiration is low
   d. its H_2O lost/CO_2 absorbed ratio is low
   e. all of the above

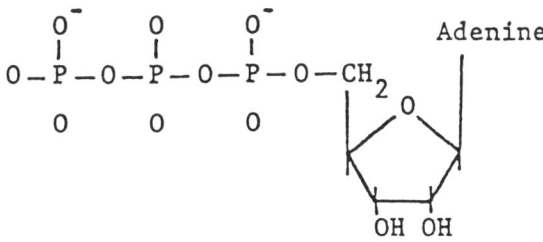

a 19. The accompanying diagram shows a molecule of
   a. ATP
   b. ADP
   c. NADPH
   d. NADH
   e. none of the above

c 20. If you identified a desert plant as belonging to the family Crassulaceae, it would very likely
   a. have thin leaves that were dropped when water was scarce
   b. exhibit C_3 photosynthesis
   c. open its stomates at night and close them during the day
   d. photosynthesize very efficiently and grow rapidly
   e. exhibit a low rate of photorespiration at all times

b 21. Which of the following is not a function or result (direct or indirect) of photosynthesis?
   a. carbon available in glucose to make other organic molecules
   b. the synthesis of ATP in the leaves to use as an energy source in the roots
   c. the conversion of light energy to chemical energy
   d. the presence of O_2 in the atmosphere
   e. the presence of O_3 in the atmosphere

d 22. In the reaction   Dye + AH_2 --------> Dye-H_2 + A
   a. only oxidation of the dye molecule occurs
   b. only reduction of the AH_2 occurs
   c. both oxidation of the dye molecule and reduction of AH_2 occurs
   d. both oxidation of AH_2 and reduction of the dye molecule occurs
   e. neither compound is oxidized or reduced

d 23. If light strikes photosystem I in an intact chloroplast, it is
   a. more likely to be absorbed by a molecule of chlorophyll b than chlorophyll a
   b. absorbed by a molecule of chlorophyll a, which passes an electron to an electron acceptor
   c. absorbed by a molecule of chlorophyll b, which passes an electron to an electron acceptor
   d. absorbed by chlorophyll a or an accessory pigment and the energy transferred to P700, which passes an electron to an electron acceptor
   e. absorbed by a molecule of chlorophyll a and the energy is then released as visible light in a process called fluorescence

b 24. The energy to synthesize ATP during photophosphorylation comes directly from
   a. ATP synthetase
   b. the flow of protons from thylakoid lumen to the stroma through ATP synthetase channels
   c. the flow of electrons from the stroma to thylakoid lumen through ATP synthetase channels
   d. the flow of electrons from phaeophytin to cytochrome b_6/f
   e. electron transport from P680 to P700

d 25. In many parts of the Amazonian rainforest, trees are being burned to open the area for agriculture. This contributes to the greenhouse effect by
   a. adding CO_2 to the atmosphere
   b. destroying trees that remove CO_2 from the atmosphere
   c. adding molecular nitrogen to the atmosphere
   d. a and b
   e. a, b and c

TRUE-FALSE QUESTIONS

T _____ 1. The energy to produce ATP in cells is provided by reactions of photosynthesis or respiration.

F _____ 2. Your eyes can detect all wavelengths of the electromagnetic radiation spectrum.

T _____ 3. Part of a plant is photoautotrophic and part is heterotrophic.

F _____ 4. Photosynthesis is an exergonic process.

T _____ 5. The process of photophosphorylation converts light energy to chemical energy.

T _____ 6. Heterotrophs could live completely in the dark.

F _____ 7. CAM plants can outcompete C_3 and C_4 plants under most conditions.

T _____ 8. RuBP carboxylase has a much lower affinity for CO_2 than PEP carboxylase.

T _____ 9. If a compound gains an electron, it has been reduced.

F _____ 10. The carbon in carbon dioxide is more reduced than the carbon in a carbohydrate.

F _____ 11. Once a molecule of NADPH is oxidized to $NADP^+$, it is broken down by a plant cell.

MATCHING

Match each term or process in List A with a process from List B. Terms in List B may be used more than once.

List A

a _____ 1. Chlorophyll b
d _____ 2. P700
a _____ 3. NADPH synthesis
d _____ 4. ATP synthetase
c _____ 5. RuBP carboxylase
a _____ 6. Splitting of H_2O
c _____ 7. 3-phosphoglyceraldehyde synthesis
b _____ 8. reduction of plastoquinone by ferredoxin
c _____ 9. Use of ATP for energy
d _____ 10. Chlorophyll a

List B

a. noncyclic electron transport
b. cyclic electron transport
c. Calvin/Benson cycle
d. noncyclic and cyclic electron transport

c _____ 9. Use of ATP for energy
d _____ 10. Chlorophyll a

Match each term or process in List A with the correct plant(s) in List B. Terms in List B may be used more than once.

	List A	List B
c	1. Stomates open at night	a. C_3 plants only
g	2. Contain RuBP carboxylase	b. C_4 plants only
f	3. Contain PEP carboxylase	c. CAM plants only
b	4. Kranz anatomy	d. C_3 and C_4 plants
g	5. Calvin/Benson cycle	e. C_3 and CAM plants
e	6. High rate of photorespiration	f. C_4 and CAM plants
f	7. Malate production	g. C_3, C_4, and CAM plants
g	8. Contain P680	

ESSAY QUESTIONS

1. Discuss the function of accessory pigments in forest understory plants.

2. Relate the Second Law of Thermodynamics to living organisms.

3. What makes the Calvin/Benson cycle a cycle?

4. What evidence supports the hypothesis that chlorophyll is the major photosynthetic pigment in plants?

5. What is the advantage to a cell of recycling molecules of ATP, NADPH, and NADH?

6. The Klingons (remember them?) have developed a ray that selectively destroys organisms containing chlorophyll. What would happen on Earth if their rays destroyed all green plants?

7. You were fortunate enough to have been accepted into NASA's space program as an astronaut and are now one of the first people to set foot on the planet Saffron, which has an atmosphere like Earth's. All of the plant life on Saffron is bright yellow due to a single pigment, saffronium.
   a. Why would it be important for some organisms on Saffron to function as plants?
   b. On a graph, plot % absorption of light by saffronium vs. wavelength; on the same graph, plot relative rate of CO_2 reduction vs. wavelength.

8. What is the greenhouse effect? What causes the greenhouse effect? How have human activities contributed to the greenhouse effect? What are the effects of global warming? What could you do to reduce global warming?

CHAPTER 11: ENERGY METABOLISM: RESPIRATION

MULTIPLE CHOICE QUESTIONS

d 1. The net gain to a cell from glycolysis is
   a. 4 ATP, 2 NADPH, plus 2 pyruvate
   b. 2 ATP, 4 NADPH, plus 1 pyruvate
   c. 6 ATP, no NADPH, and 2 pyruvate
   d. 2 ATP, 2 NADPH, and 2 pyruvate
   e. 2 ATP, 2 $FADH_2$, and 2 pyruvate

b 2. If the mitochondria were removed from a cell, which of the following would immediately stop?
   a. glycolysis
   b. the citric acid cycle
   c. the pentose phosphate pathway
   d. a and b
   e. a, b and c

b 3. The final electron acceptor in mitochondrial respiration is
   a. CO_2
   b. O_2
   c. H_2O
   d. pyruvate
   e. acetaldehyde

e 4. The synthesis of proteins from amino acids is a very endergonic process. In most organisms, the energy to drive this synthesis comes from
   a. glycolysis
   b. the citric acid cycle
   c. the electron transport chain + oxidative phosphorylation
   d. a and c
   e. a, b and c

c 5. Which of the following processes provides the most ATP used by a plant to grow?
   a. the citric acid cycle
   b. pyruvate ---> acetyl CoA
   c. the oxidation of NADH in the electron transport chain plus the accompanying chemiosmotic synthesis of ATP
   d. the oxidation of $FADH_2$ in the electron transport chain plus the accompanying chemiosmotic synthesis of ATP
   e. glycolysis

b 6. If the respiratory quotient measured for a particular plant is less than one, that plant is getting the energy it needs to grow from
   a. fermentation
   b. the aerobic respiration of lipids
   c. the aerobic respiration of carbohydrates
   d. the aerobic respiration of citric acid
   e. there is not enough information to tell

a   7. Which of the following is the most energy efficient?
     a. aerobic respiration of carbohydrates
     b. photorespiration
     c. anaerobic respiration
     d. thermogenic respiration
     e. the pentose phosphate pathway

c   8. If you grow yeast cells in a sealed container that contains a sugar solution, what begins to appear in solution is ethanol and the cells begin to die. This happens because
     a. there is not enough light for the cells to grow
     b. there is too much oxygen available
     c. there is a lack of oxygen
     d. the solution contained too much sugar initially
     e. there is too much lactate in the cells

e   9. The pathway an electron follows in the electron transport chain of mitochondria is
     a. NADH ---> FMN ---> cytochrome oxidase ---> ubiquinone ---> cytochromes ---> H_2O
     b. NADH ---> ubiquinone ---> cytochromes ---> FMN ---> cytochrome oxidase ---> O_2
     c. NADH ---> cytochromes ---> cytochrome oxidase ---> FMN ---> ubiquinone ---> O_2
     d. NADH ---> FMN ---> ubiquinone ---> cytochrome oxidase ---> cytochromes ---> H_2O
     e. NADH ---> FMN ---> ubiquinone ---> cytochromes ---> cytochrome oxidase ---> O_2

d 10. The pentose phosphate pathway is particularly common in the hyaloplasm of
     a. developing fibers because it produces the starting material for lignin
     b. meristematic cells because it produces ribose used to synthesize nucleic acids
     c. developing collenchyma cells because it produces the starting material for hemicelluloses
     d. a and b
     e. a, b and c

c 11. In chemiosmotic phosphorylation, a proton gradient develops between
     a. the mitochondrion and the hyaloplasm
     b. the cristae lumen and the hyaloplasm
     c. the cristae lumen and the mitochondrial matrix
     d. membranes of adjacent cristae
     e. the cristae membranes and the mitochondrial matrix

c 12. If you water a plant in a pot without a hole in the bottom twice a week, the plant will
     a. thrive because you are taking good care of it
     b. thrive because it has an adequate supply of water for photosynthesis
     c. die because the roots lack oxygen for respiration
     d. die because the roots have an inadequate supply of carbon dioxide
     e. die because essential minerals in the soil are washed away

a 13. Plants as a whole are
     a. obligate aerobes
     b. obligate anaerobes
     c. facultative aerobes
     d. facultative anaerobes
     e. it depends on the plant

b 14. Which of the following is not a difference between aerobic respiration and photosynthesis?
   a. site of the process
   b. function of the electron transport chain
   c. energy source for phosphorylation
   d. general type of redox reaction
   e. role of O_2

d 15. The metabolism of brown fat tissue is used to bring bats out of hibernation by increasing their body temperature. The major biochemical process involved is
   a. aerobic respiration
   b. fermentation
   c. the pentose phosphate pathway
   d. thermogenic respiration
   e. photorespiration

a 16. What is the relationship between glucose, ATP, NADH, and $FADH_2$ with respect to their energy content?
   a. glucose > NADH > $FADH_2$ > ATP
   b. ATP > glucose > $FADH_2$ > NADH
   c. NADH > $FADH_2$ > glucose > ATP
   d. $FADH_2$ > ATP > NADH > glucose
   e. glucose > $FADH_2$ > NADH > ATP

c 17. The process of photorespiration can salvage some of the carbon in glycolate, converting it to a useful form. The organelles involved are
   a. chloroplasts, endoplasmic reticululm, and dictyosomes
   b. mitochondria and peroxisomes
   c. chloroplasts, mitochondria, and peroxisomes
   d. chloroplasts and peroxisomes
   e. chloroplasts, mitochondria, and glyoxysomes

e 18. Many plants, such as cucumber, store food in their seeds in the form of fats or oils. The metabolism of the fatty acids in those molecules may
   a. be used for the synthesis of carbohydrates
   b. generate $FADH_2$ used in the production of ATP
   c. generate NADH used the production of ATP
   d. produce acetyl-CoA that enters the citric acid cycle
   e. all of the above

d 19. ATP synthetase is located in the
   a. plasma membrane
   b. hylaoplasm
   c. outer mitochondrial membrane
   d. cristae membrane
   e. mitochondrial matrix

c 20. If the amount of energy in glucose is 686 kcal/mole and in ATP is 7.3 kcal/mole, the process of aerobic respiration is how efficient?
   a. 9-11%
   b. 18-19%
   c. 38-40%
   d. 45-47%
   e. 51-53%

a 21. The major product of cellular respiration that is useful to plant cells is
   a. ATP
   b. NADH
   c. FADH$_2$
   d. NADPH
   e. O$_2$

b 22. The number of ATP molecules produced per glucose molecule in aerobic respiration is variable, either 36 or 38, because
   a. the NADH generated in the citric acid cycle yields either 2 or 3 ATP molecules in the electron transport chain
   b. of different NADH shuttle mechanisms that eventually lead to the net production of either 2 or 3 ATP molecules per NADH
   c. the number of NADH and FADH$_2$ molecules produced in the citric acid cycle varies
   d. sometimes the conversion of pyruvate to acetyl-CoA produces 1 NADH, sometimes it produces 2 NADH
   e. the amount of energy in glucose molecules varies

c 23. As electrons are passed from one compound to another in the electron transport chain
   a. electrons are pumped from the matrix out of the mitochondrion
   b. NADH is pumped from the cytosol into the matrix
   c. protons are pumped from the matrix into the cristae lumen
   d. a and b
   e. a, b and c

a 24. If plant roots stand in water until all of the dissolved oxygen has been used, the cells switch to anaerobic respiration. The main reason that the pyruvate produced in glycolysis is further metabolized is to
   a. regenerate NAD necessary for glycolysis to occur so ATP can be synthesized
   b. produce ATP using the energy in pyruvate
   c. produce lactic acid, which can be further metabolized by the root cells
   d. produce ethanol and CO$_2$, which are used in photosynthesis
   e. none of the above; pyruvate is a waste product that is secreted from the cells

e 25. Which of the following does not occur in thermogenic respiration?
   a. an electron transport chain
   b. glycolysis
   c. the citric acid cycle
   d. NADH oxidation
   e. chemiosmotic synthesis of ATP

TRUE-FALSE QUESTIONS

F _____ 1. Respiration is the exact opposite of photosynthesis.

T _____ 2. The final electron acceptor in respiration is oxygen.

T _____ 3. The net gain to a cell from glycolysis is two molecules of ATP.

T _____ 4. The rate of aerobic respiration is often not the same in all plant parts, even under identical environmental conditions.

F _____ 5. The only source of energy to synthesize ATP by aerobic respiration is carbohydrates.

F _____ 6. The biochemical pathway of respiration is isolated and not connected to other biochemical pathways in plant cells.

T _____ 7. In general, a plant can live longer without oxygen than an animal can.

F _____ 8. All of the ATP produced by cellular respiration is produced by chemiosmotic phosphorylation.

T _____ 9. Depending on the biochemical pathway, the carbon in fatty acids from a fat or phospholipid may end up in carbohydrates or carbon dioxide.

T _____ 10. ATP must be used to start the process of glycolysis.

T _____ 11. The breakdown of glucose produces organic compounds that are used to synthesize amino acids, fats and nucleic acids.

T _____ 12. The structure of a mitochondrion includes two membranes, an outer boundary membrane and an inner membrane folded into cristae. A mitochondrion must have this structure to function correctly.

MATCHING

Match each statement in List A with a biochemical pathway in List B. Answers in List B may be used more than once.

List A

e _____ 1. Occurs in mitochondria
f _____ 2. Generates ATP
a _____ 3. Occurs anaerobically
c _____ 4. Cytochromes involved
b _____ 5. Tricarboxylic acids involved
d _____ 6. Generates NADH
b _____ 7. Generates FADH$_2$
a _____ 8. Produces pyruvate

List B

a. glycolysis
b. the citric acid cycle
c. oxidative phosphorylation + electron transport chain
d. a and b
e. b and c
f. a, b and c

ESSAY QUESTIONS

1. Why do plants need the process of respiration in addition to photosynthesis?

2. Why can you kill plants by overwatering them?

3. Why do most wines contain no more than 12-14% ethanol?

4. Why do many aquatic plants have very large air channels that are continuous from leaves to stems to roots?

5 Yeast, a fungus, is a facultative aerobe. Under what conditions would yeast cell growth be optimal? Why?

6. Why is it important for mitochondria to have two membranes around the matrix?

7. Your internal organs are maintained at a temperature of approximately 37° C. What is the advantage of this?

8. Compare pyruvate, acetaldehyde, and oxygen as terminal e⁻ acceptors. What are the advantages and disadvantages of each?

9. What makes the citric acid cycle a cycle?

10. The energy released in the aerobic oxidation of glucose is 686 kcal/mole; the hydrolysis of ATP yields 7.3 kcal/mole. If 36 molecules of ATP are produced per glucose, what is the efficiency of this process? If your answer is less than 100%, what happened to the energy that does not end up in ATP?

11. Many people like to eat apple cores, seeds and all. In moderation there is no danger from this. But if consumed in **great** quantity over a short period of time, this could be dangerous because apple seeds contain cyanide. Why is it dangerous? Be as specific as possible.

CHAPTER 12: TRANSPORT PROCESSES

MULTIPLE CHOICE QUESTIONS

a 1. The source of energy driving the transport of water through the xylem is
   a. the sun
   b. ATP generated by respiration
   c. ATP produced by photosynthesis
   d. NADPH produced by photosynthesis
   e. transpiration

b 2. A basic difference between most algae and an oak tree is that
   a. the oak tree is photosynthetic, the algae are not
   b. the oak tree is capable of translocation, the algae are not
   c. the algae are capable of translocation, the oak tree is not
   d. the algae use active transport, the oak tree does not
   e. the oak tree uses active transport, the algae do not

a 3. The loss of water through stomates is called
   a. transpiration
   b. translocation
   c. transport
   d. pressure flow
   e. cohesion

b 4. Which of the following conditions promotes xylem transport of water?
   a. cool nights
   b. dry, warm days
   c. dry, warm nights
   d. humid, hot days
   e. environmental conditions do not affect xylem transport

d 5. Under experimental conditions, how could you cause stomatal openings in a plant?
   a. expose the plant to blue light
   b. lower the amount of CO_2 around the plant
   c. spray the plant with abscisic acid
   d. a and b
   e. a, b and c

c 6. Carbon dioxide molecules diffuse from the atmosphere into plant cells. This movement is due to
   a. a molecule pump powered by ATP produced in respiration
   b. a molecule pump powered by ATP produced in photosynthesis
   c. the random movement of CO_2 molecules
   d. CO_2 molecules being carried along with a stream of water
   e. wind currents

d 7. Requirement(s) for phloem transport of sugars include
   a. living sieve tube members
   b. the active transport of sugars into sieve tube members
   c. sieve tube members with no end wall
   d. a and b
   e. a, b, and c

a 8. Which of the following is true?
   a. the contents of the phloem are under pressure and the contents of the xylem are under tension
   b. sieve tube members are empty, dead cells
   c. vessel element protoplasm is unique because the vacuolar membrane disintegrates, allowing vacuolar water to mix with hyaloplasm
   d. a and c
   e. a, b and c

e 9. Companion cells regularly communicate with their associated sieve tube members. The mechanism(s) involved may be
   a. osmosis
   b. active transport
   c. by plasmodesmata
   d. b and c
   e. a, b and c

c 10. Just after the sun rises, the initial event that leads to the opening of stomatal pores is
   a. the diffusion of water into guard cells from adjacent cells
   b. the diffusion of water out of guard cells into adjacent cells
   c. the active transport of K^+ into guard cells from adjacent cells, lowering the water potential
   d. the diffusion of K^+ into guard cells from adjacent cells, lowering the water potential
   e. the synthesis of glucose, lowering the water potential in the guard cells

e 11. If an animal brushes by a plant and tears off a leaf, what prevents the plant from losing sugar from the open wound on the stem?
   a. the phloem cells die after secreting a covering of cork
   b. the phloem cells secrete a covering of suberin
   c. P-protein surges toward the sieve plates forming a P-protein plug
   d. callose plugs the sieve plates
   e. c and d

d 12. If a plant cell whose water potential is -1.0 MPa is placed in a salt solution
   a. water diffuses into the cell
   b. water diffuses out of the cell
   c. the system is in equilibrium
   d. we do not have enough information to predict what will happen

a 13. If a plant cell is placed in deionized water the water potential of that cell becomes
   a. more positive because the pressure potential becomes more positive
   b. more positive because the osmotic potential becomes more negative
   c. more negative because the pressure potential becomes more negative
   d. more negative because the osmotic potential becomes more positive
   e. more negative because the matric potential becomes more negative

c 14. Two cells are adjacent to one another. If Cell A has a water potential = -0.5 MPa and Cell B has a water potential = -1.0 MPa, then water molecules will
   a. move from A to B only
   b. move from B to A only
   c. move from A to B at a greater rate than from B to A
   d. move from B to A at a greater rate than from A to B
   e. move at equal rates in both directions

b 15. The lipid component of a lipid/protein membrane is important to
   a. prevent hydrophobic ions and molecules from leaking out of a cell
   b. prevent hydrophilic ions and molecules from leaking out of a cell
   c. pump hydrophilic ions and molecules into a cell
   d. pump hydrophobic ions and molecules into a cell
   e. b and d

c 16. Dinitrophenol (DNP) prevents oxidative phosphorylation in cells. If roots are exposed to DNP
   a. water cannot be taken into the roots
   b. water translocation stops
   c. minerals cannot be taken into the roots
   d. oxygen molecules cannot get into the roots
   e. DNP has no effect on roots

c 17. A plant cell and an animal cell, both with a water potential of -1.5 MPa, are placed in a beaker of deionized water. What happens in this system?
   a. water diffuses into both cells until equilibrium
   b. water diffuses into both cells; the plant cell bursts and the animal cell does not
   c. water diffuses into both cells; the animal cells bursts and the plant cell does not
   d. water diffuses out of both cells, shrinking them
   e. water molecules move in equal amounts in and out of both cells

b 18. Water moves from the soil into the xylem of a root via
   a. diffusion
   b. diffusion and osmosis
   c. active transport
   d. membrane vesicles
   e. translocation

d 19. Water must reach leaves of the tallest trees, such as eucalyptus. Water moves upward in one of those trees due to
   a. root pressure that builds up as water enters a root, pushing the water up
   b. pumps in the xylem cells that move the water from cell to cell
   c. capillary action in the xylem vessels
   d. the pulling of water columns up from the top
   e. we don't know how water moves through the xylem

c 20. What prevents roots from losing absorbed phosphate back into the soil?
   a. the root epidermis cuticle
   b. the Casperian strip of the pericycle cells
   c. the Casperian strip of the endodermal cells
   d. suberin in root cortical cell walls
   e. the secondary wall of phloem cells

a 21. Transport that requires energy does not occur
   a. between electrons in an atom
   b. between molecules
   c. across membranes
   d. from one plant organ to another
   e. between cells

d 22. Long distance transport systems are adaptive in land plants because
   a. most mineral nutrients are found in the soil, where light is absent
   b. they compete for sunlight, leading to taller plants
   c. air is not a very supporting medium and transport systems give a plant its only support
   d. a and b
   e. a, b and c

c 23. To move sucrose molecules from leaves to roots, the sucrose molecules must enter sieve elements. This involves
   a. moving the molecules through plasmodesmata
   b. the use of ATP to pump sucrose molecules with their concentration gradient across the plasma membrane
   c. the use of ATP to pump sucrose molecules against their concentration gradient across the plasma membrane
   d. the pumping of sucrose molecules across the plasma membrane with their concentration gradient without the use of ATP
   e. the pumping of sucrose molecules across the plasma membrane against their concentration gradient without the use of ATP

e 24. If the water potential of mature mesophyll leaf cells is -0.1 MPa, then water will diffuse
   a. to root cells
   b. to developing fruit cells
   c. to expanding leaves at the stem tip
   d. into cells of the phloem
   e. it is impossible to predict with this information

e 25. In response to the presence of an insect, the leaf of a Venus' flytrap will close, trapping the insect. The closure is due to
   a. rapid growth of leaf motor cells, bending the leaf
   b. the activation of a potassium pump, which pumps potassium out of motor cells, resulting in water diffusion out of the cells
   c. the activation of a potassium pump, which pumps potassium into motor cells, resulting in water diffusion out of the cells
   d. a rapid change in permeability of motor cell plasma membranes to potassium, causing potassium to diffuse out of and water to diffuse into the cells
   e. a rapid change in permeability of motor cell plasma membranes to potassium, causing potassium and then water to diffuse out of the cells

TRUE-FALSE QUESTIONS

T _____ 1. In a temperate area, such as Minnesota, in the winter plants are living in a physiological drought.

T _____ 2. A sugar solution is pushed through the phloem, but water is pulled through the xylem.

T _____ 3. The stage of development of many plant parts determines whether they are sources or sinks.

F _____ 4. Water molecules can move through the xylem because water molecules are not cohesive.

F _____ 5. Membranes are effective barriers against the diffusion of water and can prevent water from entering or leaving a cell.

T _____ 6. Although water normally diffuses into a root, it is possible, under the right conditions, for water to diffuse out of a root.

F _____ 7. Sources of sugars which are transported in the phloem are always leaves.

T _____ 8. Common sinks that receive transported sugars are roots, storage organs, developing fruits, meristems, growing flowers, and seeds.

T _____ 9. If a cell has an osmotic potential of -20 bars, a matric potential of -1 bar, and a water potential of -18 bars, the pressure potential is +3 bars.

F _____ 10. If the soil has a water potential of -5 bars and root cells have a water potential of -10 bars, water will diffuse into the roots.

T _____ 11. If a cell is placed in deionized water, its water potential at equilibrium will be zero.

F _____ 12. If the water potentials of two adjacent regions are equal, water molecules stop moving between the two areas.

F _____ 13. The protoplasm of adjacent cells is isolated from each other.

T _____ 14. A mineral ion, once it enters a root cell by crossing the plasma membrane, could move to the top of a plant without crossing another plasma membrane.

T _____ 15. If membrane fusion is prevented, cell plate formation will not take place.

MATCHING

Match each event in List A with an effect in List B. Answers in List B may be used more than once.

List A

e _____ 1. Water diffuses into a cell
a _____ 2. Water evaporates from a solution
b _____ 3. A cell synthesizes sugar
a _____ 4. A cell makes proteins from amino acids
d _____ 5. Translocation through the xylem begins in the morning
a _____ 6. Minerals are added to the xylem
f _____ 7. Guard cells close
g _____ 8. Guard cells open
g _____ 9. Motor cells as a Venus flytrap leaf reopens.
f _____ 10. A cell plasmolyzes

List B

a. osmotic potential becomes more positive
b. osmotic potential becomes more negative
c. pressure potential becomes more positive
d. pressure potential becomes more negative
e. a and c
f. a and d
g. b and c
h. b and d

ESSAY QUESTIONS

1. Why do most land plants have xylem and phloem?

2. Why should flower stems be cut under water, if possible?

3. What limits the height of trees? Why?

4. Discuss the relationship between environment and xylem structure.

5. Is transpiration always detrimental to a plant? Why or why not?

6. Most plants must have an adequate supply of sixteen essential elements to survive. Why is potassium one of those elements?

7. Why do people put carrot sticks in water to store them?

8. Explain the reason fertilizer solutions used to water plants must not be too strong.

9. Knowing what you do about phloem transport, how could you collect uncontaminated phloem sap for chemical analysis?

10. Why does a cell store carbohydrate as starch rather than glucose?

11. Why do many freshwater protozoa have contractile vacuoles to pump out water, while freshwater algae do not?

12. What can you say about the water potentials of two adjacent cells when they are in equilibrium? At equilibrium, is there any net movement of water from one cell to the other?

13. Explain how the leaf of a Venus' flytrap closes when stimulated by an insect and how it reopens.

14. At least one variety of tomato has been developed that can be watered with seawater. What kind of adaptation has been bred into this variety? What might be the long-term problems associated with growing tomatoes of this variety?

CHAPTER 13: SOILS AND MINERAL NUTRITION

MULTIPLE CHOICE QUESTIONS

b 1. If a plant is to grow and reproduce, it must be supplied with a large quantity of
   a. zinc
   b. phosphorus
   c. copper
   d. chlorine
   e. boron

c 2. Which is a true statement regarding element essentiality?
   a. fruits can develop without zinc, therefore it is not essential
   b. rubidium, an element similar to potassium, can substitute for the essential element potassium in a plant's growth and development
   c. the function of boron is unknown, but growth will not occur without it, therefore it is essential
   d. plants usually contain aluminum, therefore that element is essential for plant growth
   e. plants grow much better in a nutrient solution containing EDTA, a compound that keeps iron in solution, therefore EDTA is essential for plant growth

b 3. If you grow plants whose leaves start to turn yellow and become chlorotic, your plants may need to have which element added to the soil?
   a. potassium
   b. magnesium
   c. calcium
   d. sulfur
   e. boron

e 4. Essential elements would not be made available to plants by
   a. alternate freezing and thawing of rock
   b. glaciation
   c. acids released by an organism called a lichen
   d. the release of CO_2 by roots
   e. all of the above make essential elements available

d 5. In a tropical rainforest, where there is abundant vegetation and rainfall, the soil contains
   a. abundant cations due to soil acidity, but few anions due to leaching
   b. abundant cations and anions due to soil acidity
   c. abundant anions and few cations due to soil acidity and leaching
   d. few cations or anions due to soil acidity, leaching, and negatively charged soil particles
   e. high concentrations of hydroxyl ions making the soil alkaline

c 6. Which of the following would be most likely to develop deficiency symptoms?
   a. a plant growing in a tropical rainforest with associated mycorrhizae
   b. cacti growing in the desert
   c. a crop plant growing in a field that has been cultivated for many years
   d. an aquatic plant growing in lake sediments
   e. a plant growing naturally on serpentine soil

a 7. If a plant is deficient in phosphorus, a typical deficiency symptom would be
   a. older leaves that are purple
   b. young leaves that are purple
   c. older leaves that are chlorotic
   d. young leaves that are necrotic
   e. dead apical meristems

b 8. Which of the following does not perform a function of soil when plants are grown hydroponically?
   a. a hose bubbling air into the nutrient solution
   b. a greenhouse that stabilizes temperature
   c. the container holding the nutrient solution
   d. plant supports
   e. the minerals in the nutrient solution

d 9. If cubic meters of different soils are compared, in which is the total surface area of all particles greatest?
   a. coarse sand
   b. fine sand
   c. silt
   d. clay
   e. all would have the same total surface area

d 10. Organisms capable of nitrogen fixation include
   a. bacteria
   b. cyanobacteria
   c. legumes
   d. a and b
   e. a, b and c

a 11. If plants are grown in a soil deficient in molybdenum, nitrogen must be supplied in what form for the plants to survive?
   a. NH_4^+
   b. NO_3^-
   c. NO_2^-
   d. N_2
   e. any of the above

e 12. Which of the following is true concerning nitrogen fixation?
   a. organisms use little energy to fix nitrogen
   b. free-living microorganisms usually fix nitrogen at a faster rate than symbiotic nitrogen-fixers
   c. the product of nitrogen fixation is always ammonia
   d. plants, such as legumes, are capable of fixing nitrogen to satisfy their nitrogen requirement
   e. the enzyme that catalyzes nitrogen fixation is nitrogenase

c 13. From an energy-conservation standpoint, the most favorable form of nitrogen for a plant is
   a. NO_3^-
   b. NO_2^-
   c. NH_4^+
   d. N_2
   e. all are equally favorable

d 14. If seedlings of forest trees are grown in nutrient solution then transplanted to fertile prairie soils, the trees grow very poorly. This is most probably due to
   a. a lack of nitrogen in the soil
   b. a lack of phosphorus in the soil
   c. an over abundance of phosphorus in the soil
   d. a lack of the proper mycorrhizal fungi in the soil
   e. the wrong climatic conditions for the trees to grow

d 15. Acid precipitation could affect plants adversely by
   a. making cations less available because they are leached out of the soil
   b. solubilizing aluminum to toxic levels in the soil
   c. precipitating iron compounds so that iron is unavailable
   d. a and b
   e. a, b and c

c 16. The cells which give a root control over what ends up in the transpiration stream include
   a. epidermis and cortex
   b. epidermis, cortex and pericycle
   c. epidermis, cortex and endodermis
   d. epidermis, cortex, endodermis, and pericycle
   e. endodermis only

b 17. Which of the following is an incorrect match of essential element and function?
   a. carbon - structural component of organic molecules
   b. manganese - structural component of chlorophyll
   c. calcium - component of the middle lamella
   d. zinc - enzyme activator
   e. potassium - involved in stomatal opening and closing

b 18. Rock weathering does not provide which essential element to plants?
   a. phosphorus
   b. nitrogen
   c. calcium
   d. iron
   e. zinc

e 19. Successive harvests of crops from a field are likely to deplete the soil of
   a. nitrogen
   b. phosphorus
   c. potassium
   d. a and b
   e. a, b and c

c 20. Many farmers apply nitrogen in fertilizer in the form of ammonia. The form of nitrogen then taken up from the soil most often by plants is
   a. ammonia
   b. nitrite
   c. nitrate
   d. molecular nitrogen
   e. amino acids

d 21. You are growing a houseplant that begins to develop leaves that have green veins but dead tissue between the veins. This plant is most likely deficient in
   a. nitrogen
   b. calcium
   c. phosphorus
   d. manganese
   e. iron

e 22. Julius von Sachs established a nutrient solution containing calcium nitrate, potassium nitrate, potassium phosphate, and magnesium sulfate that would support plant growth. Because plants can grow in this solution, it can be concluded that
   a. calcium, nitrogen, potassium, phosphorus, magnesium, and sulfur are all essential elements
   b. calcium, nitrogen, potassium, phosphorus, magnesium, and sulfur are the only elements required by a plant
   c. phosphorus is required by plants but sulfur is not
   d. other elements are needed by plants in trace amounts
   e. further experimentation is necessary to prove or disprove plant nutrient requirements

d 23. To complete its life cycle, a grass plant requires 1 gram (gm) phosphorus, 3 gm calcium, 7 gm nitrogen, 0.25 gm iron, and 0.01 gm zinc. If the soil in a field has available 30 gm phosphorus, 600 gm calcium, 200 gm nitrogen, 5 gm iron and 1 gm zinc, then the supply of which element would most likely limit the growth of the grass population?
   a. phosphorus
   b. calcium
   c. nitrogen
   d. iron
   e. zinc

b 24. Which of the following would be the worst form of nitrogen to apply in a fertilzer if the area has acidic soil and a relatively high amount of rainfall?
   a. N_2
   b. NH_4^+
   c. NO_2^-
   d. NO_3^-
   e. all are equally good sources of nitrogen; the chemical form does not matter

a 25. Crop plants are more likely to develop mineral deficiency symptoms than native plants because
   a. of their rapid rate of growth
   b. they have lower rates of mineral transport into the root
   c. they do not transport minerals to the shoot as efficiently
   d. a and b
   e. a, b and c

TRUE-FALSE QUESTIONS

T _____ 1. The only difference between major and minor essential elements is the quantity required by a plant.

F _____ 2. The smaller the quantity of an element required, the easier it is to demonstrate its essentiality.

T _____ 3. If a plant grows normally, but does not reproduce unless supplied with a particular element, that element is considered an essential element.

T _____ 4. Once an element is found to be essential for a plant, it is assumed to be essential for most other plants.

F _____ 5. Deficiency symptoms of mobile elements show up first in young parts of a plant.

T _____ 6. If plants are grown in a nutrient solution deficient in one essential element and the stem apical meristems die, that essential element is an immobile element.

T _____ 7. A plant can use glutamate as the starting point to synthesize any required amino acid.

F _____ 8. Scientific ideas are static or unchanging.

F _____ 9. The majority of elements essential for animal growth are obtained directly from the soil.

T _____ 10. Plants contain the metabolic pathways to make every type of organic compound needed starting with carbon dioxide, water, and minerals.

T _____ 11. Nitrogen can be recycled within a plant.

F _____ 12. The essential elements for plant growth, as established at present, are required by all plants.

F _____ 13. Strontium, which is chemically similar to calcium, can substitute for that element in most plants.

F _____ 14. If an element is found in a plant, it must be essential for growth or it would be excluded from the plant.

T _____ 15. The process of respiration in root cells can make cations in the soil available for absorption into a root.

F _____ 16. The list of essential elements for plants is unlikely to change in the future.

T _____ 17. It is possible for a soil to contain concentrations of essential elements that are too high for plants to grow.

T _____ 18. Almost everyone in the city of Philadelphia has one or more azaleas growing in their yard that bloom profusely every spring. These plants do so well in this area because the soil is acidic.

MATCHING

Match each element in List A with a function in List B. Functions in List B may be used more than once.

	List A	List B
c	1. Sulfur	a. component of chlorophyll
g	2. Copper	b. integrity of membranes
d	3. Potassium	c. component of amino acids
a	4. Magnesium	d. stomatal opening and closing
f	5. Manganese	e. nitrogen reduction
b	6. Calcium	f. chlorophyll synthesis
e	7. Molybdenum	g. component of plastocyanin
f	8. Iron	

ESSAY QUESTIONS

1. What would be the major problem encountered with the genetic engineering of genes for nitrogen fixation into plants that photosynthesize?

2. Why do some farmers often rotate crops, such as corn or wheat, with legumes, such as alfalfa or clover?

3. Why might it be less expensive to raise a crop of legumes, such as soybeans, than a crop such as corn?

4. Imagine that an expedition has returned from Mars and it has a plant-like organism. Chemical analysis of the organism reveals that it contains 27 different chemical elements. Describe the experiments you would perform to determine which elements are essential for the organism. What three things would you have to prove for each element?

5. Why has it been so difficult to determine that elements such as molybdenum and chlorine are essential for plants?

6. Why do you suppose that temperate deciduous forest wildflowers bloom early in the spring rather than in the summer?

7. What problems does a plant have if it lives in an acidic soil? an alkaline soil?

8. Although acid rain is usually considered detrimental, it can also be beneficial to a plant. Why?

CHAPTER 14: DEVELOPMENT AND MORPHOGENESIS

MULTIPLE CHOICE QUESTIONS

d 1. Many plants flower in response to a change in day length. Plants with this adaptation probably have evolved in all of the following areas except
   a. Antarctica
   b. the United States
   c. Australia
   d. equatorial Africa
   e. northern Canada

b 2. The most common long-term response of a plant part to environmental stimuli is
   a. rapid movement
   b. growth
   c. hormone production
   d. turgor changes
   e. no response

c 3. The structure in mosses that produces an egg releases a chemical as it matures that attracts the flagellated sperm produced in a different structure. This is an example of
   a. positive phototropism
   b. negative phototaxis
   c. positive chemotaxis
   d. negative thigmotropism
   e. positive gravitropism

d 4. If a bulb is planted, its orientation does not matter because the
   a. root grows downward due to positive gravitropism
   b. shoot grows upward due to negative gravitropism
   c. shoot grows upward due to positive phototropism
   d. a and b
   e. a, b and c

b 5. The first plant hormone discovered was
   a. gibberellin
   b. auxin
   c. abscisic acid
   d. ethylene
   e. cytokinin

a 6. If you know in advance that a field will be temporarily flooded, an application of what hormone might help the plants survive?
   a. abscisic acid
   b. ethylene
   c. auxin
   d. cytokinin
   e. gibberellin

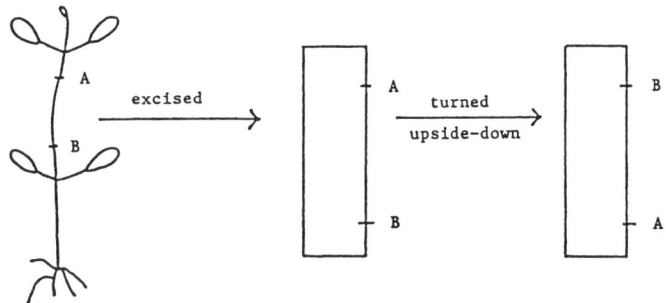

c 7. If the rate of auxin transport is measured in a stem segment that has been turned upside-down as in the accompanying diagram, it is found that the movement is
   a. from B to A and occurs at the normal rate
   b. from B to A and occurs faster than the normal rate
   c. from A to B and occurs at the normal rate
   d. from A to B and occurs slower than the normal rate
   e. stopped

d 8. If you want to produce a bushier shrub with more branches, you should
   a. water the plant with a gibberellin solution
   b. spray the plant with cytokinin
   c. spray the plant with auxin
   d. pinch off the stem tips
   e. pinch off the root tips

d 9. The only plant hormone that is a gas is
   a. cytokinin
   b. gibberellin
   c. auxin
   d. ethylene
   e. abscisic acid

b 10. The apical meristem of an actively growing woody stem is removed. If a mixture of paste and gibberellin is applied to the stem tip, cells produced by the vascular cambium will differentiate into
   a. xylem
   b. phloem
   c. parenchyma
   d. fibers
   e. the cells will stay meristematic

a 11. Many fruits are harvested before they are ripe because it is easier to transport them in that stage of development. They must be transported in a partial vacuum mainly to
   a. draw out ethylene that will hasten ripening
   b. decrease the amount of oxygen available
   c. draw out carbon dioxide produced in respiration that might harm the fruit
   d. draw out auxins that cause the fruit to produce ethylene
   e. draw out harmful cytokinins

b 12. At most plant nurseries, you can buy rooting powder that stimulates root development on stem cuttings. The rooting powder most likely contains
   a. fertilizer
   b. synthetic auxins
   c. synthetic cytokinins
   d. synthetic gibberellins
   e. abscisic acid

d 13. If all of the fruits (seeds) on the outside of a strawberry are removed when it is young, it will not develop. If the strawberry is sprayed with a hormone solution, normal development occurs. The hormone most likely to cause a response is
   a. cytokinin
   b. gibberellin
   c. ethylene
   d. auxin
   e. abscisic acid

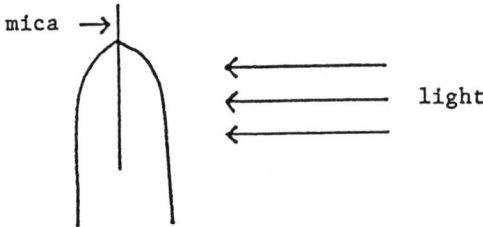

e 14. A piece of mica is inserted in the tip of a grass coleoptile and the seedling is exposed to unilateral light as shown in the accompanying diagram. In response, the coleoptile
   a. bends toward the light because of a redistribution of auxin
   b. bends away from the light because of a redistribution of auxin
   c. bends toward the light because of destruction of auxin on the lighted side
   d. bends toward the light because extra auxin is synthesized on the darker side
   e. does not bend because auxin cannot be redistributed

b 15. If plants are kept in the dark, they have elongated internodes, making them spindly. This effect is due to
   a. an overproduction of auxin
   b. phytochrome in the P_r form
   c. phytochrome in the P_{fr} form
   d. an overproduction of gibberellins
   e. a lower auxin/cytokinin ratio in the stems

d 16. Which of the following is an <u>incorrect</u> match between hormone and function?
   a. cytokinin - cell division
   b. auxin - stimulates stem elongation
   c. gibberellin - releases seeds from dormancy
   d. abscisic acid - releases buds from dormancy
   e. ethylene - latex production

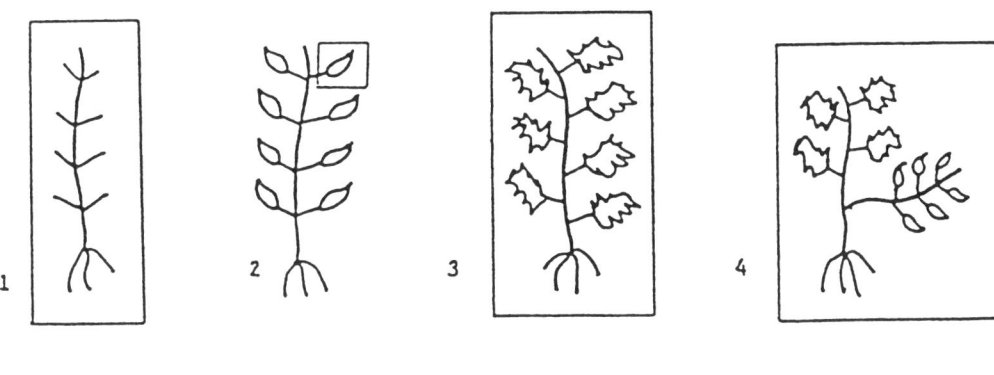

⌒⃝ = short-day plant   ⚘ = long-day plant

The blocked in portion of each plant was exposed to <u>short</u> days. In Plant 4, a branch of a short-day plant was grafted onto a long-day plant.

e 17. Which of the above plants will flower?
   a. 1,2,3, both branches of 4
   b. 2,3, short-day branch of 4
   c. 2, short-day branch of 4
   d. 2 only
   e. 2, both branches of 4

noon   midnight

e 18. As shown in the accompanying diagram, the position of bean plant leaves is different at noon and at midnight. This change in position is probably influenced by changing
   a. light conditions
   b. temperature
   c. relative humidity
   d. water availability
   e. none of the above

a 19. If a plant is placed in a horizontal position the
   a. shoot bends upward due to increased auxin on the lower side
   b. shoot bends upward due to increased auxin on the upper side
   c. root bends downward due to increased auxin on the lower side
   d. root bends downward due to an enhancer on the upper side
   e. a and c

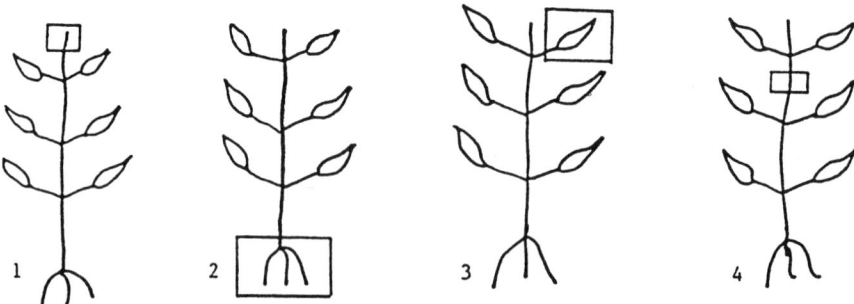

a 20. The plants in the accompanying diagram are biennial plants; the blocked in portion was exposed to cold, freezing temperatures. Which of those plants will be vernalized?
   a. 1
   b. 1 and 2
   c. 1 and 3
   d. 3 and 4
   e. 1, 2, 3 and 4

e 21. In response to external conditions, plants
   a. are more resilient than animals
   b. use less energy to build response mechanisms than animals
   c. have sophisticated response mechanisms for sexual reproduction
   d. a and b
   e. a, b and c

b 22. If a flower opens at night and closes during the day, the response is
   a. phototaxic
   b. photonastic
   c. phototropic
   d. photomorphogenic
   e. due to the presence of pollinators during the night and their absence during the day

c 23. Plant auxins
   a. affect the cells that produce them
   b. act only in moderately high concentrations
   c. bind to receptor molecules to elicit a response
   d. are carried specifically to stem cells, which respond
   e. move fairly rapidly through the phloem

d 24. If the top of a woody stem is removed
   a. axillary buds on the stem will start to grow
   b. activity of the vascular cambium will decrease
   c. cell elongation in young internodes of the stem will increase
   d. a and b
   e. a, b and c

d 25. Maple trees are unlikely to initiate an adaptive response to
   a. air temperature in autumn
   b. the presence of a boulder within its root system
   c. the early stages of a drought
   d. the leaching of phosphate from the soil
   e. day length

TRUE-FALSE QUESTIONS

T _____ 1. In general, a plant's response to external stimuli is much slower than the response of an animal.

F _____ 2. In most cases, hormones are synthesized at their place of effect.

F _____ 3. Only compounds produced naturally by plants can influence plant growth.

T _____ 4. The role of hormones is to enable plant parts to communicate with each other.

F _____ 5. If immature stem cells are exposed to high levels of auxin, they stop enlarging.

F _____ 6. Cells all respond the same way to the presence of a hormone in the environment around them.

T _____ 7. The main response elicited by auxin on a young stem cell is the weakening of its cell wall, enabling the cell to enlarge.

T _____ 8. Many conifers have their pyramidal shape due to apical dominance.

T _____ 9. To respond to a particular hormone, a cell must have the appropriate receptors.

F _____ 10. A stem responds to unilateral light at the site of auxin production.

F _____ 11. A plant grows normally if exposed only to blue light, which can be used in photosynthesis.

T _____ 12. A short-day plant will flower if given 16 hour days followed by 16 hour nights.

T _____ 13. Tomato fruits become redder if left on a windowsill to ripen rather than in a closet.

F _____ 14. The endogenous rhythms in a pine tree change from summer to winter.

T _____ 15. In nature, a circadian rhythm is exactly 24 hours long.

F _____ 16. All animal hormones are either steroids or proteins. The same is true in plants.

F _____ 17. Because plants can neither hear, see nor smell, they cannot respond to external stimuli.

T _____ 18. The different parts of a terrestrial plant live in different microenvironnments.

T _____ 19. Responses to external stimuli are faster in animals than in plants.

MATCHING

Match each process in List A with a hormone in List B. Answers in List B may be used more than once.

	List A	List B
b	1. Fruit abscission	a. auxin
a	2. Apical dominance	b. ethylene
d	3. Stomatal closure	c. gibberellin
a	4. Phototropism	d. abscisic acid
e	5. Prevention of leaf senescence	e. cytokinin
d	6. Initiation of dormancy in buds and seeds	
c	7. Release of seeds from dormancy	
e	8. Root-shoot coordination	
b	9. Initiation of root hairs	
c	10. Stimulation of pollen tube growth	

ESSAY QUESTIONS

1. What is the advantage to a biennial plant of vernalization?

2. What problem might you encounter if you emigrate to a tropical climate and intend to raise carrots, spinach, apples, peaches, and cherries?

3. Poinsettias are short-day plants. How could you get a plant to bloom at the equator?

4. Explain why gardeners do not have to worry about planting seeds upside-down.

5. Explain the difference between short-day and long-day plants. What is the critical period to induce flowering in each?

6. Design an experiment to determine whether phytochrome is involved in the response of a plant to light.

7. Would chrysanthemums set out in front of a house on a street with street lights flower? Why or why not?

8. Explain the expression "One rotten apple spoils the barrel".

9. A row of intact oat coleoptiles receives unilateral light that has first passed through a prism. Describe the appearance after 2 hours of the coleoptiles receiving each component color of white light: violet, blue, green, yellow, orange, and red.

CHAPTER 15: GENES AND THE GENETIC BASIS OF METABOLISM AND DEVELOPMENT

MULTIPLE CHOICE QUESTIONS

c 1. A cell cannot survive without
   a. a nucleus
   b. mitochondria
   c. ribosomes
   d. chloroplasts
   e. a nucleolus

d 2. At a certain stage of development of a particular plant cell, a sudden increase of one protein, Protein A, occurs. This increase could not be due directly to
   a. decreased degradation of hnRNA for Protein A
   b. increased activation of an inactive form of Protein A
   c. increased transcription of the Protein A gene
   d. increased amino acid synthesis
   e. refolding of the Protein A gene

c 3. Histone proteins
   a. act as enzymes
   b. are the basis for nucleosomes in chloroplasts
   c. protect nuclear DNA and act as the structural basis of nucleosomes
   d. are the main regulator of transcription
   e. c and d

e 4. Transcription is the synthesis of
   a. mRNA
   b. rRNA
   c. tRNA
   d. DNA
   e. a, b and c

b 5. To clone DNA, one way is to isolate mRNA and use it to produce cDNA using
   a. DNA polymerase
   b. reverse transcriptase
   c. DNA ligase
   d. RNA polymerase
   e. DNA replicase

a 6. Two different cells in a plant are unique due mainly to differences in their content of
   a. proteins
   b. lipids
   c. carbohydrates
   d. amino acids
   e. organic acids

c 7. Viruses always consist of
   a. DNA + protein
   b. RNA + protein
   c. nucleic acid + protein
   d. nucleic acid + carbohydrate
   e. nucleic acid + lipid

d 8. If a polypeptide chain contains 144 amino acids, what is the minimum number of nucleotides in the mRNA coding for this chain?
   a. 72
   b. 144
   c. 288
   d. 432
   e. 576

e 9. The type of nucleic acid found in plant viruses may be
   a. single-stranded RNA or double-stranded DNA
   b. double-stranded RNA or single-stranded DNA
   c. single-stranded RNA or DNA
   d. double-stranded RNA or DNA
   e. c and d

a 10. If you isolated a particular chromosome from the cells of a plant, isolated the DNA, melted it, then found that DNA hybridization occurred quite rapidly, that chromosome would probably contain
   a. multiple copies of rRNA and/or tRNA genes
   b. multiple copies of chloroplast protein genes
   c. a few copies of rRNA genes
   d. a few copies of tRNA genes
   e. multiple copies of glycolysis enzyme genes

d 11. If you want to insert a gene into a plant that is expressed only in root cap cells, it will most likely work best in the plant if it is
   a. inserted randomly in the plant nuclear DNA
   b. inserted into plant cells using a viral vector
   c. attached to a promotor for a glycolysis enzyme gene
   d. attached to a promotor for a specific root cap cell enzyme
   e. there is no way to insert a gene into a plant cell

a 12. The first step in the initiation of translation is the binding of
   a. initiator tRNA to the ribosomal small subunit
   b. $tRNA^{met}$ to the ribosomal large subunit
   c. mRNA to the ribosomal small subunit
   d. initiator tRNA to the start codon of mRNA
   e. initiator tRNA to an intact ribosome

c 13. Which of the following is the most energy efficient level of control of metabolism?
   a. processing of hnRNA into mRNA
   b. rate of translation
   c. rate of transcription
   d. activation/inactivation of an enzyme
   e. rate of mRNA transport from nucleus to hyaloplasm

b 14. A molecule of tRNA with the anticodon CAG binds to a codon with the sequence
   a. GUC
   b. CUG
   c. AGU
   d. UCA
   e. CTG

b 15. If a segment of DNA has the base sequence ATTGCA, the RNA transcribed from it would be
   a. TAACGT
   b. UAACGU
   c. GCCAUG
   d. GCCATG
   e. CGGTAC

c 16. The transfer of a functional gene from one organism to another is possible because
   a. DNA occurs in nucleosomes in all organisms
   b. DNA is composed of the same four bases in all organisms
   c. the genetic code is universal
   d. deoxyribonucleotides occur in the same sequence in all organisms
   e. DNA is found in nuclei in all organisms

b 17. The term "codon" refers to triplets of nucleotide bases in
   a. DNA
   b. mRNA
   c. tRNA
   d. rRNA
   e. all of the above

e 18. Which of the following is not controlled by proteins in a cell?
   a. the plane in which a cell divides
   b. the composition of a cell wall
   c. cell shape
   d. biochemical pathways of metabolism
   e. none of the above; they are all controlled by proteins

d 19. DNA molecules are protected because
   a. they are found in mitochondria and plastids
   b. the genetic code is degenerative
   c. they are covered by non-histone proteins
   d. a and b
   e. a, b and c

b 20. Which part of a gene is not transcribed?
   a. structural region
   b. TATA box
   c. introns
   d. b and c
   e. a, b and c

c 21. If a DNA segment has the base sequence TGACTCAAGCTT, then the anticodons of the tRNA molecules that bind to the mRNA transcribed from that sequence would be
   a. ACU, GAG, UUC, GAA
   b. TGA, CTC, AAG, CTT
   c. UGA, CUC, AAG, CUU
   d. UG, AC, UC, AA, GC, UU
   e. UGAC, UCAA, GCUU

d 22. The ability of a plant cell to store information in DNA accurately over the life span of the cell is ensured by all except
   a. degeneracy
   b. plastid membranes
   c. translation in the cytoplasm
   d. transcription in the cytoplasm
   e. nucleosomes

c 23. In which region would a mutation of the DNA be the least likely to affect synthesis or structure of the protein coded for by its gene?
   a. TATA
   b. exon
   c. intron
   d. enhancer element
   e. all are equally likely

b 24. Which is a correct match between a molecule and its place of synthesis in a cell?
   a. mRNA - cytoplasm
   b. RNA polymerase II - cytoplasm
   c. ribosomal subunits - cytoplasm
   d. ribosomal proteins - nucleus
   e. tRNA - cytoplasm

b 25. The major function of RNA is its participation in protein synthesis. Involved in this process are mRNA, tRNA, 28S rRNA, 18S rRNA, 5.8S rRNA and 5S rRNA. How many of these types of RNA are in their final, functional form as they are transcribed?
   a. 1
   b. 2
   c. 3
   d. 4
   e. 6

TRUE-FALSE QUESTIONS

T _____ 1. It is possible for a plant to be infected with a virus and show no symptoms of that infection.

F _____ 2. At any particular time during transcription, only two codons are being read.

F _____ 3. The ribosome merely provides a passive surface on which amino acids are linked together.

T _____ 4. Ribosomal subunits can be recycled, saving the cell energy.

F _____ 5. All genes contain the information to make a particular protein or polypeptide.

F _____ 6. Many types of plant cells can survive without a nucleolus.

T _____ 7. There are 62 different types of tRNA.

T _____ 8. tRNA molecules can be recycled.

T _____ 9. A ribosome is composed of one large subunit and one small subunit, each containing rRNA and proteins.

F _____ 10. The synthesis of the 28S, 18S, 5.8S, and 5S rRNA molecules is turned on and off at the same time.

F _____ 11. The transcription of all RNA molecules is catalyzed by the same type of RNA polymerase, therefore transcription is easier to control.

F _____ 12. If a particular protein contains 100 amino acids, then the gene that codes for that protein is 300 nucelotides long.

T _____ 13. The uniqueness of different types of cells is due mainly to different proteins, such as enzymes.

T _____ 14. If histone proteins isolated from chick brain cells are combined with DNA isolated from rose cells, functional chromosomes should form.

F _____ 15. Once a molecule of tRNA "delivers" an amino acid to a growing protein, it is degraded by the cell.

T _____ 16. The nucleotide bases in tRNA may be modified after transcription.

F _____ 17. The genetic code is totally universal.

MATCHING

Match a process in List A with an enzyme(s) in List B. Answers in List B may be used more than once.

List A

e _____ 1. Translocation of ribosome along mRNA
b _____ 2. Transcription of hnRNA
f _____ 3. DNA repair
d _____ 4. Attachment of amino acid to tRNA
g _____ 5. DNA cleavage at specific sites
e _____ 6. Termination of translation
a _____ 7. Transcription of most rRNA
e _____ 8. Formation of peptide bond
h _____ 9. Synthesis of DNA using RNA as a template
i _____ 10. DNA duplication

List B

a. RNA polymerase I
b. RNA polymerase II
c. RNA polymerase III
d. amino acid activating enzymes
e. large ribosomal subunit enzymes
f. DNA ligase
g. restriction endonucleases
h. reverse transcriptase
i. viral replicase

Match a statement from List A with a structure from List B.

List A

- e _____ 1. Used as a vector in plant genetic engineering
- a _____ 2. Excised from hnRNA
- g _____ 3. Necessary for attachment of NA polymerase II
- i _____ 4. Controls transcription
- c _____ 5. Synthesized using reverse transcriptase
- f _____ 6. Contains the complete information to make a polypeptide
- d _____ 7. Contains genes for rRNA
- b _____ 8. Codes for part of mature mRNA
- h _____ 9. Stops transcription
- j _____ 10. Found only in viruses

List B

a. intron
b. exon
c. cDNA
d. nucleolar DNA
e. ti plasmid
f. structural gene
g. TATA box
h. multiple adenine sequence
i. promotor
j. single-stranded DNA

ESSAY QUESTIONS

1. Why is protein synthesis such a complicated process?

2. Outline the flow of information from molecules of DNA to an organism's morphological characteristics.

3. How could you determine whether DNA participates directly or indirectly in protein synthesis?

4. In a tropical rainforest of South America, you discover a new species of plant. Upon analysis, 20% of the plant's DNA bases are found to be cytosine. What percentage of the bases are adenine? Guanine? Thymine?

5. If a virus enters one leaf of a plant, how could the stem or other leaves become infected?

6. What procedure(s) and precaution(s) would you use to propagate an orchid that is suspected to have a viral infection?

7. If a nucleic acid is single-stranded, how could you determine whether it is RNA or DNA?

8. What kind of experiment(s) would you do to determine how closely related two species of plants are?

9. Outline the steps of initiation, elongation, and termination of mRNA translation.

10. Why is it important that amino acid activation be catalyzed by specific enzymes?

11. Distinguish between: exons and introns; hnRNA, mRNA, tRNA, and rRNA.

12. Briefly describe how a piece of DNA can be cloned using bacteria. Why would it be important to use the same type of restriction endonuclease in the harvest phase as in the preparation phase?

13. Using Table 15.2 in your text, determine the sequence of amino acids coded for by the following segment of DNA: TACAGTCCAGTACAGCGTATGGGGATC.

14. Hybridization can occur not only between single-stranded DNAs (ssDNA), but also between RNA and ssDNA. Would the hybridization between plant mRNA and ssDNA be an exact match? Why or why not?

CHAPTER 16: GENETICS

MULTIPLE CHOICE QUESTIONS

a 1. 2-aminopurine (2AP), an analogue of guanine, is a mutagen that is incorporated into DNA in place of guanine as it replicates. When transcription occurs, the 2AP is read as adenine instead of guanine. The type of mutation involved is a(n)
   a. point mutation
   b. insertion
   c. deletion
   d. inversion
   e. none of the above

e 2. Which of the following mutations is the most likely to be harmful to a plant?
   a. a point mutation in a rRNA gene
   b. an insertion in an intron of an enzyme gene
   c. a deletion in spacer DNA
   d. an inversion in a tRNA gene
   e. a point mutation in the region of DNA coding for a start codon

a 3. Navel oranges contain no seeds, so they must be propagated vegetatively. If this seedless condition could be traced back to a single branch on a normal, seed-bearing tree, the cause of ths alteration was probably a
   a. somatic mutation
   b. reproductive mutation
   c. transposition
   d. mutation in one seed produced by the tree
   e. none of the above

d 4. Mutations usually occur frequently in a plant and may be caused by
   a. DNA copying mistakes
   b. chemicals in the environment
   c. insect pests
   d. a and b
   e. a, b and c

c 5. If plants were growing much larger than normal size in the vicinity of Chernobyl, Soviet Union, where a serious accident occurred in a nuclear reactor in 1986, a likely cause would be increased
   a. rainfall
   b. nutrient availability
   c. mutations caused by radioactivity
   d. CO_2 in the atmosphere
   e. none of the above are likely

b 6. If a plant has two identical alleles for a particular trait, then for that trait the plant is
   a. heterozygous
   b. homozygous
   c. homodominant
   d. recessive
   e. homologous

c 7. If plants with red flowers are selfed, plants of the F_1 generation produce flowers that are red or white in a ratio of 3 red:1 white. With respect to flower color, the parent plants must be
   a. homozygous with red dominant to white
   b. homozygous with white dominant to red
   c. heterozygous with red dominant to white
   d. heterozygous with white dominant to red
   e. there is not enough information given to tell

a. 8. In Jimson weed, the allele for spiny pods is dominant to the allele for smooth pods. If two plants heterozygous for this trait are crossed, their offspring would be expected to produce pods in the ratio
   a. 3 spiny pods: 1 smooth pod
   b. 3 smooth pods: 1 spiny pod
   c. 2 spiny pods: 1 smooth pod
   d. 4 spiny pods: 1 smooth pod
   e. 1 spiny pod: 2 intermediate pods: 1 smooth pod

Questions 9 through 13 refer to the following information:

A particular species of plant produces flowers of two different sizes and colors: large flowers (L) are dominant to small (l) and red flowers (R) are dominant to white (r). A plant heterozygous for the two traits is selfed, producing an F_1 population of 80 plants.

b 9. How many of the plants in the F_1 generation will breed true for flower size <u>or</u> flower color?
   a. 20
   b. 40
   c. 50
   d. 60
   e. 80

a 10. How many of the plants in the F_1 generation will breed true for <u>both</u> traits?
   a. 20
   b. 40
   c. 50
   d. 60
   e. 80

d 11. What percentage of the F_1 generation will be heterozygous for both traits?
   a. 56.25%
   b. 50%
   c. 37.5%
   d. 25%
   e. 6.25%

b 12. How many plants in the F_1 generation will produce large, red flowers?
   a. 60
   b. 45
   c. 40
   d. 15
   e. 5

c 13. What fraction of the plants that produce red flowers will be heterozygous for that trait?
   a. 1/4
   b. 1/2
   c. 2/3
   d. 3/4
   e. all of them will be heterozygous

b 14. Auxotrophic mutants in bacteria are incapable of synthesizing a particular amino acid or nucleotide base because one of the enzymes in the synthetic pathway is defective. The production of amino acids is an example of
   a. epinasty
   b. epistasis
   c. gene linkage
   d. co-dominance
   e. pleiotropy

c 15. Flower color in sweet peas is controlled by two genes, C/c and E/e. C is dominant to c and E is dominant to e. A plant produces purple flowers only if it contains at least one dominant allele for each gene, otherwise it produces white flowers. If two plants heterozygous for both genes are crossed, what will be the phenotypic ratio for purple:white flowers?
   a. 3:1
   b. 4:1
   c. 9:7
   d. 5:3
   e. 15:1

a 16. The mitochondria in some strains of *Paramecium aurelia* carry a gene which codes for a poison that kills other strains. If during sexual reproduction a poison producing cell is the mother and a non-producing cell is the father, then the F_1 progeny would most likely be
   a. all poison producers
   b. all non-poison producers
   c. 50% poison producers and 50% non-poison producers
   d. 75% poison producers and 25% non-poison producers
   e. 25% poison producers and 75% non-poison producers

c 17. A particular trait in a species of plant has two alleles, M which is dominant to m. If heterozygous plants for this trait are crossed, offspring all exhibit the dominant trait. The most logical conclusion is that
   a. cross-over has occurred
   b. the parents were not really heterozygous
   c. m is a lethal allele
   d. this gene is found only in plastids
   e. there is not enough information available to draw a conclusion

d 18. Tallness (T) is dominant to dwarfness (t) and broad leaves (B) are dominant to narrow leaves (b). To determine the genotype of a tall, broad-leaved plant, it would be best to cross it with a plant that is
   a. tall with broad leaves
   b. tall with narrow leaves
   c. dwarf with broad leaves
   d. dwarf with narrow leaves
   e. any of the above

e 19. Fruit size in a tetraploid plant is controlled by a single gene that has two alleles. The size of fruit produced by a plant depends on the number of each allele it possesses. The number of different sizes of fruit produced by a population of these plants would be
   a. 1
   b. 2
   c. 3
   d. 4
   e. 5

c 20. After one thousand generations of plant breeding, three alleles exist for a gene that originally had only one. These new alleles arose through the process of
   a. vegetative reproduction
   b. sexual reproduction
   c. mutation
   d. crossing-over
   e. all of the above

d 21. Which statement is correct about DNA replication?
   a. DNA polymerase adds nucleotides at the 5' end of the growing nucleotide chain
   b. Short strands of DNA act as primers
   c. A replicon is an enzyme for DNA replication
   d. DNA replication is semiconservative
   e. a and d are correct

b 22. Pieces of DNA that readily change their positions from one chromosome to another are
   a. hypothetically possible but not proven
   b. called transponsons
   c. usually removed from the genome by restriction enzymes
   d. of little consequence in the study of genetics
   e. not considered to be mutations

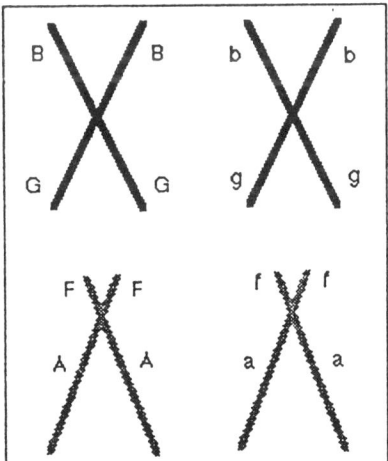

c 23 In considering the accompanying two homologous pairs of chromosomes, gametes that would form only if crossing over occurred would have the genotype
   a. BGFa
   b. BGfA
   c. BgFa
   d. bgfA
   e. bgFa

a 24. Two plants with lavender flowers are crossed. The resulting offspring include 92 white-flowered plants, 145 lavender-flowered plants, and 83 purple-flowered plants. The lavender flowered plants
   a. are heterozygous
   b. are homozygous dominant
   c. are homozygous recessive
   d. are either homozygous dominant or homozygous recessive
   e. cannot be genotypically determined

d 25. Plants that "breed true" or are "pure bred" for a trait controlled by a single pair of alleles
   a. are heterozygous
   b. are homozygous dominant
   c. are homozygous recessive
   d. are either homozygous dominant or homozygous recessive
   e. cannot be genotypically determined

c 26. Lethal alleles are often eliminated from a population because during the plant life cycle
   a. the gametophyte is diploid
   b. the sporophyte is diploid
   c. the gametophyte is haploid
   d. the sporophyte is haploid
   e. eggs and sperm are formed within tissues of the sporophyte

TRUE-FALSE QUESTIONS

F _____ 1. Nucleosomes exist only in non-dividing cells.

F _____ 2. In eukaryotic cells, DNA chains are replicated in one continuous piece from the 5' to the 3' end.

F _____ 3. It is vital that every DNA copying mistake be corrected by the appropriate DNA repair mechanism.

T _____ 4. One probable result of destruction of the ozone layer is increased mutations in plant cells, most of which will be harmful.

T _____ 5. If two traits are located at either end of a long chromosome, they undergo independent assortment during meiosis.

T _____ 6. The most accurate way to determine the percentage of crossing-over between two genes on the same chromosome would be to do a test cross using a double homologous recessive parent.

F _____ 7. The cells of a pentaploid plant can divide both mitotically and meiotically, so they are fertile.

T _____ 8. Higher plants are the evolutionary product of mutations in junk DNA.

F _____ 9. Two genes that produce 5% recombinant F_1s are farther apart than two genes that produce 15% recombinant F_1s.

T _____ 10. If there are four alleles for a particular gene, the cells of some plants may contain all four of those alleles.

MATCHING

Match each base sequence change in List A with the term in List B that best describes what has happened.

List A

d _____ 1. ATCTGAG ---> ATGTGAG
a _____ 2. GCAGTCT ---> GCACT
e _____ 3. TGACTGC ---> TGAGCCTGC
c _____ 4. ACGATGA ---> ACTAGGA
b _____ 5. ATCTGAG ---> ATGTGAG
          or GCAGTCT ---> GCACT

List B

a. deletion
b. mutation
c. inversion
d. point mutation
e. insertion

In a species of grain, herbicide resistance (H) is dominant to herbicide susceptibility (h) and large kernel size (K) is dominant to small kernels (k).

Using the accompanying information, match the crosses in List A with the ratios in List B.

List A

a _____ 1. Kk x KK
h _____ 2. Kk x Kk
b _____ 3. kk x kk
a _____ 4. HHKK x HhKk
f _____ 5. HhKk x HhKk
c _____ 6. Hh x hh

List B

a. all phenotypically dominant
b. all phenotypically recessive
c. 1:1 genotypic ratio
d. 3:1 phenotypic ratio
e. 1:2:1 genotypic ratio
f. 9:3:3:1 phenotypic ratio
g. 9:3:3:1 genotypic ratio
h. d and e

ESSAY QUESTIONS

1. In many large fields, the plants are all exactly the same variety of crop plant. What is the potential danger in this practice?

2. Considering the process of DNA replication, why isn't each molecule of DNA actually a mixture of DNA and RNA?

3. Describe how a particular plant gene could be located using transposons.

4. All mutations change the base sequence in a molecule of DNA. Explain how this might or might not have an actual effect on the plant.

5. Explain how a sporophyte could be triploid, tetraploid, or pentaploid.

6. People with Down syndrome have an extra chromosome 21. How did this arise? Why are they sterile?

7. Refer to the information in multiple choice question 15 concerning flower color in sweet peas. Explain how two white-flowered plants could produce offspring with purple flowers.

8. How could a plant produce striped or spotted leaves?

9. Describe the process of DNA replication. Why does each chromosome have thousands of replication start sites instead of just one? What are Okazaki fragments?

10. Describe the process of DNA replication. Use the terms "replicon," "primer RNA," "DNA polymerase," and "semiconservative" in your answer.

11. Why would you see wide variety in a trait within a population if the trait is controlled by a gene that displays multiple alleles?

12. What are the roles of independent assortment and crossing over in generating variety in a population?

CHAPTER 17: POPULATION GENETICS AND EVOLUTION

MULTIPLE CHOICE QUESTIONS

a   1. Speciation is the evolution of new
     a. species
     b. sub-species
     c. varieties
     d. races
     e. hybrids

c   2. Which of the following is the most likely to have evolved the most rapidly?
     a. CAM photosynthesis
     b. fleshy fruits
     c. reduced xylem in aquatic plants
     d. the development of roots
     e. woody stems

a   3. Plants
     a. that have leaves are able to intercept light for photosynthesis more efficiently than plants that lack leaves
     b. produce flowers with showy petals to attract pollinators
     c. have become taller in order to compete better for light
     d. need energy to survive and developed mitochondria and chloroplasts to fulfill that need
     e. all of the above

b   4. Natural selection would probably not occur
     a. in a dense forest
     b. in a population of plants produced by underground rhizomes
     c. in a population of plants that produces different sizes of seeds
     d. a and b
     e. a, b and c

e   5. Gene flow through a population would be enhanced by the
     a. tuft of hairs on the top of dandelion fruits
     b. fleshiness of a peach
     c. plantlets produced on the edges of *Kalanchoe* leaves
     d. co-evolution of a plant and its insect pollinator
     e. all of the above

d   6. Pollen from Species A frequently comes in contact with flower stigmas of Species B resulting in the production of hybrid plants. Species A and B remain distinct species most likely because
     a. Species A is pollinated by honey bees and Species B is pollinated by hummingbirds
     b. Species A flowers in May and Species B flowers in July
     c. Species A is native to prairies and Species B is native to temperate forest
     d. hybrid seeds fail to germinate
     e. pollen from Species A fails to germinate on stigmas of Species B flowers

d 7. In which situation would speciation probably not be phylogenetic?
   a. a plant is pollinated by a species of migrating birds
   b. a fungus produces billions of spores that easily travel on wind currents
   c. individuals of a particular species of tree in tropical rainforests are widely separated, but are pollinated by bats
   d. a particular species of tree produces very heavy seeds that are poisonous to animals
   e. any fragment of an aquatic plant may produce a new plant

c 8. An example of a postzygotic isolating mechanism between closely related species would be
   a. one species flowers in April, the other in August
   b. the two species are pollinated by different species of bees
   c. a zygote produced by syngamy of sperm and egg of the two species contains an uneven number of chromosomes
   d. the pollen from one species will not germinate on the stigma of the other
   e. one species is terrestrial, the other totally aquatic

e 9. Which of the following observations could have led scientists to start considering the concept of evolution?
   a. the close resemblance of European and South American orchid flowers
   b. the erosion of mountains
   c. the remarkable vegetative similarity of unrelated desert plants in Africa and North America
   d. a and b
   e. a, b and c

b 10. Which of the following is not an example of natural selection?
   a. only those seeds produced by a plant that land on moist sand germinate
   b. red and pink flowers produced in a population of one species are pollinated equally well by bees
   c. some plants in a population produce more efficient mitochondria than others
   d. plants of a particular species with a more extensive root system absorb more minerals from the soil
   e. within a particular population, the taller plants intercept more light

a 11. A gene has two alleles, Y and y. If you introduce YY individuals and yy individuals into a new area, the frequency of these two alleles will remain the same as long as
   a. there is random sexual mating
   b. there is sexual reproduction where one genotype is preferred over another
   c. one or both of the alleles mutate
   d. a portion of the population is randomly lost
   e. one genotype survives better than another

c 12. Many domesticated plants do not closely resemble their ancestors. This is due mainly to
   a. natural selection
   b. mutations
   c. artificial selection
   d. accidents
   e. sexual reproduction

a 13. The name most commonly associated with the concept of natural selection is
   a. Charles Darwin
   b. Alfred Russel Wallace
   c. G.H. Hardy
   d. G. Weinberg
   e. Jean Baptiste de Lamarck

b 14. If two species exhibit convergent evolution, the result will be
   a. one species
   b. two distinct species that resemble one another very closely
   c. a variety of species produced over a short period of time
   d. two distinct species that are very different from each other
   e. it is impossible to predict what might happen

e 15. In a large population of plants, which of the following is least likely to cause a change in allele frequencies within the population?
   a. a forest fire that destroys part of the population
   b. radioactive fallout from an accident at a nuclear power plant
   c. microhabitats within the range of the population where certain phenotypes have a better chance of surviving
   d. the preference of a pollinator for a certain flower color
   e. wind pollination of the flowers

b 16. Which of the following is not capable of evolving?
   a. a population of fruit flies
   b. your professor
   c. a culture of bacteria
   d. the collective cats of a city
   e. a particular species of plant

b 17. One of the biggest problems with saving endangered species, plant or animal, by captive breeding is
   a. mutation
   b. genetic drift
   c. natural selection
   d. geographic barriers
   e. sexual reproduction

c 18. A circular petri dish containing nutrient agar has a disc of filter paper saturated with an antibiotic in the center. If bacteria are put in the petri dish, they grow until they form three zones: A zone (A) right next to the disc where there is no growth; a zone (B) to the outside of A where bacteria resistant to the antibiotic grow; and a zone (C) to the outside of B where bacteria susceptible to the antibiotic grow. Which of the following best explains why bacteria can grow in Zone B?
   a. some bacteria already had an allele specifically to protect them from the antibiotic
   b. the antibiotic caused a mutation, producing an allele in some bacteria giving them resistance
   c. some bacteria, by chance, had an allele giving them resistance
   d. during the experiment, some bacteria produced an allele specifically to give them resistance
   e. this experiment cannot be explained

a 19. According to the chemosynthesis theory of the origin of life on Earth
   a. the atmosphere was exposed to ultraviolet and gamma radiation from the sun
   b. the atmosphere was oxidizing
   c. organic chemicals that formed abiotically were quickly oxidized
   d. b and c
   e. a, b and c

e 20. The second atmosphere around Earth lacked
   a. hydrogen

b. water
c. ammonia
d. hydrogen sulfide
e. molecular oxygen

c 21. In a lawn, it is observed that nearly all of the dandelions have tall, upright flower stalks. After several years of regular mowing where the tall flower stalks were cut off, it is found that nearly all dandelions in the lawn have short, curled flower stalks that are not readily cut off by the lawn mower. This is an example of:
a. artificial selection
b. accidents
c. natural selection
d. competition
e. isolating mechanisms

a 22. Artificial selection and natural selection differ in that
a. artificial selection involves purpose; natural selection does not involve purpose
b. natural selection results in changes in allele frequency; artificial selection does not
c. factors involved in artificial selection are accidents; those in natural selection are purposeful
d. natural selection strives to improve a species; artificial selection results in more useful plants
e. artificial selection results in changes in allele frequency; natural selection does not

e 23. Current studies are focusing on _____ as the earliest heritable information molecule.
a. DNA
b. protein
c. polysaccharide
d. peptidoglycan
e. RNA

e 24. The buildup of oxygen in the atmosphere of the early, evolving earth likely resulted in
a. the evolution of the mitochondrion
b. rust
c. reduction in UV light irradiation levels at the surface of the earth
d. evolution of terrestrial life
e. all of the above

TRUE-FALSE QUESTIONS

F _____ 1. Natural selection not only causes differential survival within a population, it also causes mutations.

F _____ 2. The evolutionary gain of a plant structure usually proceeds more rapidly than the evolutionary loss of a structure.

F _____ 3. The only outcome of divergent speciation is the development of one subpopulation into a new species while the other subpopulation remains the original species.

T _____ 4. If you pick a flower, it is possible that you are eliminating an allele from that plant's population.

F _____ 5. Alleles produced in a population by mutation always have an immediate effect on the survival of the individuals possessing them.

T _____ 6. Alleles that are disadvantageous may become advantageous and vice versa.

F _____ 7. Natural selection can occur in a population of individuals homozygous dominant for every trait.

T _____ 8. The importance of random changes in allele frequencies is higher in small populations.

T _____ 9. The driving force behind the evolution of aerobic respiration was photosynthesis.

T _____ 10. One reason we should be concerned today about massive deforestation in tropical areas is a possible decrease of molecular oxygen in the atmosphere.

ESSAY QUESTIONS

1. Distinguish between phyletic and divergent speciation and the causes of each.

2. What two conditions are necessary for natural selection to occur? Why must they occur?

3. Which type of speciation, phyletic or divergent, do you think occurs more commonly? Justify your answer.

4. One solution to the problem of species extinction is captive breeding in zoos or gardens. What are some of the problems associated with this solution?

5. If an insecticide is applied to a field to decrease a population of pest insects, not all of the pest insects are killed. Explain this in terms of genetics. What problems can result from this?

6. Explain why, in the absence of living organisms, moon rocks may contain amino acids.

7. Discuss the reasons why the evolution of photosynthesis was so important to life on Earth.

CHAPTER 18: CLASSIFICATION AND SYSTEMATICS

MULTIPLE CHOICE QUESTIONS

e 1. If you were a scientist trying to study the taxonomy of a group of plants, you might study the plants'
   a. phytochemistry
   b. plastid DNA sequences
   c. microscopic anatomy using a scanning electron microscope
   d. rRNA base sequences
   e. all of the above

b 2. In which of the following situations would a particular plant species be more likely to evolve into a large family?
   a. an environment where only a few phenotypes are fit enough to survive and almost all mutations are selectively disadvantageous
   b. an environment where numerous phenotypes are successful and many mutations are advantageous
   c. an environment that is very stable
   d. the species is pollinated by one species of insect
   e. the species lives in only one particular type of habitat

d 3. Which of the following is most likely to be a phylogenetic classification system?
   a. trees, shrubs, vines, and herbs
   b. aquatic and terrestrial plants
   c. bee-pollinated, hummingbird-pollinated, beetle-pollinated, and wind-pollinated plants
   d. plants producing bilateral flowers with 5 sepals, 5 petals, 10 stamens, and one hypogynous gynoecium; plants producing radial flowers with 3 sepals, 3 petals, 3 stamens, and one epigynous gynoecium composed of 3 fused carpels
   e. bilateral or radial flowers

b 4. Thorns, spines, and prickles are all elongated structures that are relatively hard with a sharp point at one end. Thorns are modified stems, spines are modified leaves, and prickles are outgrowths from a stem surface. These three structures are
   a. homologous
   b. analogous
   c. aggregate structures
   d. phylogenetic
   e. harmonious

e 5. The kingdom which contains organisms which are not as closely related evolutionarily to one another as are organisms in other kingdoms is
   a. Monera
   b. Plantae
   c. Fungi
   d. Animalia
   e. Protista

d 6. Two species of plants have succulent leaves covered by similarly shaped hairs. These similarities could be the product of
   a. a common ancestor
   b. convergent evolution
   c. coevolution
   d. a and b
   e. a, b and c

b 7. The correct way to write the species name of corn is
   a. Zea mays
   b. *Zea mays*
   c. *Zea Mays*
   d. mays
   e. *mays*

e 8. Why have many plants not been described or given a scientific name?
   a. many live in inaccessible places
   b. many have been overlooked
   c. many closely resemble species already described
   d. a and b
   e. a, b and c

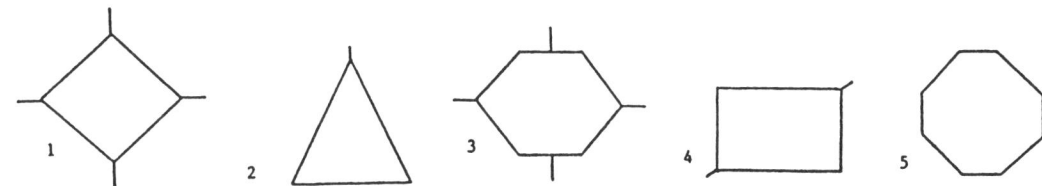

b 9. If each geometric figure in the accompanying diagram represents a species, which two are probably the most closely related and may have a common ancestor?
   a. 1 and 2
   b. 1 and 3
   c. 2 and 4
   d. 3 and 5
   e. 1 and 5

c 10. A classification system that shows evolutionary relationships is
   a. taxonomic
   b. artificial
   c. phylogenetic
   d. binomial
   e. analogous

a 11. Which of the following would be the best criterion for placing two separate species in the same genus?
   a. they occasionally interbreed, producing a sterile hybrid
   b. both produce red colored flowers
   c. both are trees
   c. both exhibit C_4 photosynthesis
   e. both have exactly the same growth requirements

e 12. The correct sequence of taxonomic catagories, from smallest to largest, is
   a. species ---> genus ---> order ---> division ---> class ---> family ---> kingdom
   b. genus ---> species ---> family ---> class ---> order ---> division ---> kingdom
   c. species ---> family ---> class ---> genus ---> division ---> order ---> kingdom
   d. family ---> species ---> genus ---> order ---> class ---> division ---> kingdom
   e. species ---> genus ---> family ---> order ---> class ---> division ---> kingdom

c 13. The scientist who contributed most to plant classification before Christ was
     a. Pliny the Elder
     b. Dioscorides
     c. Theophrastus
     d. Carolus Linnaeus
     e. Gaspard Bauhin

d 14. The foundation of modern scientific classification of organisms can be traced to
     a. ancient China
     b. the Roman Empire
     c. Renaissance Italy
     d. ancient Greece
     e. eighteenth century Sweden

b 15. The idea that the children of Arnold Schwarzenegger would have large muscles like their father without working out to get them would be attributed to
     a. Gregor Mendel
     b. John Baptiste Lamarck
     c. Charles Darwin
     d. Alfred Russel Wallace
     e. Carolus Linnaeus

d 16. Organisms are given scientific names because
     a. Latin is the language of scholars
     b. it frustrates amateur botanists
     c. it gives taxonomists something to do
     d. it makes communication about organisms easier
     d. all of the above

a 17. The basis of scientific classification is
     a. evolutionary relationships
     b. morphology
     c. the opinion of the classifier
     d. the area where an organism lives
     e. all of the above

b 18. Two organisms are considered the same species if they
     a. are morphologically similar
     b. can interbreed
     c. live in the same habitat
     d. a and b
     e. a, b and c

c 19. Fossils are often classified in artificial taxa referred to as
     a. fossil genera
     b. past genera
     c. form genera
     d. morph genera
     e. paleogenera

b   20. A truly natural, phylogenetic classification scheme for plants was first developed and widely used by
   a. Theophrastus
   b. Engler and Prantl
   c. Darwin
   d. Lamarck
   e. Linnaeus

e   21. *Acer rubrum*, *Acer saccharinum*, and *Acer saccharum* are all classified as members of the same
   a. family
   b. genus
   c. species
   d. subspecies
   e. a and b

a   22. Homologous features
   a. are similar to each other due to descent from a common ancestor
   b. are similar to each other due to convergent evolution
   c. generally show little similarity in morphology
   d. are of little use in taxonomy and systematics
   e. b and d

d   23. Evolutionarily, the earliest kingdom was
   a. Animalia
   b. Protista
   c. Myceteae
   d. Monera
   e. Plantae

TRUE-FALSE QUESTIONS

F _____ 1. The science of classification of organisms is called nomenclature.

T _____ 2. The current classification system of plants is an approximation.

F _____ 3. The concept of the genus was first established by Carolus Linnaeus.

F _____ 4. Sub-species of a particular plant must be very similar morphologically.

T _____ 5. Characteristics used to determine plant species may be anatomical or physiological.

T _____ 6. Plants are still being described today.

T _____ 7. A place for the storage of dried plant specimens is called an herbarium.

T _____ 8. The process of naming a new species of plant has a series of checks and balances to lessen the chance that the plant has been previously described.

F _____ 9. The discovery of a new species of plant, while interesting in its own right, usually has very little affect on our knowledge of plants already described.

T _____ 10. The taxonomy of many organisms depends on the opinions of the taxonomer.

F _____ 11. Once a particular species has been recognized, there is no question that is a separate species.

T _____ 12. It is much easier to classify taxonomically and to establish phylogenetic relationships for some plants than for others.

T _____ 13. The ancestors of all present-day organisms were primitive bacteria.

F _____ 14. Each couplet of a taxonomic key divides plants into phylogenetic catagories.

F _____ 15. If two plants exhibit C_4 metabolism, those metabolic pathways must be homologous and the plants must be very closely related.

MATCHING

Match each characteristic in List A with a kingdom in List B. Answers in List B may be used more than once.

	List A	List B
c _____	1. Cell walls contain chitin	a. Monera
a _____	2. No true nucleus	b. Protista
f _____	3. Photosynthesis	c. Fungi
d _____	4. Produce tissues of very specialized cells; no photosynthesis	d. Animalia
b _____	5. Very simple eukaryotic organisms	e. Plantae
e _____	6. Produce tissues of very specialized cells; photosynthesis	f. Two or more kingdoms
f _____	7. Descended from Protista-like organisms	

ESSAY QUESTIONS

1. Distinguish between: homology and analogy; artifical and phylogenetic classification; genus and form genus.

2. What criteria would you use to establish that a particular plant species is a new species?

3. Give your students a set of objects, such as different types of nails, pins, or clothes fasteners, and have them write a taxonomic key to identify them.

4. What is a species? How does a taxonomist decide whether two plants are members of the same species or of two closely related species? One genus or two closely related genera?

5. How would you determine whether two similar structures are homologous or analogous?

CHAPTER 19: KINGDOM MONERA: PROKARYOTES

MULTIPLE CHOICE QUESTIONS

b 1. The study of algae is
   a. physiology
   b. phycology
   c. bacteriology
   d. pteridology
   e. none of the above

c 2. Which of the following statements about prokaryotes is <u>incorrect</u>?
   a. they make minerals reavailable to plants through the process of decay
   b. they cause numerous plant and animal diseases
   c. they divide, sometimes quite rapidly, by the process of mitosis
   d. some can convert atmospheric nitrogen to a form which can be used by plants
   e. many produce antibiotics that we use to fight diseases

d 3. The main reason(s) a botanist might isolate a specific plant gene and insert it into bacteria would be to study
   a. the isolated gene
   b. the protein that the isolated gene codes for
   c. the effect of that gene on the bacteria
   d. a and b
   e. a, b and c

a 4. *Escherichia coli*, a common intestinal bacterium, is unicellular with cells that are rod-shaped. They are
   a. bacilli
   b. cocci
   c. spirilla
   d. pleomorphic
   e. filamentous

d 5. If some filamentous cyanobacteria are exposed to adverse environmental conditions, they form resistant structures called
   a. heterocysts
   b. endospores
   c. sclerotia
   d. akinetes
   e. no cyanobacteria can form resistant structures

d 6. If you compared a plant leaf cell and a cyanobacterial cell, which of the following would be found in both?
   a. ribosomes
   b. plasma membrane
   c. chloroplasts
   d. a and b
   e. a, b and c

e 7. Penicillin is probably the most common antibiotic used to fight bacterial infections in people. It is most effective against
   a. gram-negative bacteria because it prevents the synthesis of N-acetylmuramic acid
   b. gram-negative bacteria because it prevents the synthesis of diaminopimelic acid
   c. gram-positive bacteria because it inhibits protein synthesis
   d. gram-positive bacteria because it prevents the synthesis of hemicellulose molecules
   e. gram-positive bacteria because it prevents the cross-linking of the acetylglucosamine/acetylmuramic acid polysaccharide molecules

e 8. A flagellum produced by a gram-negative bacterium must pass through which layers to reach the environment around the cell?
   a. plasma membrane only
   b. plasma membrane, peptidoglycan layer, and glycocalyx
   c. peptidoglycan layer, LPS layer, and glycocalyx
   d. peptidoglycan layer and glycocalyx
   e. plasma membrane, peptidoglycan layer, LPS layer, and glycocalyx

a 9. A bacteriophage is a type of virus that invades a bacterial cell. If, in addition to the viral DNA, a bacteriophage injects some bacterial DNA into a cell, that DNA might be incorporated into the DNA of the cell. This process is called
   a. transduction
   b. transformation
   c. conjugation
   d. translocation
   e. replication

b 10. The Red Sea is due to an abundance of cyanobacteria that contain an abundance of
   a. phycocyanin
   b. phycoerythrin
   c. chlorophyll a
   d. chlorophyll b
   e. carotenoids

c 11. The early atmosphere around Earth contained no free oxygen; free oxygen was added due to photosynthesis. The most likely type of organisms to initially add O_2 to the atmosphere resembled present day
   a. green bacteria
   b. purple bacteria
   c. cyanobacteria
   d. a and b
   e. a, b and c

a 12. If you try to grow bacteria as autotrophs in culture, the carbon should be supplied in the form of
   a. carbon dioxide
   b. sugars
   c. amino acids
   d. fats
   e. any of the above

c 13. A characteristic that is not important for classifying prokaryotes is
   a. metabolism
   b. wall chemistry
   c. internal structure
   d. ability to carry out photosynthesis
   e. sensitivity to O_2

b 14. You have isolated a parasitic organism from a plant. Upon closer examination, you find that the organism is prokaryotic, has little DNA, and no cell wall. You would feel confident classifying that organism as a(n)
   a. cyanobacterium
   b. mycoplasma
   c. archaebacterium
   d. green bacterium
   e. purple bacterium

d 15. The main way that archaebacteria differ from eubacteria is in
   a. their lack of a true nucleus
   b. the lack of a cell wall
   c. their lack of internal organelles
   d. their metabolism
   e. the presence of 70S ribosomes

e 16. Marsh plants often live in stagnant water, where the sediments may contain H_2S, which is toxic to plant roots. One way these plants survive is by forming a mutually beneficial association with
   a. *Cytophaga*
   b. *Agrobacterium*
   c. *Rhizobium*
   d. *Azotobacter*
   e. *Beggiatoa*

a 17. To maximize energy conservation with respect to nitrogen metabolism within a plant, which soil bacteria would be most advantageous?
   a. *Azotobacter*
   b. *Azotobacter* and *Nitrosomonas*
   c. *Azotobacter*, *Nitrosomonas*, and *Nitrobacter*
   d. *Nitrosomonas* and *Nitrobacter*
   e. *Azotobacter* and *Pseudomonas*

c 18. If you were trying to filter-sterilze a solution, which type of organisms would give you the most problems?
   a. eubacteria
   b. cyanobacteria
   c. mycoplasmas
   d. archaebacteria
   e. fungi

a 19. The main photosynthetic pigment in cyanobacteria is
   a. chlorophyll a
   b. chlorophyll b
   c. bacteriochlorophyll
   d. phycocyanin
   e. phycoerythrin

d 20. The strongest evidence that suggests a close relationship between chloroplasts and prochlorophytes is
   a. the presence of chlorophyll a and b in both
   b. the presence of photosystem II in both
   c. their lack of phycobilins
   d. their 16S rRNA base sequence
   e. their O_2 production

b 21. Exchange of DNA between compatible bacterial cells via a pilus is
   a. transduction
   b. conjugation
   c. replication
   d. transformation
   e. asexual reproduction

e 22. For an experiment, you need a source of bacteriochlorophyll. To obtain this pigment, you could culture
   a. any prokaryote
   b. any cyanobacteria
   c. purple bacteria
   d. green bacteria
   e. c or d could be used as sources

c 23. The photosynthetic apparatus of green and purple bacteria lack photosystem (PS)II. This means that these autotrophs cannot produce
   a. sugars
   b. ATP
   c. O_2
   d. a and c
   e. a, b, or c

d 24. The putrid smell of rotten eggs comes from
   a. anaerobic fermentation of lipids by bacteria
   b. breakdown products from death and decay of gram-positive bacteria
   c. breakdown products from death and decay of gram-negative bacteria
   d. anaerobic fermentation of proteins by bacteria
   e. byproducts of peptidoglycan synthesis that accumulate inside the egg shell

a 25. You have found a prokaryote in a salt pool. Tests reveal that this organism's wall lacks peptidoglycan but contains glycoproteins and polysaccharides, has unusual lipid chemistry, and is not affected by drugs that would normally inhibit the actions of ribosomes. You could conclude that this prokaryote is most likely a type of
   a. archaebacteria
   b. nitrogen-fixing bacteria
   c. cyanobacteria
   d. purple bacteria
   e. facultatively anaerobic eubacteria

c 26. The aquatic fern *Azolla* is often grown in rice paddies. The most likely reason for this is that *Azolla*
   a. is a source of carbohydrates for the rice plants
   b. is a source of fish food in the paddy
   c. is in a symbiotic relationship with a nitrogen-fixing cyanobacterium, *Anabaena*, which also fertilizes the rice with fixed nitrogen
   d. releases anti-bacterial compounds into the paddy, protecting the rice plants
   e. none of the above

TRUE-FALSE QUESTIONS

T _____ 1. Genetic engineering involves the insertion of a foreign gene into an organism.

F _____ 2. Plants can use atmospheric nitrogen as a source of nitrogen by converting it into nitrates.

F _____ 3. In some special cases, bacteria may group together to form a tissue.

T _____ 4. Cells of bacteria and archaebacteria are never connected by intercellular cytoplasmic strands, but cells of cyanobacteria may be connected.

F _____ 5. One form of reproduction in cyanobacteria is the formation of akinetes.

F _____ 6. Because prokaryotes contain no true nucleus, they do not contain DNA.

F _____ 7. If a flagellum seen in an electron micrograph is covered by a plasma membrane, it must be part of a prokaryotic cell.

F _____ 8. Conjugation usually increases the amount of DNA in the recipient cell permanently.

T _____ 9. The most common fate of foreign DNA taken into a bacterial cell is for it to be depolymerized.

T _____ 10. As far as we know, cyanobacteria have no mechanism for genetic exchange.

T _____ 11. The respiration of organic compounds in bacteria is basically the same process as respiration in plant cells.

T _____ 12. Energy sources for lithotrophic bacteria include ammonia, hydrogen, hydrogen sulfide, and ferrous iron.

T _____ 13. The base sequence of 16S rRNA has been used to determine phylogenetic relationships among prokaryotes.

T _____ 14. Many antibiotics that are effective against bacteria have no effect on archaebacteria.

F _____ 15. Flagella provide the principal means of motility for gliding bacteria.

MATCHING

Match each characteristic in List A with a group of organisms in List B. Organisms in List B may be used more than once.

List A

b _____ 1. Lack of a cell wall
a _____ 2. Oxygenic photosynthesis
f _____ 3. Contain bacteriochlorophyll
a _____ 4. Contain phycobilins
f _____ 5. Anoxygenic photosynthesis
g _____ 6. Convert Fe^{++} to Fe^{+++} to produce ATP
e _____ 7. Oxidize NH_4^+ tp NO_2^- to produce ATP
c _____ 8. Cell wall lacks peptidoglycan
d _____ 9. Important in recycling the elements in cellulose and chitin
a _____ 10. Often form water blooms that can contribute to fish kills

List B

a. cyanobacteria
b. mycoplasmas
c. archaebacteria
d. gliding bacteria
e. nitrifying bacteria
f. purple bacteria
g. lithotrophic bacteria

ESSAY QUESTIONS

1. Some mitochondrial and chloroplast proteins are made in those organelles and some are made in the hyaloplasm and then transported into the organelles. How could you find out where a particular organelle protein is made in a eukaryotic cell?

2. What are the advantages and disadvantages of being an autotroph? A heterotroph?

3. The availablity of nitrogen or phosphorus often limits the growth of natural plant populations. Of these two elements, phosphorus is more often limiting than nitrogen. Why?

4. Describe the genetic system of prokaryotes. How do they exchange genetic material? Divide?

5. Explain how motile bacteria react to external stimuli.

6. Heat from the Earth and pressure in some deep ocean vent systems convert sulfate to hydrogen sulfide. What type of organisms use the hydrogen sulfide for energy? How?

CHAPTER 20: KINGDOM MYCETEAE: FUNGI

MULTIPLE CHOICE QUESTIONS

d 1. A major reason fungi are classified in a separate group from the eubacteria is that they
   a. are heterotrophic
   b. lack cellulose in their cell walls
   c. are decomposers
   d. are eukaryotic
   e. are mainly filamentous

c 2. In the past, fungi were classified in the plant kingdom mainly because they
   a. are photosynthetic
   b. may produce large, macroscopic structures
   c. produce cell walls
   d. produce recognizable tissues
   e. are eukaryotic

c 3. *Rhizopus stolonifera* is a zygomycete sometimes called black bread mold because it often uses bread as a food source. This fungus is a
   a. biotroph
   b. necrotroph
   c. saprotroph
   d. parasite
   e. lithotroph

e 4. Biotrophs sometimes insert a small, peg-like portion of a cell into a host cell. The portion of the fungal cell inserted is a(n)
   a. plasmodesma
   b. rhizoid
   c. mycelium
   d. sporangium
   e. haustorium

a 5. Which of the following is not a response of some plants to fungal attack?
   a. the production of alternaric acid that kills the fungal cells
   b. the production of phytoalexins
   c. the death of plant cells in immediate contact with the fungal cells
   d. the formation of wound cork to limit the spread of the fungus
   e. a high vacuolar concentration of phenolic compounds

c 6. The major type of fungal vegetative body is
   a. unicellular and eukaryotic
   b. unicellular and prokaryotic
   c. filamentous and eukaryotic
   d. filamentous and prokaryotic
   e. parenchymatous, macroscopic, and eukaryotic

a 7. If a fungal filament grows 2 mm per day and you mark a hypha with dye near its base, how far will the dye have moved after 10 days?
   a. 0 mm
   b. 2 mm
   c. 10 mm
   d. 20 mm
   e. 30 mm

e 8. All of the following can be substrates or environments for some type of fungi except
      a. jams and jellies
      b. frozen meat
      c. compost piles
      d. hot springs
      e. fungi can live in or on all of the above

e 9. Slime molds are placed in the Kingdom Myceteae along with the fungi, but differ from them because the vegetative body
      a. lacks distinct cells
      b. lacks a cell wall
      c. engulfs its food
      d. b and c
      e. a, b and c

d 10. The plasmodium of a true slime mold contains
      a. a single haploid nucleus
      b. a single diploid nucleus
      c. many paired haploid nuclei
      d. many diploid nuclei
      e. a mixture of haploid and diploid nuclei

b 11. Many botanists believe that the ancestors of fungi diverged very early from the ancestors of plants and animals. One piece of evidence that supports this hypothesis is that
      a. fungi contain no plastids
      b. fungal mitosis differs from that of plants and animals
      c. fungi are very simple in organization
      d. fungi have no sexual reproduction
      e. fungal cells have a cell wall

b 12. If you examine the stalk of a mushroom, you would find
      a. parenchymatous tissue
      b. compact hyphae only
      c. millions of individual cells
      d. specialized tissues such as xylem and phloem
      e. it cannot be generalized because mushroom structure varies from species to species

a 13. Chlamydospores in fungi are analogous to what structure in cyanobacteria?
      a. akinetes
      b. heterocysts
      c. endospores
      d. cyanophycean granules
      e. there is no analogous structure in cyanobacteria

a 14. The main way oomycetes differ from ascomycetes and basidiomycetes is that they
      a. contain cellulose in their cell walls
      b. are only unicellular
      c. are all parasites on other organisms
      d. are mainly terrestrial
      e. do not cause any plant diseases

b 15. Many people of Irish ancestry live in the United States primarily because of the great potato famine caused by
   a. *Plasmopara viticola*
   b. *Phytophthora infestans*
   c. *Rhizopus stolonifera*
   d. *Saprolegnia* sp.
   e. *Achlya* sp.

d 16. The basidium produced by basidiomycetes is analogous to what structure produced by ascomycetes?
   a. oogonium
   b. antheridium
   c. ascogonium
   d. ascus
   e. zygosporangium

a 17. Why are fungi classified in the Subdivision Deuteromycotina?
   a. sexual reproduction has not been observed
   b. sexual reproduction is different than in other subdivisions
   c. hyphal structure is different than in other subdivisions
   d. spore structure is different than in other subdivisions
   e. b, c and d

d 18. The algal component of a lichen may benefit from this association because it
   a. occupies habitats that would be too harsh without the fungus
   b. is protected from desiccation by the fungus
   c. grows more rapidly when combined with the fungus
   d. a and b
   e. a, b and c

c 19. Which of the following is correct about *Puccinia graminis* f. sp. *tritici*?
   a. urediniospores are produced mainly in the summer and infect barberry plants
   b. teliospores are produced mainly in the late summer or early autumn and infect wheat plants
   c. basidiospores are produced mainly in the spring and infect barberry plants
   d. spermatia are produced mainly in the late summer and infect wheat plants
   e. aeciospores are produced mainly in the spring and infect barberry plants

e 20. Hyphae of different mating types of a specific fungus fuse prior to sexual reproduction. Ultimately, spores are produced by meiosis. How many different types of spores could be produced by this process?
   a. 1
   b. 2
   c. 4
   d. 100
   e. it is impossible to tell exactly, but the number is probably very large

c 21. The hyphae of zygomycetes differ from hyphae of basidiomycetes in that hyphae of
   a. basidiomycetes do not have septa
   b. zygomycetes do not have septa
   c. zygomycetes have much smaller cells
   d. zygomycetes do not form a mycelium
   e. basidomycetes do not form a mycelium

a 22. The cells of basidiomycetes form clamp connections that
   a. allow dikaryotic (binucleate) cells to divide
   b. permit the hypha to separate more readily for asexual reproduction
   c. attach cells to each other with a stronger connection for forming a mycelium
   d. develop only between cells that will undergo meiosis
   e. prevent cells from becoming dikaryotic (binucleate)

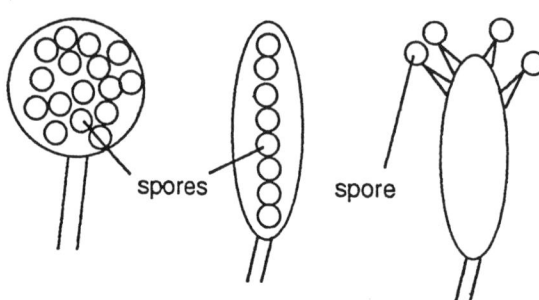

c 23. Which of the following correctly identifies the accompanying reproductive structures to their respective subdivision?

	I	II	III
a.	basidiomycetes	zygomycetes	ascomycetes
b.	basidiomycetes	ascomycetes	zygomycetes
c.	zygomycetes	ascomycetes	basidiomycetes
d.	oomycetes	zygomycetes	basidiomycetes
e.	zygomycetes	oomycetes	ascomycetes

b 24. If you wanted to locate a diploid nucleus in an ascomycete, you would look in the
   a. ascospore
   b. ascus
   c. conidiospore
   d. sclerotium
   e. none of the above

a 25. Chytrids differ from ascomycetes and basidiomycetes in that chytrids
   a. have flagella
   b. have chitin in their cell walls
   c. are all saprotrophic
   d. are unknown in aquatic systems
   e. all of the above

c 26. An ascocarp that is a small simple disk is a(n):
   a. cleistothecium
   b. perithecium
   c. apothecium
   d. conidothecium
   e. sclerotium

e 27. Within which of the following groups would you find fungi that have cells that are coenocytic?
        I. Zygomycetes
        II. Ascomycetes
        III. Myxomycota
  a. I only
  b. II only
  c. III only
  d. I and II only
  e. I and III only

b 28. You are studying brown rot. In order to induce the fungi to reproduce sexually, you would
  a. place a mesh bag over infected fruit while still on the tree
  b. knock the fruit off the tree onto the ground
  c. spray the fruit with gibberellins
  d. leave the fruit on the tree
  e. none of the above would induce sexual reproduction

TRUE-FALSE QUESTIONS

T _____ 1. Some fungi produce cells that have flagella, but the majority of fungi do not.

F _____ 2. The Kingdom Myceteae represents a natural, monophyletic group of organisms.

T _____ 3. Some fungi probably evolved from ancestors that were true algae.

F _____ 4. To classify a filamentous fungus with certainty, all you need are a few filaments because filament structure differs between subdivisions.

T _____ 5. Fungi are completely heterotrophic, with no exceptions.

F _____ 6. A plant that produces phytoalexins as a defense against pathogenic fungi is protected from attack by any type of fungus.

T _____ 7. Filaments produced by fungi in the Subdivision Mastigomycotina are multinucleate coenocytes.

T _____ 8. Water enters xerophilous fungi by the process of osmosis.

F _____ 9. While the cap of a mushroom is composed only of hyphae, the stalk contains tissues such as sclerenchyma for support.

F _____ 10. The flagella of mastigomycetes are composed of the protein flagellin.

T _____ 11. An alternative to pesticides for mosquito control would be to spread the chytrid *Coelomomyces*.

T _____ 12. The Subdivision Deuteromycotina is an artificial catagory and all members of this subdivision should belong to other subdivisions.

MATCHING

Match each characteristic in List A with a group or groups in List B. Answers in List B may be used more than once.

	List A	List B
i	1. Heterotrophic	a. myxomycetes
a	2. Engulf their food	b. oomycetes
g	3. Cell walls contain chitin	c. zygomycetes
f	4. Vegetative body may consist of coenocytic hyphae	d. ascomycetes
a	5. Produces a plasmodium	e. basidiomycetes
i	6. Reproduce sexually	f. b and c
d	7. May produce a perithecium	g. c, d and e
e	8. May produce a mushroom	h. b, c, d and e
h	9. Include forms that are economically important	i. a, b, c, d and e
c	10. Nutrients absorbed through rhizoids	
d	11. Commonly reproduce asexually by condia	
e	12. Clamp connections	

ESSAY QUESTIONS

1. The vegetative hyphae of zygomycetes and oomycetes are nonseptate. Why do these fungi form septa at the base of reproductive structures?

2. A friend brings you a log that contains fungal hyphae, but no reproductive structures. How could you determine what subdivisions of fungi are represented in the log?

3. You are studying a population of cells using a light microscope and do not see nuclei in the cells. Does that mean they are prokaryotic? Why or why not?

4. Compare and contrast fungal and plant cell mitosis.

5. What is the advantage to a fungus of producing upright conidiophores or fruiting bodies on stalks?

6. What are the similarities and differences between heterokaryosis and diploidy?

7. One of the only ways to control the disease wheat rust, which is caused by a basidiomycete, is through the production of resistant varieties of wheat. Unfortunately, it usually does not take long for the fungus to ovecome the wheat's resistance. Why does it take take such a short time?

8. What type of relationship exists between the components of a lichen? How do lichens reproduce? Do they reproduce sexually? Why or why not?

9. Distinguish between: heteroecious and autoecious; ascus and basidium; conidium and ascospore; lichen and mycorrhiza.

10. Concerning the life cycle of *Puccinia graminis* f. sp. *tritici*, describe the conditions necessary for the germination of each of the following: urediniospore, basidiospore, teliospore, aeciospore, spermatium.

11. What is a parasexual cycle in fungi?

12. Why are some fungi that are normally not pathogenic to humans beginning to cause problems in people?

13. In 1958, Steve McQueen made his screen debut in a film called The Blob. In that film, an organism that resembled a huge shapeless slime mold moved around devouring people. Could a slime mold like the large Blob actually exist? Why or why not?

CHAPTER 21: ALGAE AND THE ORIGIN OF EUKARYOTIC CELLS

MULTIPLE CHOICE QUESTIONS

c 1. Which type of general life cycle found in algae is basically the same type found in most animals?
   a. dibiontic and isomorphic
   b. dibiontic and heteromorphic
   c. monobiontic, where the diploid phase represents the individual
   d. monobiontic, where the haploid phase represents the individual
   e. there is no algal life cycle with an animal equivalent

e 2. Which of the following is an example of isogamy?
   a. gametes are both the same size and shape and both are motile
   b. two non-motile cell protoplasts that fuse
   c. gametes are both motile, one larger than the other
   d. one gamete is large and non-motile, the other is smaller and motile
   e. a and b

d 3. The Chlorophyta, as a group, shows remarkable diversity with respect to
   a. body construction
   b. habitat
   c. photosynthetic pigments
   d. a and b
   e. a, b and c

a 4. There are many different species in the Chlorophyta that are motile colonies. In all of those colonies, the cells are held together by
   a. a gelatinous matrix
   b. a middle lamella
   c. plasmodesmata
   d. a and b
   e. a, b and c

c 5. If all cells divide in two planes and do not separate after cytokinesis, the resulting organism is
   a. an unbranched filament
   b. a branching filament
   c. a sheet of cells one cell thick
   d. a sheet of cells two cells thick
   e. parenchymatous

b 6. The major characteristic used to identify species of diatoms is
   a. their shape
   b. the pattern of ridges, depressions, and pores in the cell wall
   c. their photosynthetic pigments
   d. their type of cell division
   e. the number and position of their flagella

c 7. Sexual reproduction is triggered in diatoms by
   a. hormones
   b. environmental chemicals
   c. cell size
   d. presence of a compatible mating type
   e. day length

b   8. The group of algae thought to be ancestral to land plants is the
       a. Charophyceae
       b. Chlorophyceae
       c. Euglenophyta
       d. Phaeophyta
       e. Rhodophyta

c   9. A characteristic of the Chrysophyta unifying its three classes is the presence of
       a. two anterior, unequal flagella
       b. silica in the cell wall
       c. chlorophylls a and c
       d. laminarin as a food storage product
       e. there are no unifying characteristics and the three classes should each be a separate division

d   10. When cell division occurs in a diatom
       a. the nuclear envelope persists
       b. the frustules separate, releasing a naked protoplast that divides; each daughter cell produces two new frustules
       c. one daughter cell uses the parent outer frustule as its outer frustule and one daughter cell uses the parent inner frustule as its inner frustule; the result is two cells of equal size
       d. both daughter cells use one parent frustule as their outer frustule; the result is two cells of unequal size
       e. c or d depending on the species

b   11. *Euglena*
       a. has cells that are a fixed, non-changing shape
       b. stores food as a unique carbohydrate, paramylon, outside the chloroplasts
       c. has two anterior flagella and uses both to swim
       d. has one cup-shaped chloroplast
       e. may divide by meiosis

e   12. Euglenoids and dinoflagellates are similar in that both
       a. have the same photosynthetic pigments
       b. have two anterior flagella
       c. store food reserves as oil
       d. reproduce sexually
       e. have a nuclear envelope and nucleolus that persist during mitosis

c   13. All dinoflagellates lack
       a. chloroplasts
       b. a cell wall
       c. histones
       d. microtubules and a spindle
       e. starch

d   14. Dinoflagellates are more dissimilar to other algae than other divisions. This is due to their
       a. photosynthetic pigments
       b. lack of sexual reproduction
       c. almost universal motility
       d. nuclear structure and cell division
       e. all of the above

e 15. Algae whose cytokinesis is most similar to your own cells are algae that produce
   a. a phragmoplast and a cleavage furrow
   b. a phragmoplast and a cell plate that grows radially outward
   c. a phycoplast and a cleavage furrow
   d. a phycoplast and a cell plate that grows radially inward
   e. a cleavage furrow only

a 16. You have discovered a new organism. Which would be the most important characteristic causing you to classify it in Kingdom Protista rather than Kingdom Plantae?
   a. reproductive structures that are completely converted to spores or gametes
   b. reproductive structures with an outer layer of sterile cells that do not become spores or gametes
   c. the presence of photosynthesis
   d. the presence of well-defined tissues
   e. none of the above

b 17. Many motile, photosynthetic organisms can perceive light gradients and swim toward or away from the light. The internal structure responsible for light detection in this process is a
   a. chloroplast
   b. eyespot
   c. mitochondrion
   d. liposome
   e. flagellum

e 18. Large brown algae called kelps often inhabit the intertidal zone. What structural feature(s) enable them to survive in this habitat?
   a. food reserves of laminarin
   b. a body that is firm and thick, yet elastic
   c. a holdfast
   d. intercalary meristems
   e. all of the above

c 19. In which division of algae does the most cell specialization occur?
   a. Rhodophyta
   b. Chlorophyta
   c. Phaeophyta
   d. Chrysophyta
   e. Pyrrophyta

b 20. The pigments of the red algae most closely resemble the pigments of
   a. true plants
   b. Cyanobacteria
   c. Pyrrophyta
   d. Phaeophyta
   e. none of the above

b 21. Eukaryotic organelles that are thought to have originated by endosymbiosis include
   a. the nucleus
   b. plastids
   c. endoplasmic reticulum
   d. 80s ribosomes
   e. 70s ribosomes

a 22. Green, red, and brown algae may have originated by separate endosymbiotic associations of chloroplasts with eukaryotic cells. Evidence for this includes
   a. pigments in the chloroplasts of each group
   b. membrane lipids associated with the chloroplasts of each group
   c. differences in associations of the chloroplasts with endoplasmic reticulum
   d. DNA sequences that clearly show this
   e. interactions of each group with parasites

c 23. Which of the following correctly pairs the algal taxon with its characteristic food storage molecule?
   a. green algae - oils
   b. euglenoids - laminarin
   c. red algae - floridean starch
   d. dinoflagellates - glycogen
   e. chrysophytes - paramylon

d 24. Chlorophyll c is found in
   a. euglenoids
   b. green algae
   c. red algae
   d. brown algae
   e. none of the above

c 25. Algae with isomorphic generations
   a. are unknown
   b. produce gametes by meiosis within the spore wall
   c. have morphologically indistinguishable diploid and haploid generations
   d. have morphologically different diploid and haploid generations
   e. b and c

e 26. Sulfated galactans are
   a. used as stabilizing agents for ice cream
   b. derived from the cell walls of red algae
   c. mucilages
   d. b and c
   e. a, b, and c

TRUE-FALSE QUESTIONS

T _____ 1. Unlike the diatoms, which have frustules, the golden-brown algae have silaceous scales that lie either on the cell surface or just beneath the plasma membrane.

F _____ 2. Red tides are an interesting scientific novelty, but have no effect on people.

T _____ 3. Cytokinesis in red algae resembles most other plants except in the direction of cell wall growth between the new daughter cells.

T _____ 4. The chromosomes of dinoflagellates retain some prokaryotic characteristics, such as a lack of histones.

T _____ 5. Most of the non-pigmented algae are euglenoids or dinoflagellates.

F _____ 6. Prokaryotic cells may divide by either mitosis or meiosis.

T _____ 7. More DNA is found per cell in eukaryotes than in prokaryotes.

F _____ 8. According to the endosymbiont theory, chloroplasts probably arose before mitochondria because chloroplasts produce the oxygen needed by mitochondria.

F _____ 9. All true algae contain chlorophyll a and are photosynthetic.

F _____ 10. Eukaryotic flagella, like chloroplasts, probably evolved several times in several separate evolutionary lines of organisms.

T _____ 11. Most motile cells can respond to external stimuli, such as light or chemical gradients.

T _____ 12. Photosynthetic euglenoids may also be heterotrophic.

T _____ 13. A common ancestor probably gave rise to both the Charophyceae and true plants.

F _____ 14. Most brown algae are multicellular, but a few are unicellular.

T _____ 15. Red algae lack plasmodesmata, but do have distinctive pit connections whose function is unknown.

MATCHING

Match each characteristic in List A with a group of organisms in List B. Answers in List B may be used more than once.

List A

f _____ 1. No motile cells produced
g _____ 2. Contain chlorophyll a
a _____ 3. Store food as paramylon
a _____ 4. Produce a pellicle
c _____ 5. Cell wall contains silica
e _____ 6. May produce air bladders
g _____ 7. Produce unicellular or multicellular sporangia where all cells are fertile
a _____ 8. Entirely unicellular
e _____ 9. Never unicellular
d _____ 10. Contain chlorophyll b and starch
f _____ 11. Life cycle with three generations
b _____ 12. Usually have two flagella, one flat and ribbon-like
f _____ 13. Cell wall contains sulfated galactans which are important commercially as stabilizers
b _____ 14. Cause red tides
d _____ 15. Ancestral to true plants

List B

a. Euglenophyta
b. Pyrrophyta
c. Bacillariophyceae
d. Chlorophyta
e. Phaeophyta
f. Rhodophyta
g. all of the above

ESSAY QUESTIONS

1. Starting with a motile, unicellular organism, explain the possible evolutionary origin of each body type found within the Chlorophyta.

2. What is the major characteristic used to separate true algae from true plants? Why is this characteristic significant? Would any alga be classified as a plant on this basis?

3. Draw diagrams that illustrate all types of monobiontic and dibiontic life cycles.

4. Using labeled diagrams, show the life cycle of *Ulothrix*; indicate where mitosis, meiosis, and fertilization occur.

5. Is it more likely that chloroplasts had a single origin or multiple origins? Defend your answer.

6. List ten characteristics of prokaryotes and then compare each to the situation in eukaryotes.

7. Distinguish between the autogenous and endosymbiont theories. Include evidence to support one of those theories.

8. *Euglena*, *Chlamydomonas*, and *Gonyaulax* look similar using low power on a light microscope, yet they are classified in three separate divisions. To which division does each belong? Why are they classified in separate divisions?

CHAPTER 22: NONVASCULAR PLANTS: MOSSES, LIVERWORTS, AND HORNWORTS

MULTIPLE CHOICE QUESTIONS

a 1. Nonvascular plants have probably not adapted and evolved as well as vascular plants because
   a. the dominant generation in their life cycle is haploid
   b. they contain no vascular tissue
   c. the gametophyte is usually small in size
   d. the gametophyte is perennial and photosynthetic
   e. their life cycle is dibiontic

c 2. The accompanying diagram is a mature
   a. microgametangium
   b. antheridium
   c. archegonium
   d. sporangium
   e. oogonium

b 3. One characteristic that nonvascular and vascular plants have in common is
   a. a dibiontic life cycle where the same generation is dominant
   b. the retention and initial development of the zygote within the megagametangium
   c. their size
   d. a and b
   e. a, b and c

d 4. Moss spores are released from a capsule when
   a. cells of the seta respond to humidity, twisting and turning to shake the spores out of the capsule
   b. cells of the operculum respond to humidity by changing shape, opening holes
   c. cells of the capsule wall respond to humidity by changing shape and splitting open the capsule wall
   d. cells of the peristome respond to humidity, bending the peristome teeth outward
   e. elaters uncoil, pushing the spores out

e 5. If you cut a cross-section through a moss stem, you might find
   a. cells that are fairly uniform from outer and inner areas
   b. slight differences between cell structure in outer and inner areas
   c. large differences in cell structure between outer and inner areas
   d. a and b
   e. a, b and c

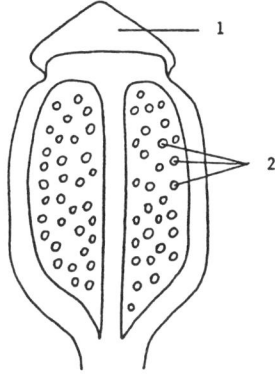

Questions 6 through 9 refer to the accompanying diagram.

c 6. The structure shown in the accompanying drawing is a
    a. liverwort sporangium
    b. liverwort antheridium
    c. moss sporangium
    d. moss antheridium
    e. hornwort sporangium

c 7. The structure labeled number 1 is(are)
    a. a columella
    b. spores
    c. an operculum
    d. a layer of sporogenous cells
    e. a peristome

a 8. The cells labeled number 2 are
    a. spores
    b. sporogenous cells
    c. part of the columella
    d. sperm cells
    e. egg cells

b 9. The cells labeled number 2 were produced by
    a. mitosis
    b. meiosis
    c. binary fission

d 10. The cells responsible for water and mineral transport in some mosses are called
    a. tracheids
    b. vessel elements
    c. sieve-tube elements
    d. hydroids
    e. leptoids

e 11. The cells that transport sugars in some mosses are similar to sieve-tube elements because
   a. they are elongate
   b. they lack nuclei at maturity
   c. they have prominent cytoplasmic connections with other transporting cells
   d. their metabolism may be controlled by adjacent cells
   e. all of the above

e 12. Examples of moss plants could be collected from
   a. between the bricks of a patio
   b. along roadsides
   c. the soil in pots in a greenhouse
   d. attached to rocks in a stream
   e. all of the above

c 13. Moss leaves are usually
   a. one cell thick with a cuticle on the upper surface and stomates along the midrib in the lower surface
   b. three cells thick
   c. one cell thick with a cuticle on the upper surface and no stomates
   d. one cell thick with a cuticle on both upper and lower surfaces and no stomates
   e. two cells thick with no cuticle and somates in both upper and lower surfaces

a 14. The gametophyte of leafy liverworts resembles the gametophyte of a moss because both have
   a. a stem with three rows of leaves
   b. pointed leaves with a midrib
   c. leaves with conducting tissue
   d. a stem with thick walled cells on the outside
   e. all of the above

b 15. The most common way that sperm cells travel from antheridia to archegonia in *Marchantia* is by
   a. swimming from one to the other
   b. being splashed by raindrops
   c. being carried by wind currents
   d. being carried by insects
   e. by the antheridiophore growing to the archegoniophore

c 16. The gametophyte of thallose liverworts has
   a. mucilage chambers that eventually develop into air chambers
   b. air chambers that contain the cyanobacterium *Nostoc*
   c. air chambers that open to the upper surface by large pores
   d. conducting cells
   e. all of the above

c 17. One of the major ways hornworts differ from liverworts is their
   a. general gametophyte shape
   b. alternation of generations
   c. lack of oil bodies in their cells
   d. dependence of the sporophyte on the gametophyte
   e. distinct stems and leaves

e 18. Sporophytes of mosses and hornworts are similar because both
   a. are permanently photosynthetic
   b. consist of a foot, seta, and sporangium
   c. produce sporangia containing elaters
   d. both produce spores over a long period due to a basal meristem
   e. none of the above

b 19. Hornworts almost universally form a symbiotic relationship with the cyanobacterium *Nostoc*. The probable advantage to the hornwort of this relationship is that they
   a. are protected from desiccation
   b. gain fixed nitrogen
   c. gain absorbed phosphorus
   d. receive sugars for growth
   e. receive hormones that permit the gametophyte to grow

e 20. Which of the following is an hypothesis concerning the origin of nonvascular plants?
   a. they all evolved from early vascular plants by becoming simpler and losing their vascular tissue
   b. they evolved from green algae, as did the vascular plants, but at a different time
   c. hornworts evolved from an early vascular plant, but mosses and liverworts did not
   d. mosses, liverworts, and hornworts all had separate evolutionary origins
   e. all of the above

e 21. Which of the following is a true moss?
   a. reindeer moss
   b. Spanish moss
   c. club moss
   d. pond moss
   e. none of the above

a 22. Rhizoids of mosses
   a. provide anchorage to the gametophore
   b. absorb water for the rest of the plant
   c. absorb minerals for the rest of the plant
   d. have chloroplasts
   e. all of the above

c 23. In the life cycle of a typical moss, the protonema
   a. covers the sporangiophore
   b. is the diploid phase of the gametophore
   c. is a branched system of filaments composed of long, slender cells
   d. produces antheridia and archegonia
   e. transport sperm cells to the egg

a 24. Cells of the calyptra are derived from the archegonium and are
   a. haploid
   b. diploid
   c. triploid
   d. similar in structure to leptoids
   e. a and d

b 25. According to the fossil record, mosses have been as morphologically complex as present-day mosses since the
    a. Devonian Period, about 380 million years ago
    b. Permian Period, about 250 million years ago
    c. Late Cretaceous Period, about 80 million years ago
    d. Eocene Epoch, about, 50 million years ago
    e. Pleistocine Epoch, about 1 million years ago

TRUE-FALSE QUESTIONS

F _____ 1. Scientists, in general, agree about the classification of mosses, liverworts, and hornworts.

T _____ 2. Although all mosses, liverworts, and hornworts lack true xylem and phloem, some do possess cells that transport water and sugars.

T _____ 3. In all nonvascular plants, the sporophyte must obtain minerals and sugars from the gametophyte.

T _____ 4. What you recognize and identify as a moss is the gametophyte.

F _____ 5. The classification system where mosses, liverworts, and hornworts are placed in separate divisions suggests that all of these plants have a common ancestor.

T _____ 6. It is possible that all of the gametophytes in a dense mound of moss plants are genetically identical.

F _____ 7. Moss sperm usually locate mature eggs by swimming randomly.

T _____ 8. A moss sporophyte is basically parasitic on its gametophyte.

F _____ 9. The spores in a liverwort capsule are produced by mitosis.

F _____ 10. Mosses are generally confined to habitats that are warm and moist.

T _____ 11. Stems of leafy liverworts are not homologous to vascular plant stems.

F _____ 12. Spores are usually released from sporangia when the air is humid or it is rainy because they will survive better under those conditons.

T _____ 13. A hornwort sporophyte can produce spores continuously due to its basal meristems, while mosses and liverworts cannot.

MATCHING

Match each characteristic in List A with a group or groups of plants in List B. Answers in List B may be used more than once.

List A

g _____ 1. Sporophyte dependent on gametophyte
b _____ 2. Leaves with two rounded lobes and no midrib
f _____ 3. Gametophyte may be flat and ribbonlike
c _____ 4. Sporophyte has a basal meristem
c _____ 5. Most susceptible to drying
a _____ 6. May contain leptoids
g _____ 7. Totally lack xylem and phloem
f _____ 8. Sporangium contains elaters
e _____ 9. Sporangium has a columella
c _____ 10. Archegonia greatly reduced
d _____ 11. Capsule often covered by a calyptra
a _____ 12. Spore release controlled by a peristome
g _____ 13. Produce a protonema
a _____ 14. Important in soil formation
d _____ 15. Many disc-shaped chloroplasts per cell

List B

a. mosses
b. liverworts
c. hornworts
d. mosses and liverworts
e. mosses and hornworts
f. liverworts and hornworts
g. mosses, liverworts, and hornworts

ESSAY QUESTIONS

1. Why have nonvascular plants stayed fairly small in size?

2. Most mosses do not retain water well. What adaptations have evolved to permit them to inhabit terrestrial areas?

3. In what way(s) do mosses enable vascular plants to become established in an area?

4. How do water and minerals move in most mosses? How fast does it occur?

5. Why do you think that the fingerlike cap of a mature *Marchantia* archegoniophore is at the top of a long stalk?

6. Compare sporophyte structure in mosses, liverworts, and hornworts.

7. When would be the best time of year to try to find hornworts? Why?

CHAPTER 23: VASCULAR PLANTS WITHOUT SEEDS

MULTIPLE CHOICE QUESTIONS

b 1. The major differnce between *Cooksonia* and *Rhynia* is the
   a. absence of leaves in *Rhynia*
   b. absence of rhizomes in *Cooksonia*
   c. presence of a cuticle in *Cooksonia*
   d. presence of a pith in *Rhynia*
   e. presence of terminal sporangia in *Rhynia*

e 2. You find a fossil of an early vascular plant and believe it belongs in either the Rhyniophyta or the Zosterophyllophyta. No sporangia are present. The most reliable characteristic to use would be the
   a. pattern of leaves on the stem
   b. presence or absence of a cuticle
   c. presence or absence of stomates
   d. pattern of stem branching
   e. position of protoxylem and metaxylem

a 3. A derived structural characteristic found in many present-day plants is
   a. xylem vessel elements
   b. xylem tracheids
   c. primary growth
   d. parenchyma tissue
   e. spores

d 4. Fossils of early vascular plants suggest that vascular plants can survive without
   a. leaves
   b. roots
   c. stems
   d. leaves and roots
   e. leaves and stems

a 5. The height many plants reach today probably evolved as a result of competition for
   a. light
   b. water
   c. space
   d. carbon dioxide
   e. minerals

c 6. The height that many plants exhibit was made possible principally by the evolution of
   a. xylem
   b. phloem
   c. xylem and phloem
   d. sclerenchyma
   e. collenchyma

d 7. Features present in living groups of plants that are similar to very early species are said to be
   a. primitive
   b. advanced
   c. original
   d. relictual
   e. derived

d 8. Structural characteristics have developed in vascular plants that are advantageous for coping with drought. These adaptations include
   a. the formation of a cuticle that covers all aerial plant parts
   b. a low surface-to-volume ratio
   c. rhizoids
   d. a and b
   e. a, b and c

c 9. The greatest advantage initally for the movement of plants into terrestrial habitats was the
   a. greater amount of light available
   b. the ability to be metabolically active during short dry periods
   c. lack of organisms that eat plants
   d. greater ease of reproduction
   e. greater availability of water

c 10. The most important structural characteristic(s) that suggests that the group ancestral to the ferns was the Rhyniophyta is the
   a. pattern of stem branching
   b. type of xylem secondary walls
   c. position of the sporangia
   d. a and b
   e. a, b and c

d 11. Although *Psilotum* and *Rhynia* are sometimes classified in the same division because of their similar structure, they are not exactly alike because *Rhynia* lacks
   a. horizontal rhizomes
   b. a protostele
   c. terminal sporangia
   d. enations
   e. upright, dichotomously branched stems

e 12. *Psilotum* is unique among living vascular plants because
   a. the gametophyte produces rhizoids
   b. the gametophyte is heterotrophic
   c. the sporophyte has an endarch siphonostele
   d. the sporophyte lacks stomates
   e. the gametophyte contains vascular tissue

c 13. Early lycophytes are thought to have evolved from
   a. *Psilotum*-like plants
   b. *Rhynia*-like plants
   c. *Zosterophyllum*-like plants
   d. *Cooksonia*-like plants
   e. the evolutionary origin is unknown

b 14. A necessary precondition for the evolution of seeds is
   a. homospory
   b. heterospory
   c. isogamy
   d. anisogamy
   e. b and d

d 15. The now extinct lycophytes *Lepidostrobus* and *Lepidophloios* share many characteristics with seed plants except
   a. heterospory
   b. megaspore development within the megaspore wall
   c. megaspore retention within the sporangia
   d. indehiscence of the sporangia
   e. all of the above

b 16. Circinate vernation in ferns is
   a. the dichotomous branching of vascular tissue in the leaves
   b. the developmental process that produces a fiddlehead
   c. a fiddlehead
   d. the developmental process that produces spores
   e. the development of compound leaves

b 17. A sorus is a
   a. group of sporangia on the upperside of a fern leaf
   b. group of sporangia on the underside of a fern leaf
   c. group of sporangia along the edge of a fern stem
   d. single sporangium on the underside of a fern leaf
   e. group of fern spores

a 18. The sequence of events summarized in the telome theory is
   a. totally dichotomous branching ---> overtopping ---> planation ---> webbing ---> megaphylls
   b. totally dichotomous branching ---> overtopping ---> planation ---> webbing ---> microphylls
   c. overtopping ---> totally dichotomous branching ---> planation ---> webbing ---> megaphylls
   d. totally dichotomous branching ---> planation ---> overtopping ---> webbing ---> microphylls
   e. totally dichotomous branching ---> overtoppping ---> webbing ---> planation ---> megaphylls

c 19. The most distinctive characteristic of the Arthrophyta is their
   a. epidermal structure
   b. homospory
   c. strobilus structure
   d. xylem structure
   e. gametophyte structure

a 20. Which group of characteristics best describes sporophytes of the Arthrophyta?
   a. jointed, herbaceous, aerial stems; whorled reduced megaphylls; subterranean rhizomes; roots
   b. jointed, woody, aerial stems; whorled reduced microphylls; subterranean rhizomes; roots
   c. dichotomous, herbaceous, aerial stems; alternate reduced megaphylls; roots
   d. herbaceous, aerial stems; opposite reduced megaphylls; subterranean rhizomes; rhizoids
   e. woody, aerial stems; whorled reduced megaphylls; subterranean rhizomes; rhizoids

e 21. The evolution of phloem permitted
   a. development of leaves in early plants
   b. development of deep roots in early plants
   c. development of rhizoids in early plants
   d. invasion of drier environments by early plants
   e. b and d

c 22. The evolution of seeds and pollen eliminated the requirement in those plants for
   a. protection of the gametophyte phase from predators
   b. production of sperm
   c. an environment wet enough for transport of sperm to the egg
   d. the gametophyte phase of the life cycle
   e. the sporophyte phase of the life cycle

a 23. According to the transformation hypothesis
   a. plants with isomorphic alternation of generations evolved by elaboration of the sporophyte phase and reduction of the gametophyte phase
   b. plants with isomorphic alternation of generations evolved by elaboration of the gametophyte phase and reduction of the sporophyte phase
   c. plants without a multicellular sporophyte evolved by elaboration of the zygote into the sporophyte by mitotic divisions
   d. plants without a multicellular gametophyte evolved by elaboration of the zygote into the gametophyte by mitotic divisions
   e. gametophytes were transformed into sporophytes, completing the life cycle

d 24. Homosporous plants produce spores that
   a. are all of the same size
   b. germinate into gametophytes bearing antheridia and archegonia
   c. germinate into gametophytes bearing either antheridia or archegonia, not both
   d. a and b
   e. a and c

d 25. Features in which lycophytes differ from rhyniophytes and zosterophyllophytes include the presence of _____ in lycophytes
   a. sporangia
   b. true leaves, stems and roots
   c. vascular tissues
   d. cuticle and stomata
   e. rhizoids

c 26. Cones (strobili) occur in
   a. ferns
   b. rhyniophytes
   c. arthrophytes
   d. *Psilotum*
   e. all homosporous vascular cryptogams

TRUE-FALSE QUESTIONS

F _____ 1. The fossil record indicates that seed-producing vascular plants definitely evolved according to the transformation theory.

F _____ 2. The earliest fossils of vascular plants are exclusively sporophytes because most have attached sporangia.

T _____ 3. A sterile jacket layer of cells in sporangia and gametangia probably evolved in response to dry periods.

T _____ 4. The advantage of a plant producing enations is increased surface area for photosynthesis.

T _____ 5. Some living lycophytes have roots, leaves, and a small amount of secondary growth.

F _____ 6. Microphylls are always small as their name implies.

F _____ 7. Microphylls have the same evolutionary origin as megaphylls.

T _____ 8. Lycophytes probably evolved from *Zosterophyllum*-like plants.

F _____ 9. The fact that lycophytes produce roots is one piece of evidence supporting the theory that they were ancestral to ferns and seed plants.

T _____ 10. The division of seedless vascular plants that contains the largest number of living species is the Pteridophyta.

F _____ 11. The trunk of a tree fern would be a good source of lumber.

F _____ 12. Because arthrophyte leaves are small with a single trace of vascular tissue, they are microphylls.

T _____ 13. At one time, some arthrophytes produced secondary tissue and were the size of trees.

MATCHING

Match each characteristic in List A with a leaf type in List B. Answers in List B may be used more than once.

List A

a _____ 1. May be produced by a plant with a protostele
b _____ 2. May be produced by a plant megaphylls
c _____ 3. Have a leaf trace
b _____ 4. Often associated with a leaf gap
c _____ 5. May be small in size
c _____ 6. May be large in size
a _____ 7. Originated as outgrowths from a stem
b _____ 8. Represent an entire branch system

List B

a. microphylls
b. megaphylls
c. microphylls and with a siphonostele

Match each characteristic in List A with a group of organisms in List B. Answers in List B may be used more than once.

List A

a _____ 1. Lack roots
c _____ 2. Produce sporangiophores
e _____ 3. May produce rhizomes
b _____ 4. Have microphylls
c _____ 5. Have silica in their epidermis
a _____ 6. Gametophyte never photosynthetic
e _____ 7. May produce bisexual gametophytes
d _____ 8. Exhibit circinate vernation

List B

a. Psilophyta
b. Lycophyta
c. Arthrophyta
d. Pteridophyta
e. all of the above

ESSAY QUESTIONS

1. Distinguish between the concepts of: primitive and relictual; advanced and derived.

2. Compare the evolutionary origin of moss leaves, microphylls, and megaphylls. Do microphylls or megaphylls more closely resemble moss leaves? Why?

3. A friend, knowing that you have taken a botany course, asks for advice about a "sick" fern that has brown spots all over the undersides of its leaves. What would be your advice?

4. Antheridia and archegonia produced by an individual fern gametophyte do not mature at the same time. What are the advantages of this timing? Disadvantages?

5. What is the difference between a genus and a form genus?

6. Which must develop first in an evolutionary line: an impervious cuticle or stomata and guard cells; vascular tissue or roots?

7. Distinguish between the transformation hypothesis and the interpolation hypothesis for the origin of life cycles in early terrestrial plants.

CHAPTER 24: SEED PLANTS I: GYMNOSPERMS

MULTIPLE CHOICE QUESTIONS

a 1. Which of the following is the <u>same</u> in heterosporous seedless vs seed-producing vascular plants?
   a. basic type of life cycle
   b. place of development of the megagametophyte
   c. place of development of the microgametophyte
   d. source of nourishment for the embryo
   e. all of the above

d 2. The group directly ancestral to the Division Coniferophyta was the
   a. cycads
   b. ginkos
   c. gnetophytes
   d. progymnosperms
   e. trimerophytes

c 3. Fossil material of progymnosperms, a now extinct group, <u>never</u> contains
   a. megaphylls
   b. two kinds of spores
   c. seeds
   d. secondary xylem and phloem
   e. cork cambia

d 4. The major function(s) of cupules in fossil plants was to
   a. protect the megasporangia
   b. increase the possibility of pollination by creating favorable wind patterns
   c. nourish the developing embryo
   d. a and b
   e. a, b and c

a 5. Pycnoxylic wood is strong because is has few or little
   a. parenchyma
   b. rays
   c. tracheids
   d. ray tracheids
   e. secondary xylem

e 6. A major way pine differs from other conifers is in the
   a. structure of its stems
   b. structure of its seed cones
   c. lack of seed cones
   d. presence of a cork cambium
   e. production of two types of leaves

d 7. Adaptations found in pine needles that would help the plants survive a dry climate or period include
   a. sunken stomates
   b. cylindrical shape
   c. long length
   d. a and b
   e. a, b and c

a 8. The function of a suspensor is to
   a. push an embryo deeper into its food source, the megagametophyte
   b. nourish a developing embryo by absorbing food
   c. provide cells that develop into an embryo
   d. connect a developing ovule or seed to the parent sporophyte
   e. store food reserves for a developing embryo

d 9. Wood of *Archaeopteris* contained
   a. tracheids
   b. rays
   c. resin canals
   d. a and b
   e. a, b and c

c 10. A common ornamental shrub is yew (*Taxus*), belonging to the order Taxales. The principal way this order differs from the Coniferales is the
   a. absence of secondary growth
   b. structure of the leaves
   c. lack of seed cones
   d. a and b
   e. a, b and c

d 11. Seed fern wood was softer than progymnosperm wood because
   a. its rays were not uniseriate
   b. its tracheids were longer and wider than progymnosperm tracheids
   c. it contained a lot of axial parenchyma
   d. a and b
   e. a, b and c

e 12. Most support in cycads is provided by
   a. collenchyma
   b. pycnoxylic wood
   c. manoxylic wood
   d. turgor pressure in the abundant parenchyma
   e. persistent leaf bases

a 13. Which of the following statements is not correct?
   a. cycads produce large, frond-like simple leaves
   b. cycads differ from ferns because they produce seeds
   c. cycads differ from palms because they lack flowers
   d. cycad megasporophylls evolved from leaves
   e. cycads produce large multiflagellated sperm cells

c 14. Cycadeoids produced
   a. separate seed and pollen cones on the same plant
   b. separate seed and pollen cones on different plants
   c. cones containing both microsporophylls and megasporophylls
   d. seed cones only; pollen was not produced in cones
   e. pollen cones only; seeds were not produced in cones

e 15. Ginkos look remarkably like dicots, but are not because they have
   a. unprotected ovules
   b. dichotomously branched veins in their leaves
   c. pycnoxylic wood
   d. a and b
   e. a, b and c

b 16. Which characteristic of *Gnetum* is least like angiosperms? Its
   a. leaves
   b. reproductive structures
   c. basic structure
   d. tracheary elements
   e. wood

e 17. Paper recycling
   a. saves trees
   b. reduces acid rain
   c. reduces use of bisulfites and their release
   d. reduces use of sulfurous acid
   e. all of the above

b 18. You have just discovered a new living plant and upon examination it has the following characteristics: secondary growth, simple leaves, compound seed cones, and compound pollen cones. This plant should be classified in the division
   a. Cycadophyta
   b. Gnetophyta
   c. Pteridospermophyta
   d. Ginkophyta
   e. Coniferophyta

a 19. What is the nutritive tissue in a conifer seed?
   a. female gametophyte
   b. male gametophyte
   c. integuments
   d. endosperm produced by union of polar nuclei and a sperm cell
   e. cotyledons

c 20. The ancestors of the ginkos were one of the
   a. Taxales
   b. Coniferales
   c. Pteridospermophyta
   d. Gnetophyta
   e. the origin is completely unknown

e 21. In the evolution of the seed
   a. telomes surrounded the microsporagium to protect it
   b. telomes surrounded the megasporangium then fused together to form the integument
   c. the megasporangium became free-sporing to form the seed
   d. a pollen chamber developed as a holding area for pollen
   e. b and d

d 22. Conifers are
   a. vines
   b. herbs
   c. annuals
   d. shrubs or trees
   e. all of the above

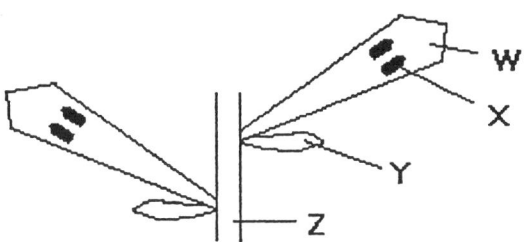

e 23. The accompanying diagram shows the parts of a conifer cone. Which of the following correctly pairs the name of the structure with the letter labeling it?
   a. X - bract
   b. Y - seed
   c. Y - ovuliferous scale
   d. Z - bract
   e. W - ovuliferous scale

b 24. The earliest conifers are known from the
   a. Devonian Period
   b. Carboniferous Period
   c. Cretaceous Period
   d. Eocene Epoch
   e. Pliocene Epoch

d 25. The seeds of yew (*Taxus*) are surrounded by
   a. a protective whorl of leaves
   b. a dense sclerenchymatous husk
   c. a protective whorl of branches
   d. a fleshy, red aril
   e. a whorl of sterile pollen cones

a 26. Cycadeoids and cycads differed from each other in that
   a. the stomata of each group is different
   b. cycad cones have both ovules and pollen organs
   c. the leaves looked very different
   d. cycadeoids were tall, woody trees
   e. all of the above

c 27. Vessels are found in the wood of
   a. conifers
   b. seed ferns
   c. gnetophytes
   d. cycadeoids
   e. cycads

TRUE-FALSE QUESTIONS

T _____ 1. The term gymnosperm literally means naked ovule or seed.

F _____ 2. If you see a tree with no leaves in the winter, it is definitely not a conifer.

F _____ 3. The wood of pine contains only tracheids which are of uniform diameter.

T _____ 4. A pine pollen tube is parasitic on the megasporangium.

T _____ 5. The only difference between embryo structure in conifers and flowering plants is the number of cotyledons.

T _____ 6. The main function of an aril is to attract animals that will eventually distribute the seeds.

T _____ 7. Without any attached reproductive structures, it would be difficult to tell a fossil seed fern leaf and fossil true fern leaf apart.

F _____ 8. The stem of a cycad contains a protostele.

F _____ 9. Cycad seed cones and pollen cones may be found on the same plant.

F _____ 10. Cycadeoids are commonly used as ornamentals in states such as Florida.

T _____ 11. The main structural characteristic that groups *Gnetum*, *Ephedra*, and *Welwitschia* together in the same division is their reproductive structures.

T _____ 12. Adaptations of pine needles suggest that pines may have evolved in a habitat short of water all or part of the year.

MATCHING

Match each characteristic in List A with a group of plants in List B. Answers in List B may be used more than once.

List A

h _____ 1. Heterosporous
h _____ 2. Produce secondary growth
g _____ 3. Vessels
f _____ 4. Simple, dichotomously veined leaves
a _____ 5. Lack seeds
b _____ 6. Resin canals
e _____ 7. Bisexual cones
b _____ 8. Most important economically
h _____ 9. Leaves megaphylls
g _____ 10. Male cones compound
c _____ 11. Seeds produced on the foliage leaves
a _____ 12. Protostele

List B

a. Progymnospermophyta
b. Coniferophyta
c. Pteridospermophyta
d. Cycadophyta
e. Cycadeoidophyta
f. Ginkophyta
g. Gnetophyta
h. all of the above

ESSAY QUESTIONS

1. Distinguish between pollination and fertilization.

2. Starting with the rhyniophytes, diagram the evolution of all of the divisions of gymnosperms.

3. How do cycads differ from seed ferns?

4. Both ginkos and taxads produce single seeds that are not in a seed cone. How could you tell the difference between them?

5. The Gnetophyta exhibit several features which are characteristic of the Anthophyta rather than gymnosperms. What are those characteristics? Why are gnetophytes considered gymnosperms?

6. You're walking in the park and pass several female ginko trees covered with mature reproductive structures. Your friend remarks that the ginko trees have produced a lot of fruit. Would you agree or disagree with your friend's statement? Defend your answer.

7. Compare the advantages and disadvantages of two life cycles: one in which alternate generations are completely independent of the preceding generation and one where they are not.

8. The description of aneurophytes and trimerophytes are very similar. Why are they placed in separate divisions?

9. Describe the evolution of the ovule integument and the selective advantage of its development.

10. How could you distinguish between fossil microspores and pollen grains?

11. In what ways do cycads differ from cycadeoids?

12. Describe and give the effect on the wood of each treatment used in paper making.

a 8. In most situations where you do not have a microscope, you could recognize a plant as a monocot if the plant had
   a. parallel venation in the leaves
   b. reticulate venation in the leaves
   c. flowers with five petals
   d. vascular bundles found throughout the stem
   e. secondary growth

d 9. The direct ancestors of monocots very likely
   a. were woody
   b. did not produce flowers
   c. had a gynoecium composed of many fused carpels
   d. had flowers with separate, not fused, petals and sepals
   e. had bilateral flowers

e 10. The line of evolution in the monocots accepted by a majority of taxonomists is
   a. Alismatidae ---> Arecidae ---> Commelinidae ---> Zingiberidae ---> Liliidae
   b. Alismatidae ---> Commelinidae ---> Zingiberidae ---> Liliidae ---> Arecidae
   c. Arecidae ---> Alismatidae ---> Zingiberidae ---> Liliidae ---> Commelinidae
   d. Liliidae ---> Alismatidae ---> Commelinidae ---> Arecidae ---> Zingiberidae
   e. none of these subclasses was ancestral to any other

b 11. Which subclass of Liliopsida has the greatest number of relictual features?
   a. Liliidae
   b. Alismatidae
   c. Zingiberidae
   d. Commelinidae
   e. Arecidae

a 12. Grasses are
   a. wind pollinated, so they have a very reduced perianth
   b. bee pollinated, so they have a very reduced perianth
   c. beetle pollinated, so they have showy petals
   d. butterfly pollinated, so they have showy petals and sepals
   e. moth pollinated, so their flowers are open mainly at night

d 13. A flower characteristic that is quite derived, a product of considerable evolutionary modification, is
   a. petals fused into a tube
   b. stamens fused to the petals
   c. sepals very small
   d. an inferior gynoecium
   e. a compound gynoecium

c 14. The Orchidaceae supply us with
   a. cinnamon
   b. nutmeg
   c. vanilla
   d. sage
   e. tarragon

e 15. Why are members of the subclass Magnoliidae thought to resemble early angiosperms?
   a. the lack of xylem vessels
   b. numerous separate carpels
   c. uniaperaturate pollen grains
   d. spirally arranged stamens
   e. all of the above

e 16. You encounter a plant that is the size of a small tree, has a single sturdy trunk, and has a crown of pinnately compound leaves, but no reproductive structures are present. Without further investigation, this plant could belong to the
   a. Magnoliophyta, Sublcass Arecidae
   b. Pteridophyta
   c. Cycadophyta
   d. a and c
   e. a, b and c

b 17. Flowers in the Family Araceae are embedded in a
   a. spathe
   b. spadix
   c. spadon
   d. spandex
   e. spatula

c 18. Almost all civilizations have depended primarily on which subclass of Magnoliophyta for food?
   a. Arecidae
   b. Magnoliidae
   c. Commelinidae
   d. Rosidae
   e. Dilleniidae

d 19. A unifying characteristic of the Caryophyllidae is
   a. the shape of their petals
   b. the pattern in vessel element secondary walls
   c. their leaf shape and size
   d. the presence of perisperm in the seeds
   e. the type of pollinators

b 20. Economically, the Asteridae is extremely important because some plants in the subclass produce
   a. morphine
   b. potatoes
   c. corn
   d. apples
   e. spinach

d 21. Which of the following is unique to the angiosperms?
   a. vessels in the wood
   b. deciduous leaves
   c. seed- and pollen-producing structures in the same reproductive organ
   d. double fertilization leading to an embryo and endosperm
   e. herbaceous habit

a 22. C.E. Bessey developed a hypothesis of generalized, ancestral flowers. Bessey's hypothetical flowers had
   a. sepals, petals, stamens, and carpels
   b. carpels in an inferior position
   c. either stamens or carpels but not both
   d. either sepals or petals but not both
   e. no petals

c 23. The earliest definite angiosperm leaf fossils are from the
   a. Eocene Epoch and are common
   b. Pleistocene and are rare
   c. Lower Cretaceous and are rare
   d. Carboniferous Period and are rare
   e. Devonian Period and are common

a 24. Which of the following plants is a member of the most derived group of monocots?
   a. an orchid
   b. *Dieffenbachia*
   c. ginger
   d. rice
   e. date palm

c 25. A pigment uniquely characteristic of the Caryophyllaceae is
   a. anthocyanin
   b. chlorophyll b
   c. betalain
   d. B-carotene
   e. xanthophyll

b 26. A specialized character of the Hamamelidae is
   a. fly pollination
   b. wind pollination
   c. perisperm
   d. an aquatic habit
   e. iridoid compounds

e 27. The largest family of dicots is
   a. Caryophyllaceae
   b. Solanaceae
   c. Scrophulariaceae
   d. Hamamelidaceae
   e. Asteraceae

TRUE-FALSE QUESTIONS

F _____ 1. For any specific flower, all of the features must be either relictual or derived, not a mixture.

T _____ 2. Complete, radially symmetrical flowers are considered more relictual than incomplete, bilateral flowers.

F _____ 3. Most characteristics of a plant evolve at the same rate.

T _____ 4. If a flowering plant is woody, it must be a dicot.

F _____ 5. If a plant is evergreen, is must be a gymnosperm.

T _____ 6. Many angiosperms produce chemicals that are useful medicinally.

F _____ 7. The fact that Hammelidae is a fairly restricted subclass suggests that this group was never large and did not diversify well.

F _____ 8. While taxonomists do not all agree about the classification of many groups of plants, they do agree about classification within the Magnoliophyta.

F _____ 9. In the Rosidae, simple leaves are a relictual characteristic.

T _____ 10. The largest family of dicots is the composites.

T _____ 11. A species with bilateral flowers is usually pollinated by one specific type of pollinator.

MATCHING

Match each plant or plant product in List A with a subclass in List B. Answers in List B may be used more than once.

	List A		List B
i	1. Sunflower seeds	a.	Arecidae
a	2. Coconut	b.	Commelinidae
f	3. Beets	c.	Zingiberidae
i	4. Tomato	d.	Liliidae
g	5. Watermelon	e.	Magnoliidae
e	6. Curare, a muscle relaxant	f.	Caryophyllidae
d	7. Onion	g.	Dilleniidae
b	8. Wheat	h.	Rosidae
i	9. Vincristine, an anticancer drug	i.	Asteridae
h	10. Rubber		
b	11. Sugar cane		
h	12. Beans		

Match each plant or plant product product in List A with a family in List B.

	List A		List B
d	1. Vanilla	a.	Araceae
e	2. Avocado	b.	Poaceae
k	3. Coffee	c.	Bromeliaceae
b	4. Bamboo	d.	Orchidaceae
g	5. Tea	e.	Lauraceae
a	6. Houseplant *Philodendron*	f.	Papaveraceae
c	7. Pineapple	g.	Theaceae
h	8. Chocolate	h.	Sterculiaceae
i	9. Grapes	i.	Vitaceae
f	10. Morphine, a strong analgesic	j.	Rosaceae
j	11. Cherry	k.	Rubiaceae
l	12. Digitalis, a heart medication	l.	Scrophulariaceae

ESSAY QUESTIONS

1. What characteristics did early flowers most likely have? In what ways have they evolved since then?

2. Explain how parallel venation in monocot leaves might have evolved.

3. In some members of the Alismatidae, flowers lack a perianth completely. Under what conditions does this occur most often? What is the advantage to the plant? Why can the flowers function without a perianth?

4. What do you think pollinates many epiphytic members of the Order Bromeliales? Why?

5. Several members of the subclass Magnoliidae produce compounds that we use medicinally. What compounds are produced? Of what advantage are they to the plants?

6. Name four drugs produced by members of the Asteridae and explain how each is used medicinally.

7. Discuss the role of germplasm banks in maintaining genetic diversity of plants.

8. Describe the floral characteristics of wind pollinated angiosperms. Why are these adaptations important in the reproduction of these plants?

CHAPTER 26: POPULATIONS AND ECOSYSTEMS

MULTIPLE CHOICE QUESTIONS

b   1. Which best illustrates a population?
   a. a single plant in the eastern United States
   b. all of the sugar maple trees in the eastern United States
   c. all of the poison ivy and spring beauty plants in the eastern United States
   d. all of the bloodroot plants and the physical environment in the eastern United States
   e. the individuals of all species + the physical environment in the eastern United States

c   2. If you dig the soil to plant a garden in Columbus, Ohio, you often hit yellow clay, which is part of the
   a. A horizon
   b. zone of rock fragments
   c. zone of deposition
   d. C horizon
   e. D horizon

d   3. A major problem for organisms found only in high altitude habitats is
   a. high winds
   b. poor soil
   c. a narrow temperature range
   d. intense ultraviolet radiation
   e. a short growing season

a   4. Certain populations of *Agrostis tenuis*, a grass, grow successfully on mine tailings enriched in either copper or lead, heavy metals which are normally toxic to plants in high concentrations. These populations are known as
   a. ecotypes
   b. subspecies
   c. species
   d. varieties
   e. races

c   5. Many birds eat berries and deposit the seeds (plus a small amount of fertilzer) at a site removed from the parent plant. This is an example of
   a. interspecific competition
   b. intraspecific competition
   c. mutualism
   d. parasitism
   e. commensalism

d   6. Tulips and daffodils overwinter as bulbs below the soil surface, so they would be catagorized as
   a. phanerophytes
   b. hemicryptophytes
   c. therophytes
   d. geophytes
   e. chamaephytes

CHAPTER 25: SEED PLANTS II: ANGIOSPERMS

MULTIPLE CHOICE QUESTIONS

d 1. The most recently evolved group of plants are the
   a. ferns
   b. conifers
   c. gnetophytes
   d. flowering plants
   e. cycads

d 2. Probably the most important advantage of carpels is that
   a. the ovules are enclosed by tissue of the sporophyte
   b. the ovules are protected from other organisms
   c. the ovules do not dry out as easily as 'naked' ovules
   d. fewer ovules are wasted because the sporophyte can test pollen
   e. the ovules are distributed better when mature

c 3. What is the most probable order of evolution?
   a. sieve tubes ---> vessels ---> flowers ---> herbaceousness
   b. flowers ---> herbaceousness ---> sieve tubes ---> vessels
   c. flowers ---> sieve tubes ---> vessels ---> herbaceousness
   d. herbaceousness ---> flowers ---> sieve tubes ---> vessels
   e. vessels ---> flowers ---> sieve tubes ---> herbaceousness

a 4. Which of the following is more likely to be a derived feature in angiosperms?
   a. deciduous leaves
   b. being a small tree
   c. little axial parenchyma
   d. tracheary elements tracheids
   e. all rays narrow and tall

e 5. Which characteristic of flowers is considered to be the most relictual?
   a. complete flowers
   b. parts arranged in spirals
   c. superior ovary
   d. no fusion of parts
   e. all of these features are relictual and you cannot rank them

a 6. The angiosperms are most probably the result of how many lines of evolution from the gymnosperms?
   a. one
   b. two
   c. three
   d. four
   e. many

c 7. The focus is on what group as possible direct ancestor of the flowering plants?
   a. pteridospermophytes
   b. conifers
   c. cycadophytes
   d. progymnosperms
   e. gnetophytes

e   7. The 'gateway' by which energy enters the organisms in an ecosystem is through the
- a. herbivores
- b. primary consumers
- c. decomposers
- d. carnivores
- e. autotrophs

c   8. You would expect which of the following plants to be an r-selected species?
- a. oak trees in a mature forest
- b. understory trees in a tropical rainforest
- c. desert herbs that grow quickly and flower only when water is available
- d. an avocado plant, which produces a few, very large seeds
- e. the creosote bush, whose leaf litter releases chemicals which are toxic to other plants

d   9. The term 'trophic level magnification' refers to an increase in the concentration of chemicals from plants to herbivores to carnivores. This occurs primarily because
- a. the chemicals are not metabolized well
- b. the chemicals are not excreted well
- c. biomass increases
- d. a and b
- e. a, b and c

a   10. Which two types of organisms are absolutely vital to ecosystems?
- a. autotrophs and decomposers
- b. autotrophs and herbivores
- c. herbivores and carnivores
- d. herbivores and decomposers
- e. carnivores and decomposers

e   11. The limiting factor for cacti living in a desert would be
- a. available light
- b. available CO_2
- c. available minerals
- d. high temperatures
- e. available water

b   12. Eastern red cedar grows best on less acidic soils, which are commonly found on limestone outcrops in its range. The type of geographic distribution of this plant would be
- a. sparse
- b. clumped
- c. random
- d. uniform
- e. allelopathic

a   13. The carrying capacity for a plant population would not be influenced by
- a. periodic floods
- b. the concentration of minerals in the soil
- c. the total annual rainfall
- d. the number of herbivores in the area
- e. the amount of space available for growth

a 14. The habitat of a plant species would include all of the following except
   a. a person who picks some of the species' flowers once
   b. nearby plants of other species
   c. the composition of the soil
   d. insects that lay their eggs in the plant's leaves
   e. length of the growing season

b 15. An abiotic component(s) of a plant's habitat would be
   a. insect pollinators
   b. a fungal pathogen
   c. other plants of the same species
   d. temperature extremes
   e. a, b and c

e 16. If two species, A and B, occupy the same ecological niche in an area, what are possible outcomes?
   a. Species A is better adapted and Species B is ultimately excluded from the ecosystem
   b. both species diverge from their original niches, reducing competition
   c. Species B diverges from its original niche reducing competition
   d. b and c
   e. a, b and c

b 17. An example of commensalism is
   a. a mycorrhizal relationship
   b. epiphytic orchids in a tree
   c. mistletoe plants in a tree
   d. the pollination of a flower by a hummingbird
   e. wheat rust

e 18. The boundaries of a plant species' geographic range might be determined by
   a. day length
   b. temperature extremes during the year
   c. yearly precipitation pattern
   d. existence of seed dispersing animals
   e. all of the above

b 19. Which of the following populations exhibit the fastest growth in an unlimited environment?
   a. a species with a generation time of one year and each individual produces ten seeds
   b. a species with a generation time of one year and each individual produces one hundred seeds
   c. a species with a generation time of two years and each individual produces one hundred seeds
   d. a species with a generation time of two years and each individual produces fifty seeds
   e. a species with a generation time of ten years and each individual produces one hundred seeds

c 20. A volcanic eruption would not
   a. alter the soil in the area
   b. change the species of plants living in the area
   c. affect yearly temperature patterns in the area
   d. affect the amount of water available to plants in the area
   e. affect the amount of wind plants in the area would be exposed to

e 21. The niche of a forest wildflower would include
  a. soil moisture
  b. the annual temperature cycle
  c. insect pollinators
  d. a and b
  e. a, b and c

a 22. Cacti are restricted to desert regions primarily because of the
  a. climate
  b. soil type
  c. flash floods
  d. animal species present
  e. fungal species present

c 23. An example of mutualism is
  a. a bird's nest in a tree
  b. the HIV virus in humans
  c. species of nitrogen-fixing *Rhizobium* living in legume root nodules
  d. the growth of *Phytophthora infestans*, which causes late blight of potato, on potato plants
  e. barnacles on a whale

b 24. Individuals in a desert are often evenly spaced from their neighbors. This would not be caused by
  a. competition between roots for water
  b. seed dispersal
  c. germination inhibitors released by the plants
  d. growth inhibitors released by the plants
  e. herbivores living in the plants

d 25. r conditions would not be produced by
  a. a high wind blowing down a large tree in a forest
  b. a drought
  c. the abandonment of a field by a farmer
  d. the closing of the tree canopy in late spring or early summer
  e. a volcanic eruption that produces a new island

a 26. If a few seeds of a species blow into an area, the population will at first increase exponentially and then increase further, but not exponentially, until it reaches its carrying capacity. What will be the relationship between birth rate (germination) and death rate as time progresses?
  a. birth rate >>> death rate ---> birth rate > death rate ---> birth rate = death rate
  b. birth rate = death rate ---> birth rate >>> death rate ---> birth rate > death rate ---> birth rate = death rate
  c. birth rate > death rate ---> birth rate >>> death rate ---> birth rate = death rate
  d. birth rate >>> death rate indefinately
  e. there is no way to predict how birth rate and death rate will compare

TRUE-FALSE QUESTIONS

T _____ 1. Ecology is the study of organisms in relation to their physical environnment and other organisms.

F _____ 2. It is always advantageous for species of plants to live in populations.

T _____ 3. Two species of plants that receive the same amount of precipitation live in different habitats if the yearly pattern of that precipitation is different.

T _____ 4. Most species of temperate plants have a broader tolerance range for temperature than many tropical plants because they are exposed to a wider range of temperatures.

T _____ 5. Plants in tropical areas or deserts that survive the dry period during the year as seeds are therophytes.

F _____ 6. Tropical plant species that are mycorrhizal are more probably r-selected than K-selected.

F _____ 7. If a plant needs 1 unit of iron, 3 units of phosphorus, and 5 units of calcium to survive, then the growth of a population of that plant in an area that contains 20 units of iron, 50 units of phosphorus, and 100 units of calcium would be limited by the amount of iron available.

T _____ 8. While all populations have the potential for exponential growth, the actual rate of increase depends, in part, on generation time.

F _____ 9. A plant's habitat would not include mycorrhizal fungi in the soil.

T _____ 10. Those aspects of the habitat that definitely affect a plant constitute its operational habitat.

F _____ 11. In tropical ecosystems, the temporal structure is constant because there are no yearly temperature changes.

T _____ 12. Ecosystems need a continual input of energy because much of the energy in each trophic level is lost as heat.

F _____ 13. Limiting factors are always nutritional.

F _____ 14. Populations can grow exponentially indefinitely under natural conditions.

MATCHING

Match each example in List A with one type of relationship in List B. Answers in List B may be used more than once.

List A

a _____ 1. Ants living on acacias
e _____ 2. Late blight of potato, a fungal disease of potato
b _____ 3. Epiphytic bromeliads on a tree
c _____ 4. Crowded maple seedlings on the forest floor
d _____ 5. Mealybugs that suck material out of the phloem of a plant
c _____ 6. Two species of plants that produce flowers of the same size, shape, and color at the same time of year
b _____ 7. Algae living on turtle shells
d _____ 8. Rabbits eating carrots in a garden
a _____ 9. Yucca moths lay their eggs in the gynoecium of the yucca flowers and also pollinate them

List B

a. mutualism
b. commensalism
c. competition
d. predation
e. parasitism

ESSAY QUESTIONS

1. Relate what you know about the effects of fire on ecosystems and the debate about putting out fires in Yellowstone National park.

2. One element often lost from grassland ecosystems during fires is nitrogen. What group of flowering plants (besides grasses) would you expect to find in grasslands? Why?

3. If two species occupy exactly the same ecological niche in an ecosystem, what are the possible outcomes of this situation? Why?

4. Distinguish between generation time and biotic potential. Describe how each affects population growth.

5. Diagram the typical growth curve exhibited by most species when they invade a new habitat. Explain what is happening in each part of the growth curve.

6. In lakes, excessive amounts of algae are usually caused by excess nutrients in the water. Often the limiting factor for growth of the algae is the amount of either nitrogen or phosphorus available to those organisms. To control growth, some communities try to eliminate as much as possible the amount of a specific element entering the lake, usually the element that is the limiting factor. Which would be easier to control- nitrogen or phosphorus? Why.

7. Draw a pyramid of energy and explain why it has that particular shape.

8. Sycamores and oaks often grow in the same general area, but oaks tend to be restricted to well drained soils on hills, while sycamores are mainly found on wetter soils on floodplains. Each type of tree is capable of growing on either well drained or wet soil. Why do these two species exhibit this distribution pattern?

9. In 1859, a farmer in Australia imported 12 pairs of rabbits from England and released them for hunting. What do you think happened to the rabbit population in Australia? To native species?

CHAPTER 27: BIOMES

MULTIPLE CHOICE QUESTIONS

a 1. The North Pole points as directly toward the sun as possible at the
    a. summer solstice
    b. autumnal equinox
    c. winter solstice
    d. vernal equinox
    e. the orientation of the North Pole does not change during the year

b 2. The least amount of seasonal change occurs
    a. at the North Pole
    b. at the equator
    c. at the South Pole
    d. between 23.5° N and the North Pole
    e. between 23.5° S and the South Pole

b 3. Which present day continent was not part of Gondwanaland?
    a. Africa
    b. Asia
    c. South America
    d. Australia
    e. Antarctica

c 4. As North America collided with Eurasia, forming the Appalachian Mountains, what group of plants was first evolving?
    a. Aneurophytales
    b. Trimerophytophyta
    c. Rhyniophyta
    d. Archaeopteridales
    e. Medullosales

d 5. Conifers evolved
    a. as Gondwanaland formed and the climate became cooler and drier
    b. as Gondwanaland collided with Laurasia and climate became hotter and drier
    c. as Pangaea formed with its diverse climate
    d. as North America and Eurasia moved northward and climate became cooler and moister
    e. during the Triassic, when the climate was dry

e 6. Which continent has had the shortest period of time for plants to evolve due to its position in Pangaea?
    a. Australia
    b. Europe
    c. Asia
    d. South America
    e. Africa

d 7. In the past, North America has been part of
    a. Gondwanaland
    b. Gondwanaland and Laurasia
    c. Pangaea
    d. Pangaea and Laurasia
    e. none of the above; it has always been separate

a 8. The Gulf of Mexico coast is humid in the summer mainly due to
   a. westerly equatorial ocean currents
   b. easterly equatorial ocean currents
   c. prevailing westerlies
   d. prevailing easterlies
   e. a and c

e 9. What is the main similarity between temperate and tropical rainforests?
   a. the yearly temperature pattern
   b. the dominant trees
   c. species diversity
   d. daily fog
   e. a large amount of precipitation each year

c 10. The largest organisms alive today are
   a. blue whales
   b. Douglas fir trees
   c. sequoias
   d. western hemlock trees
   e. eucalyptus trees

b 11. The United States contains a wide variety of biomes, but lacks
   a. tropical rainforest
   b. tropical savannah
   c. subalpine forest
   d. boreal coniferous forest
   e. desert

d 12. The biome with the greatest species diversity is
   a. temperate deciduous forest
   b. boreal coniferous forest
   c. tropical grassland
   d. tropical rainforest
   e. temperate rainforest

a 13. The most distinctive feature(s) of a biome that allows you to quickly recognize each type is(are)
   a. the dominant plants
   b. the dominant animals
   c. the climate
   d. a and b
   e. a, b and c

e 14. Why did pioneers traveling west in the 1800s in the United States find prairie soils especially rich for farming?
   a. the burrowing of small mammals
   b. the eating habits of large mammals
   c. enough, but not too much, rainfall
   d. the lack of trees
   e. all of the above

b 15. The country containing the widest variety of biomes is
   a. Australia
   b. the United States
   c. Brazil
   d. the Soviet Union
   e. China

a 16. All of the following biomes are dominated by conifers. In which is fire a major factor in that domination?
   a. southern evergreen forest
   b. temperate rainforest
   c. boreal coniferous forest
   d. subalpine forest
   e. montane forest

c 17. In tropical rainforests, most small shrubs and herbs occur as epiphytes, high in the canopy, because
   a. the soil is poor and they receive nutrients from the trees
   b. temperatures are warmer
   c. there is more available light
   d. there is more available water
   e. all of the above

d 18. What type of biome most often supports herds of large herbivores?
   a. tropical rainforest
   b. shrublands
   c. temperate deciduous forest
   d. savannah
   e. boreal coniferous forest

e 19. A similarity between deserts and arctic tundra is
   a. hot summers and cold winters
   b. abundant lichens
   c. thin, rocky soil
   d. low mineral availability
   e. low yearly precipitation

c 20. The major difference between a grassland and a savannah is
   a. species diversity
   b. the presence of trees in grassland, none in savannah
   c. the presence of trees in savannah, none in grassland
   d. the types of trees present
   e. their location

d 21. Deciduous forest is present in both North America and Europe because they both have
   a. the same climate
   b. similar soil factors
   c. exactly the same species
   d. a and b
   e. a, b and c

d 22. The entire United States would have a uniform distribution of rainfall if it
   a. had mountains in the west and the rest was flat
   b. had mountains in the east and in the west
   c. had mountains in the center but none on the coasts
   d. was fairly flat
   e. none of the above

e 23. During the Jurassic period, which of the following did not occur?
   a. the North American and Eurasian continents became more temperate
   b. North America separated from Eurasia
   c. Australia separated from Antarctica
   d. the Himalayan Mountains formed
   e. The Ural Mountains formed

b 24. If you were a botanist interested in the study of lichens, which biome would be the best source of those organisms?
   a. tropical rainforest
   b. arctic tundra
   c. temperate deciduous forest
   d. desert
   e. grassland

a 25. A biome that does not exhibit climate differences during the year that produce a growing season alternating with a dormant season would be
   a. tropical rainforest
   b. temperate deciduous forest
   c. grasslands
   d. boreal coniferous forest
   e. all biomes alternate growing and dormant seasons

TRUE-FALSE QUESTIONS

F _____ 1. In the northern hemisphere, the prevailing westerlies occur farther south than the northeast trade winds.

T _____ 2. The area on the landward side of a mountain range, where there is low rainfall, is called a rain shadow.

F _____ 3. Biome physiognomy refers to the metabolism of organisms in the biome.

T _____ 4. Redistribution of heat from equator to poles is due mainly to winds and ocean currents.

T _____ 5. The biome on the northwest coast of the United States is temperate rainforest, a result of mountain ranges that force the prevailing westerlies to rise.

F _____ 6. In the United States, subalpine forest occurs only in some western states.

F _____ 7. The most important information used to determine continental position through time has come from the study of fossils.

F _____ 8. From Minnesota to Louisiana, temperate deciduous forest is practically identical with respect to species competition.

T _____ 9. It is the dominant plants that most strongly characterize a biome.

T _____ 10. Cacti in a North American desert and euphorbs in an African desert are very similar in appearance due to convergent or parallel evolution.

T _____ 11. It sometimes snows in some deserts.

F _____ 12. No species of trees grow in the arctic tundra biome.

F _____ 13. Boreal coniferous forests are dominated by species of pine.

T _____ 14. In a tropical rainforest, the soil itself is very poor in essential elements because those elements are almost exclusively found in the organisms.

F _____ 15. The structure of biomes is a result of climate only.

T _____ 16. In the summer, arctic tundra soil can be very marshy due to the permafrost.

T _____ 17. The taiga is dominated by conifers because the growing season is too short for broad-leaved, deciduous plants.

F _____ 18. The majority of small, herbaceous plants in a temperate deciduous forest bloom in the summer.

T _____ 19. The majority of animals living in a tropical rainforest would be tree-dwellers.

MATCHING

Match each characteristic in List A with a biome in List B. Answers in List B may be used more than once.

List A

a _____ 1. Greatest species diversity
c _____ 2. Richest soil in essential elements
e _____ 3. Permafrost
f _____ 4. May be dominated by one or two species
d _____ 5. Soil thin and rocky
e _____ 6. Lowest species diversity
a _____ 7. Highest rainfall per year
b _____ 8. Winters cold, summers long and warm
d _____ 9. Vegetation patchy due partly to drainage
b _____ 10. Usually has a sub-canopy contributing to species diversity

List B

a. tropical rainforest
b. temperate deciduous forest
c. grasslands
d. desert
e. arctic tundra
f. taiga

ESSAY QUESTIONS

1. Explain how the the western part of Washington state can support lush vegetation, while the eastern part is grassland.

2. For their latitude, which is basically the same as Canada, Ireland and England are warmer than they should be. Why?

3. Explain why the central part of the United States is grassland.

4. Explain in terms of meteorology why: tropical rainforests receive so much rainfall; the Sahara desert is so hot and dry.

5. Describe the origin of the northeast trade and prevailing westerly winds. How does each influence climate in the United States?

6. Explain why there are so few native species of plants in India.

7. Why are the only two continents that have temperate rainforests North and South America?

8. Why are many biologists concerned today about the loss of tropical rainforests?

9. Many tropical rainforests are being cut down or burned by people to grow crops. Knowing what you do about soils, predict their success. Justify your prediction.

10. Discuss why your continent experiences seasonality.

Bio-Art

The following pages contain Bio-Art, a unique testing and studying resource. These are 37 important pieces of text art rendered in black and white, and generally without labels. You can tear them out, add whatever labels you want, and photocopy them onto clear acetates for overhead projector use. You can also copy them and give them to your students as part of a test or to help them study for exams. You do not need to request permission to copy these pieces for classroom use.

1. evolution of cell type
2. cut away of mitochondrion
3. cut away of chloroplast
4. cut away of flagellum
5. cell wall rosettes
6. diagram of mitosis
7. meiosis
8. leaf shapes
9. leaf shapes
10. leaf margins
11. mycorrhize
12. cambium formation
13. angiosperm life cycle
14. inflorescence types
15. Z scheme embedded in membrane
16. glycolysis
17. Krebs cycle
18. guard cells opening and closing
19. phloem loading and unloading
20. mRNA translation
21. monohybrid cross
22. gram negative and positive walls
23. life cycle of rhizopus
24. life cycle of ascomycete
25. life cycle of basidiomycete
26. body form evolution in green algae
27. cell division in algae
28. cell division in algae
29. algal life cycle, theory
30. algal life cycle
31. life cycle of Ulva
32. life cycle of Fucus
33. life cycle of moss
34. life cycle of liverwort
35. branching patterns
36. life cycle of fern
37. conifer ovule and megagametophyte

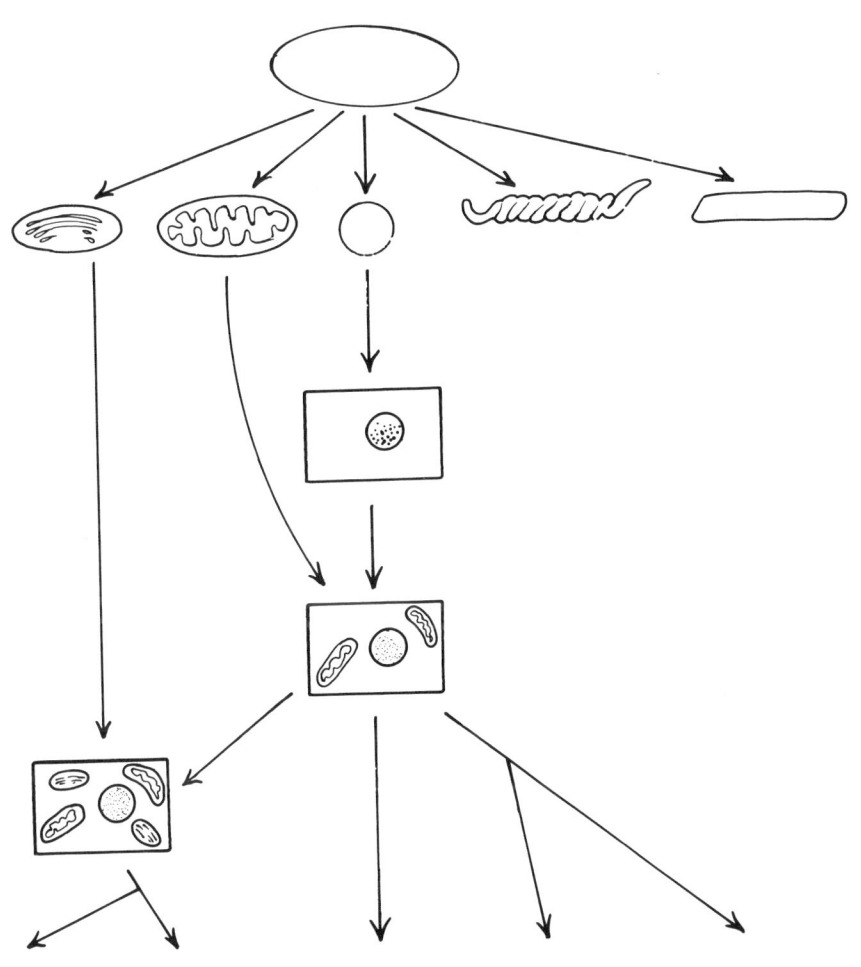

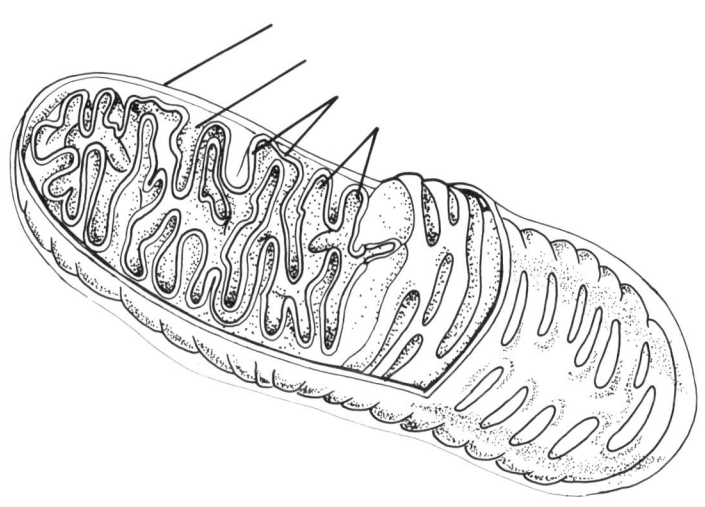

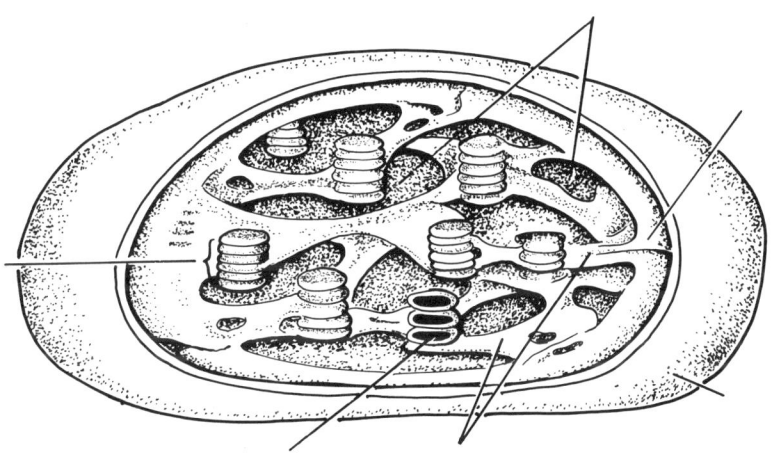

3

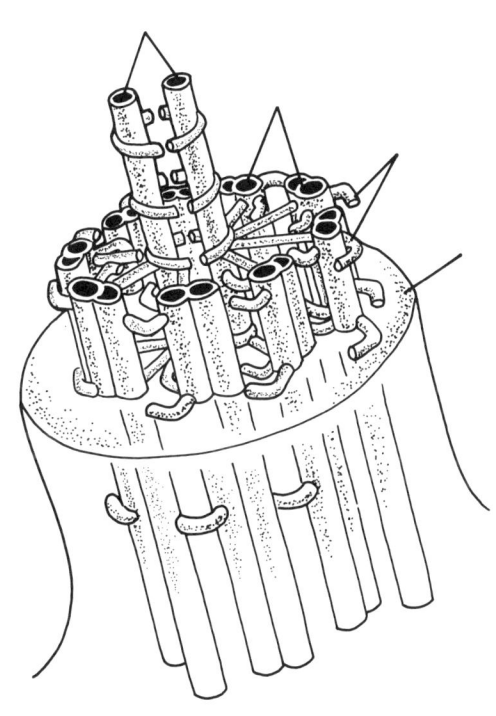

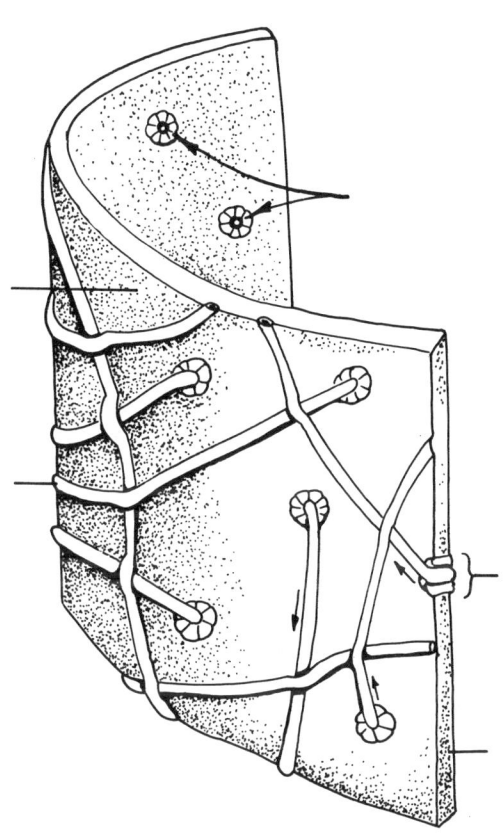

 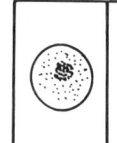 

 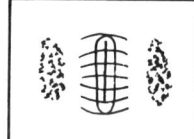

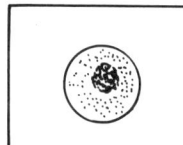 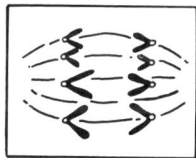

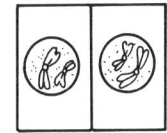 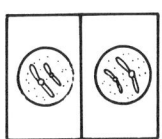 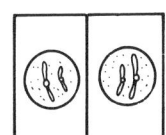

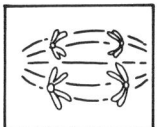 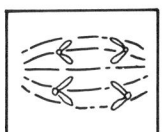 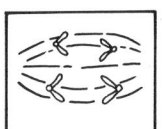

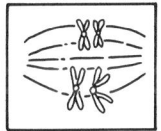 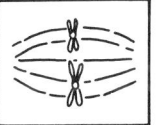 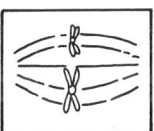

  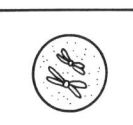

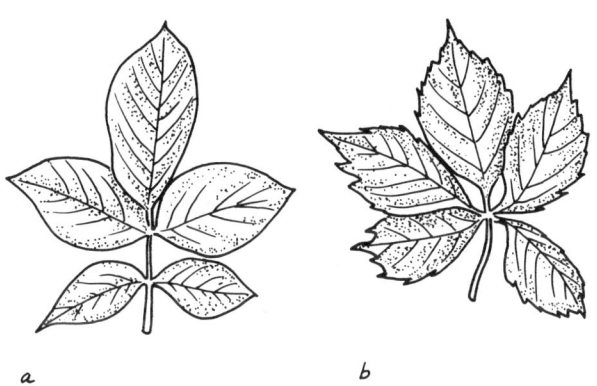

a b

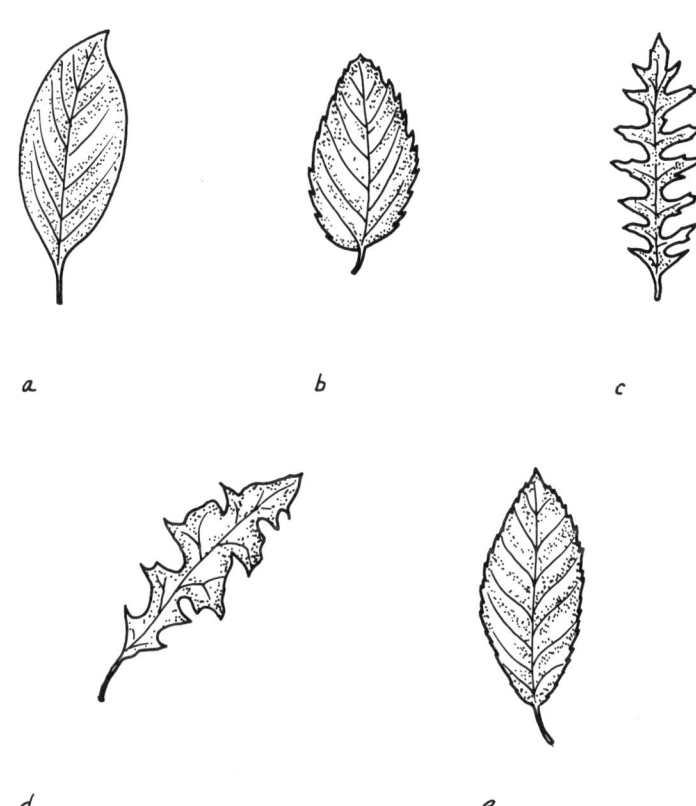

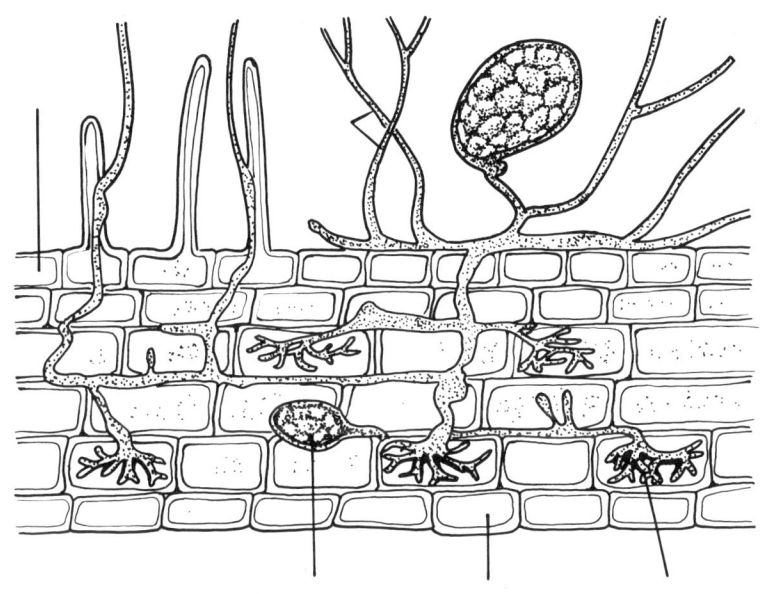

11

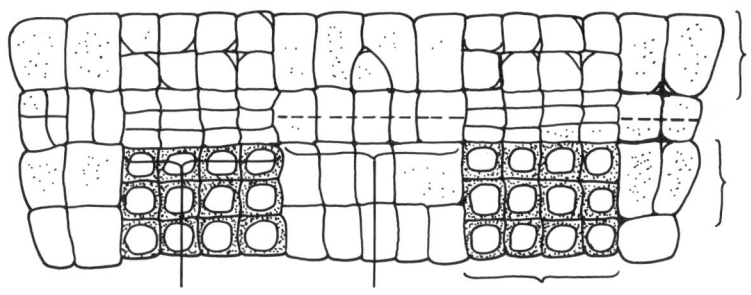

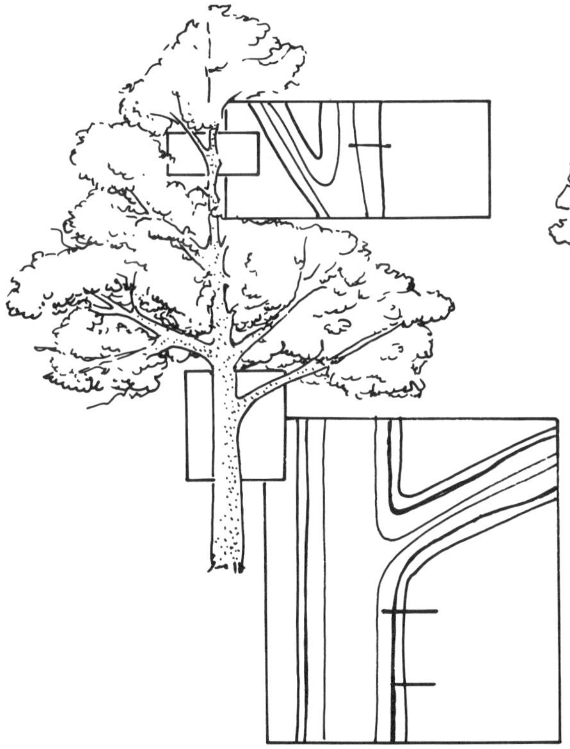

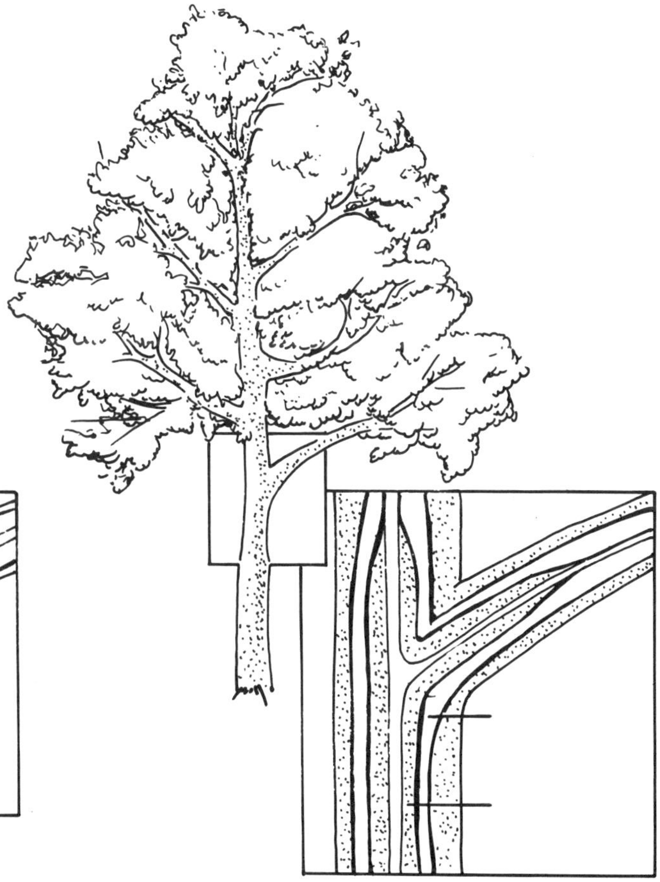

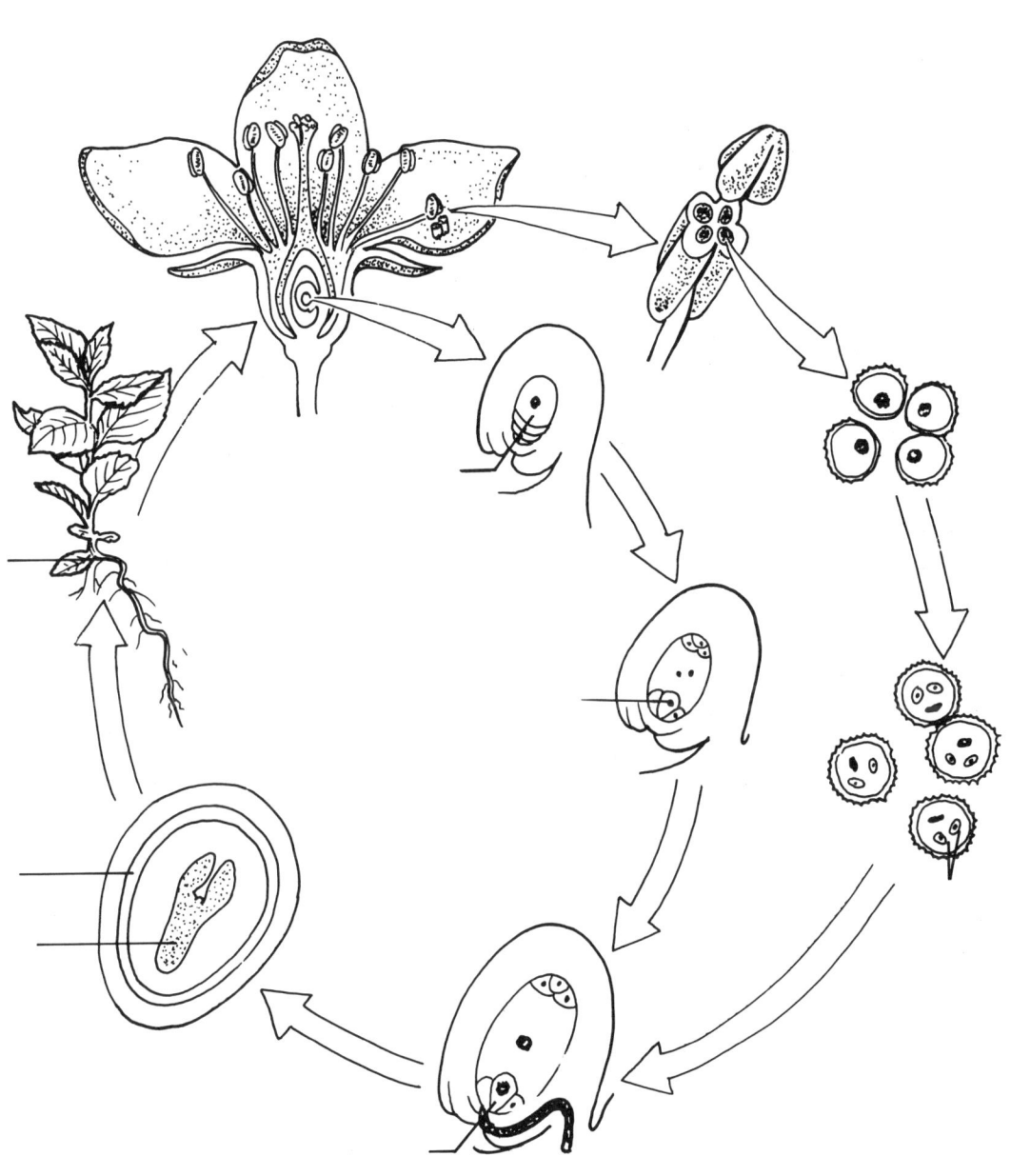

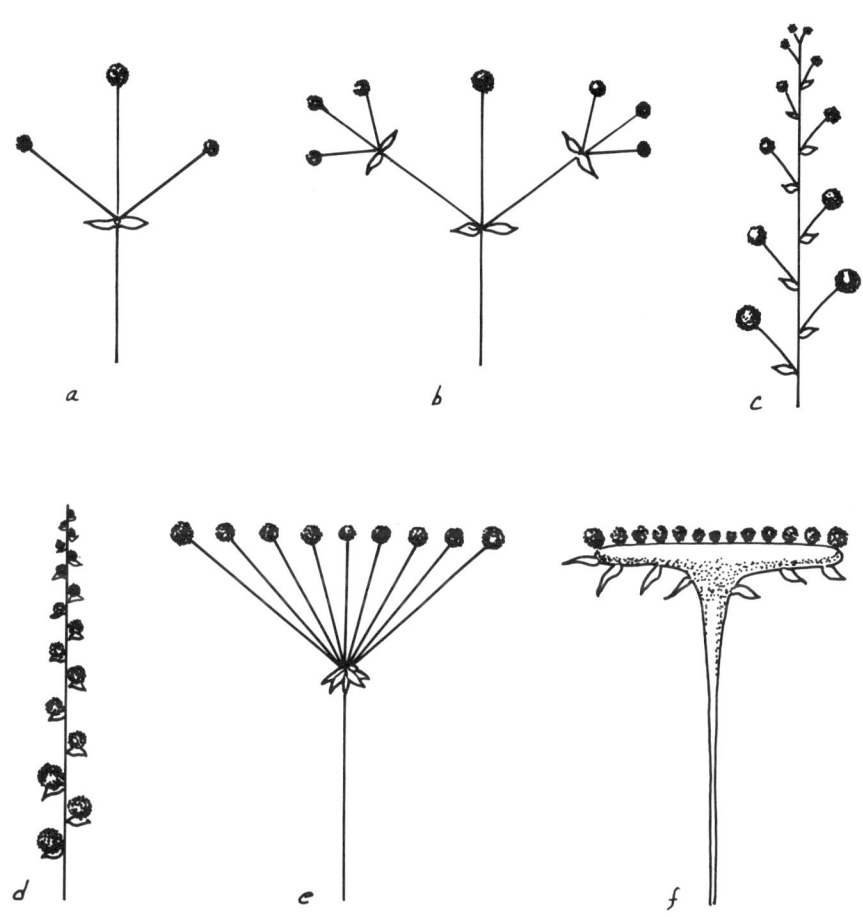

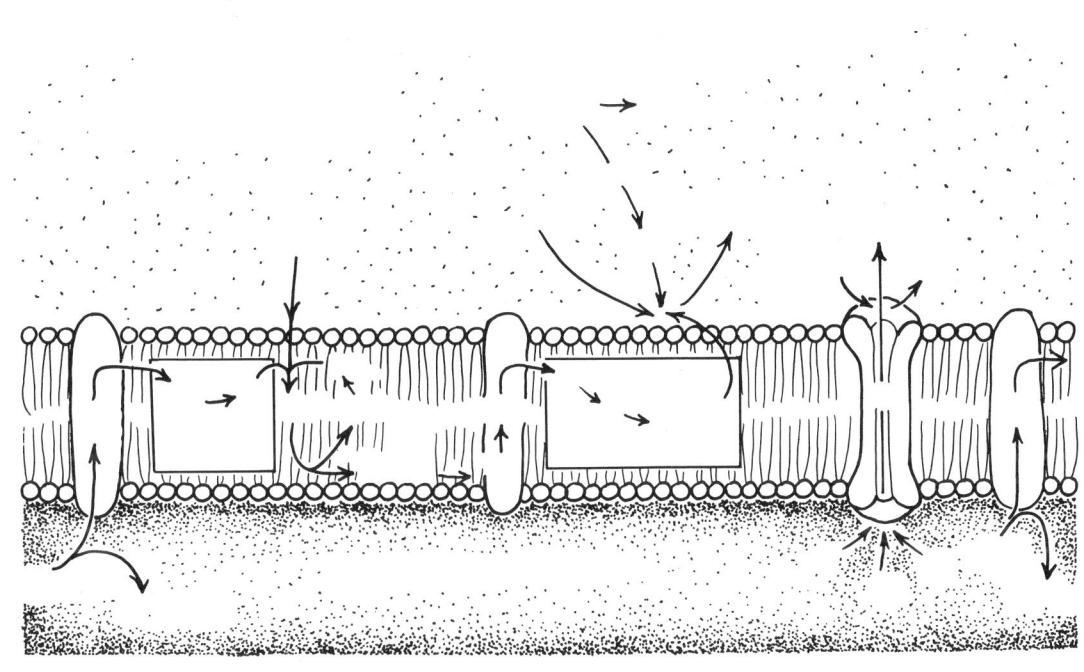

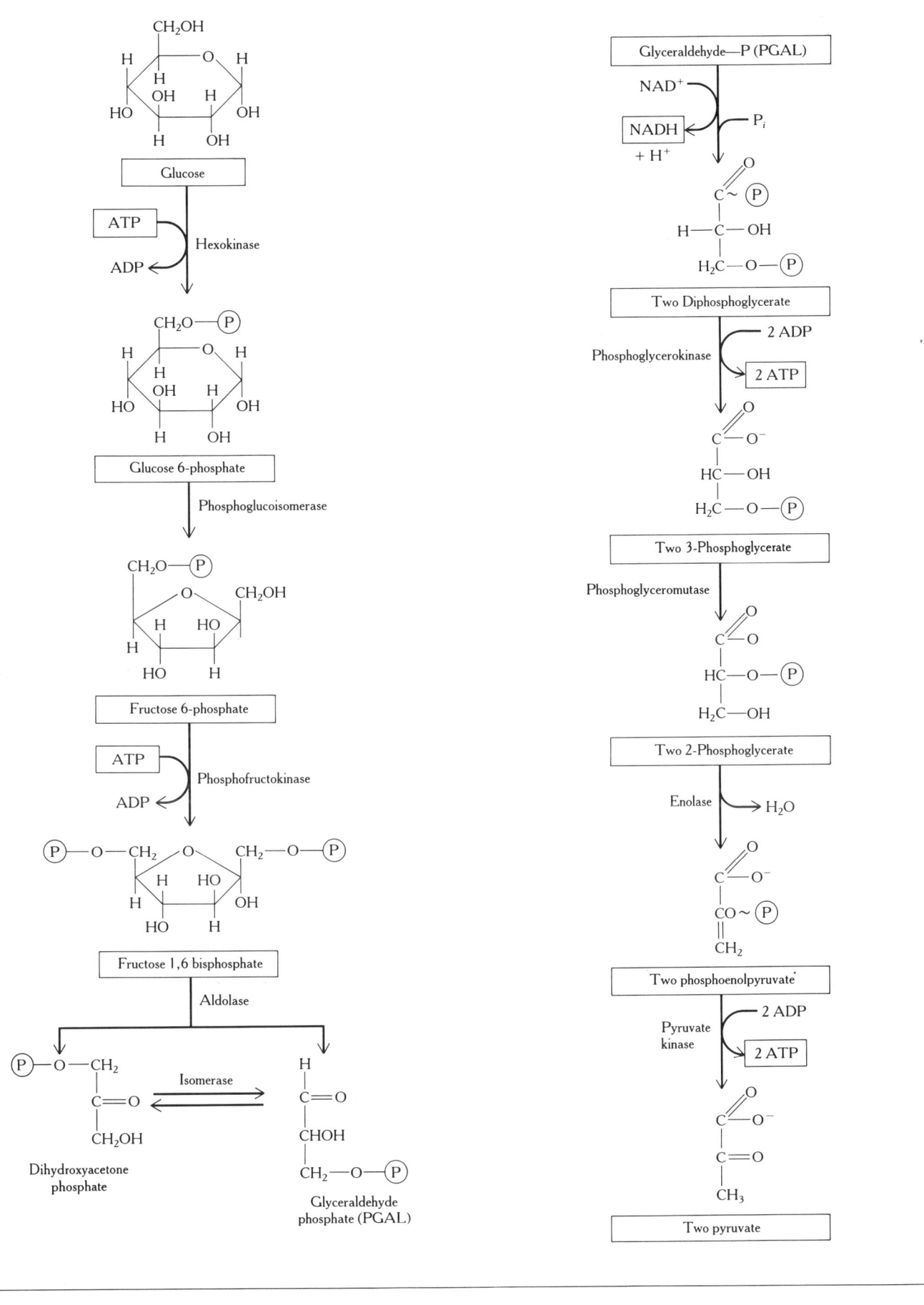

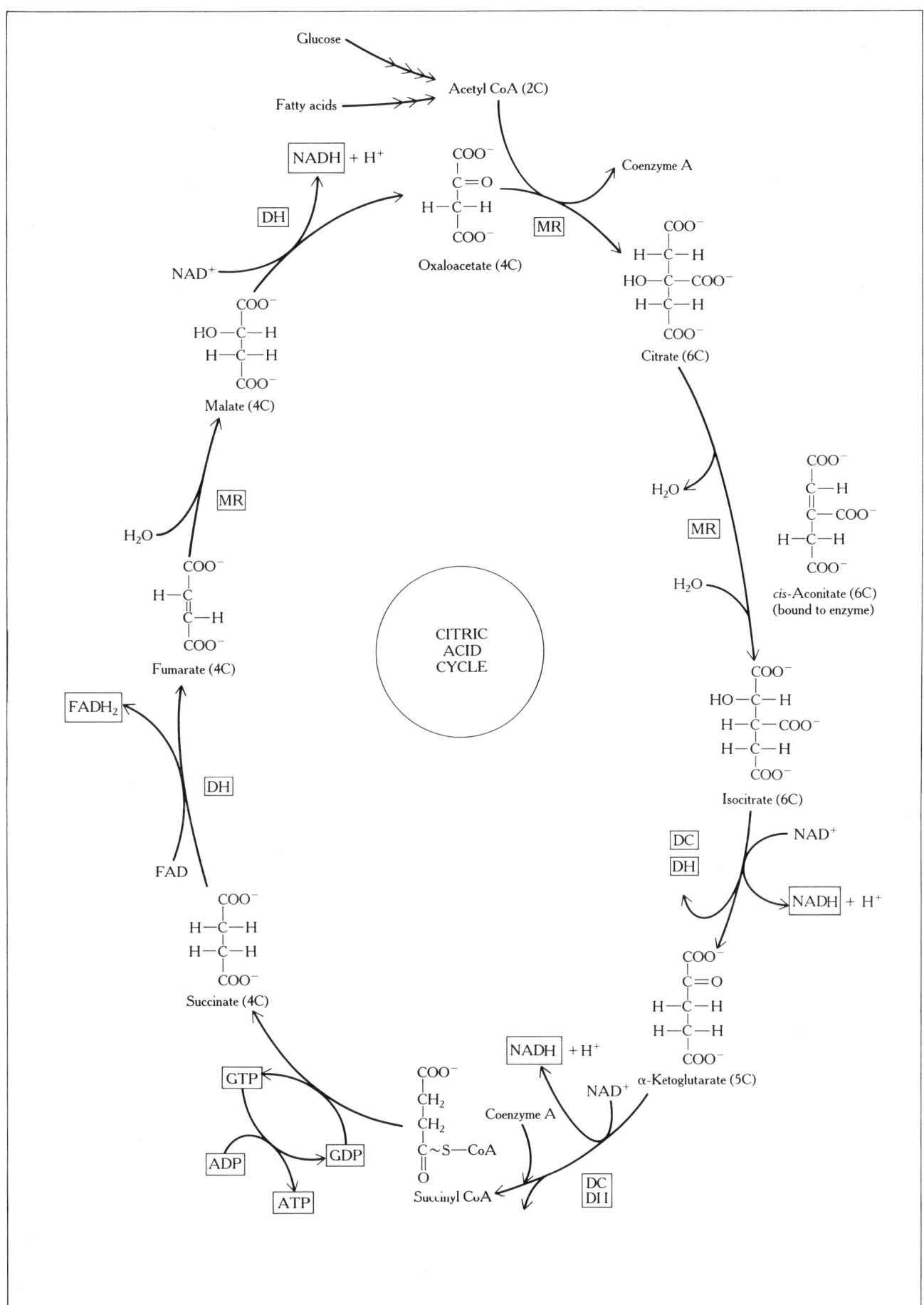

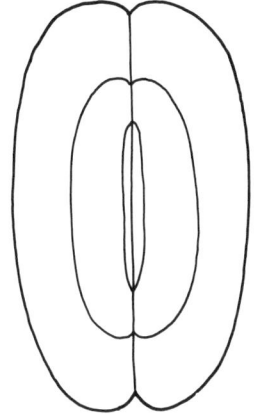

(a) Night

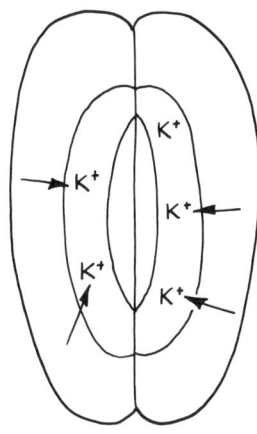

(b) Morning

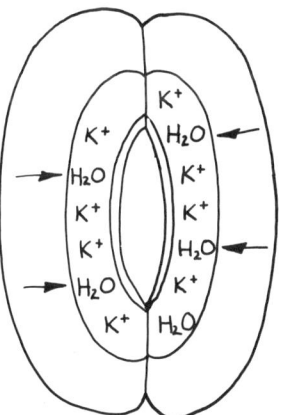

(c) Morning

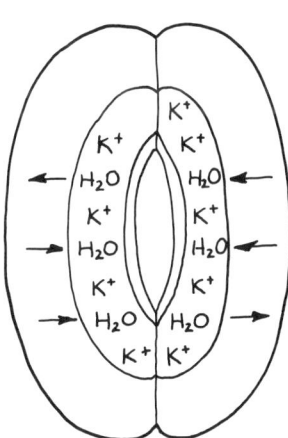

(d) Morning

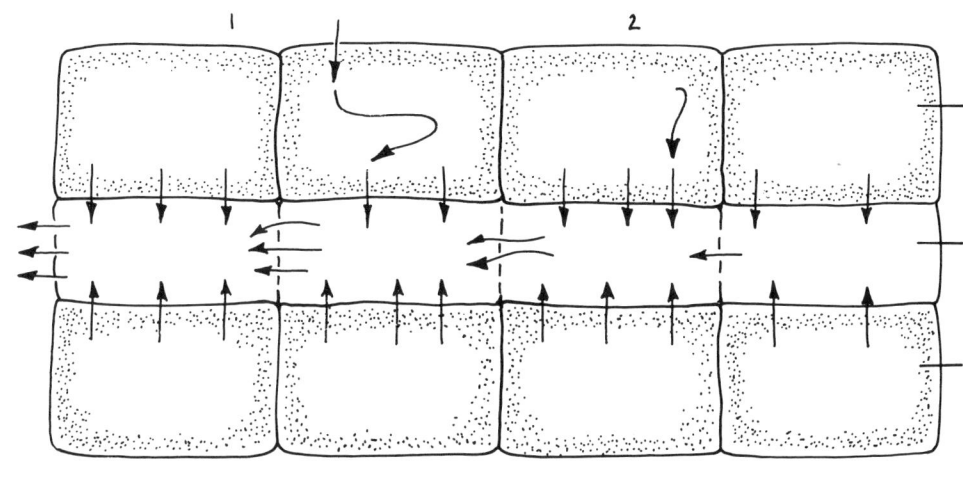

(a)

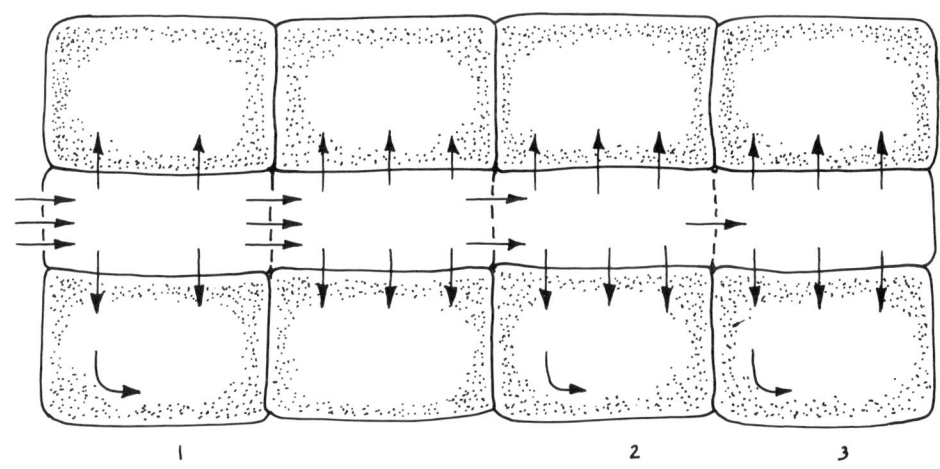

(b)

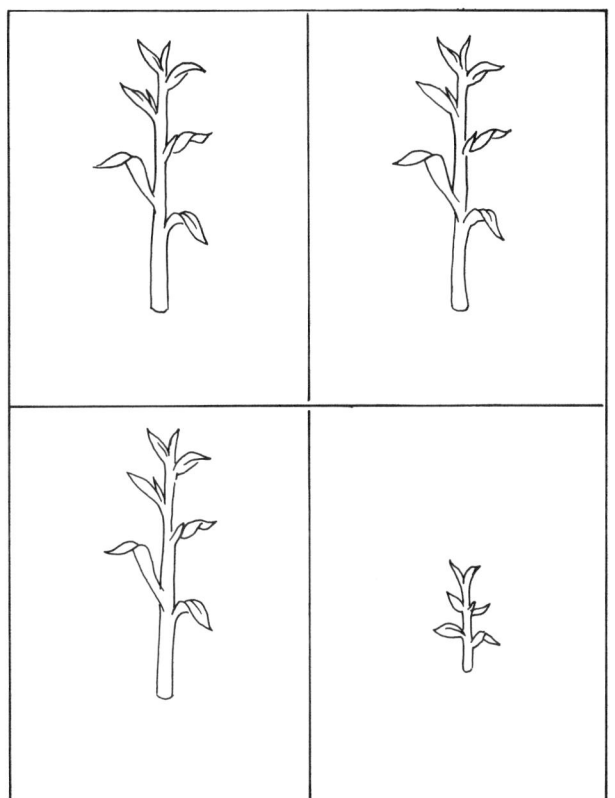

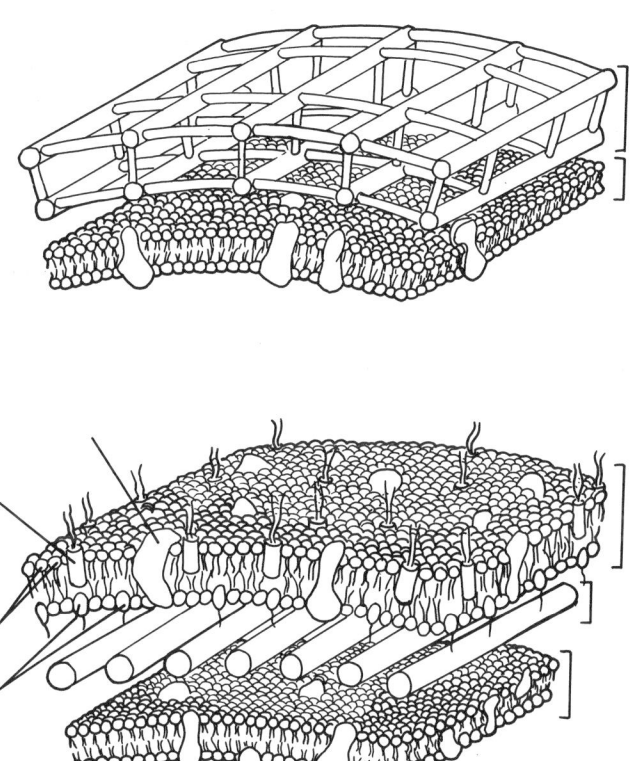

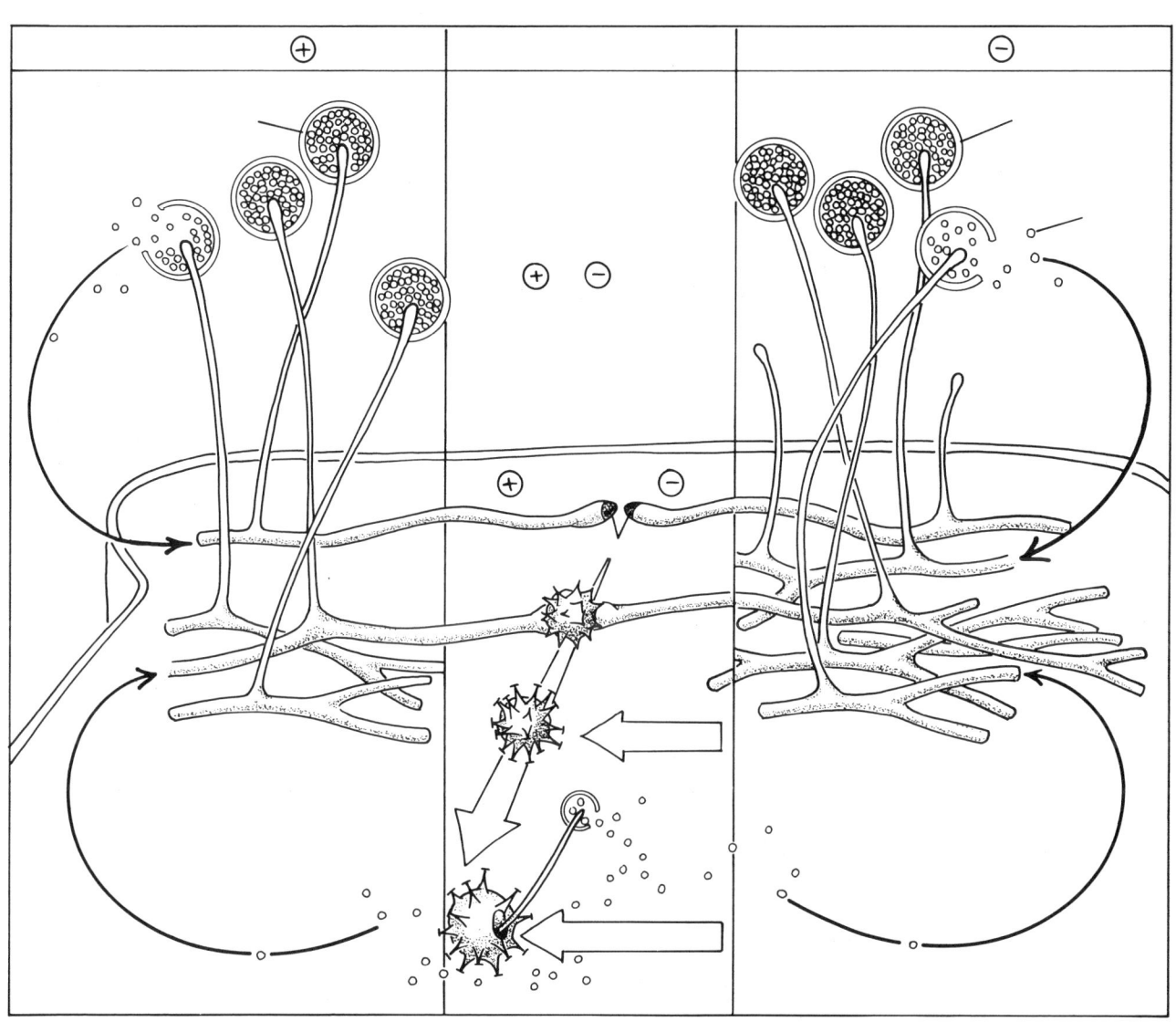

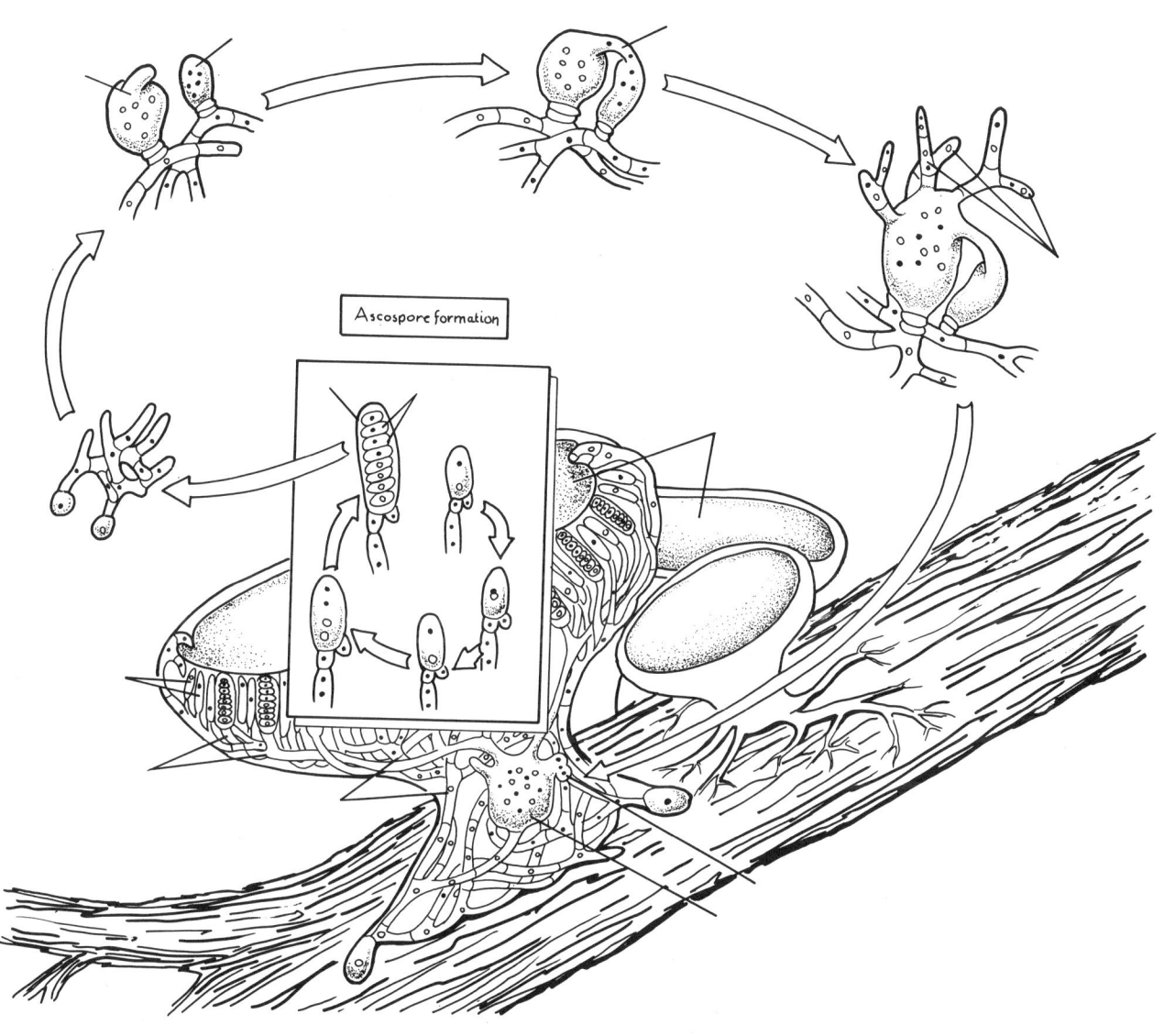

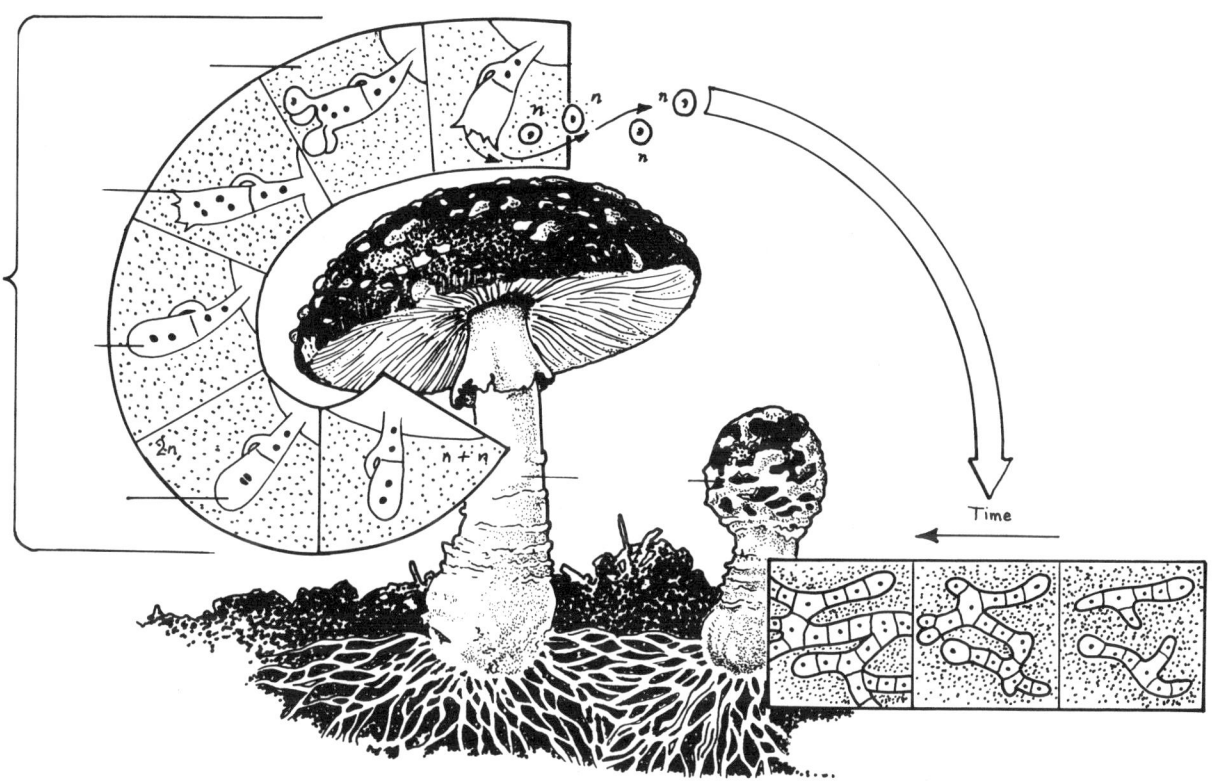

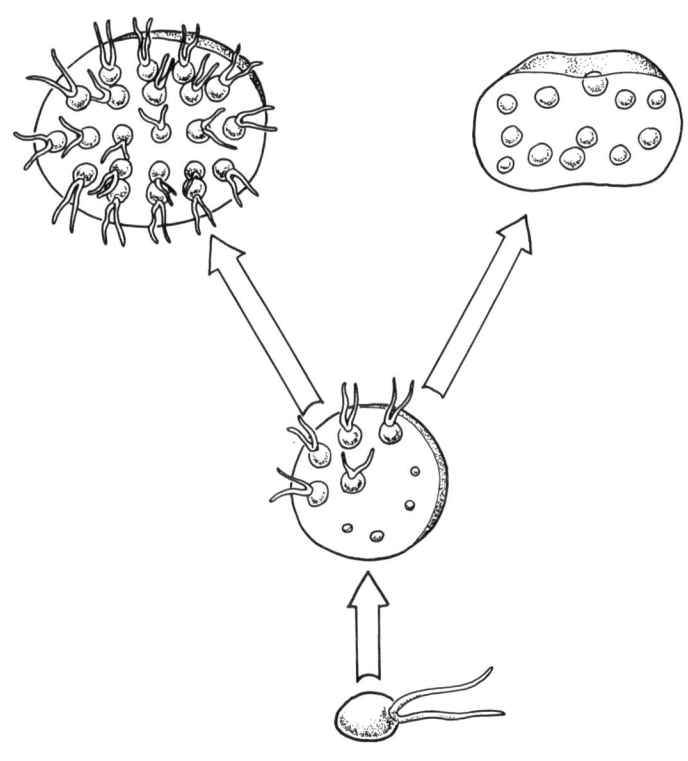

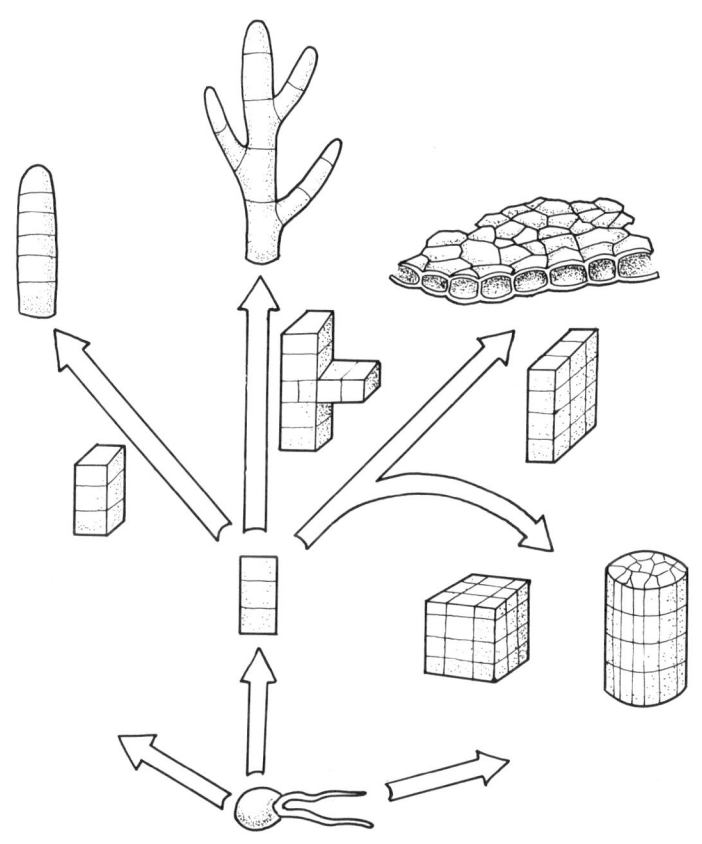

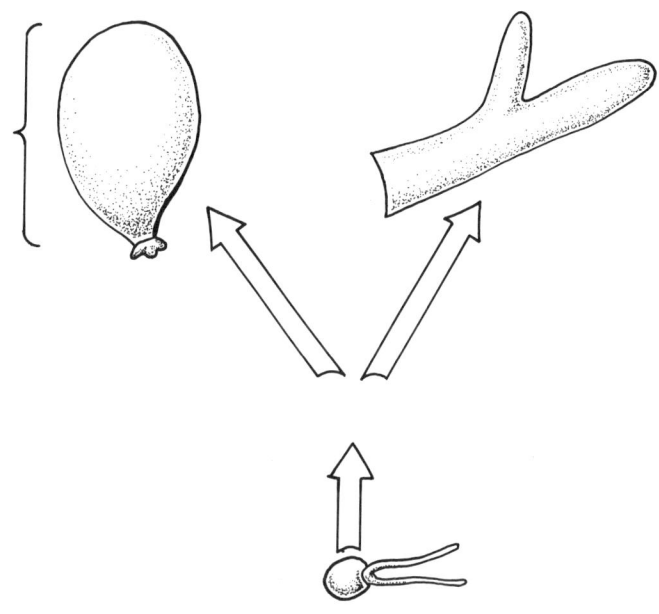

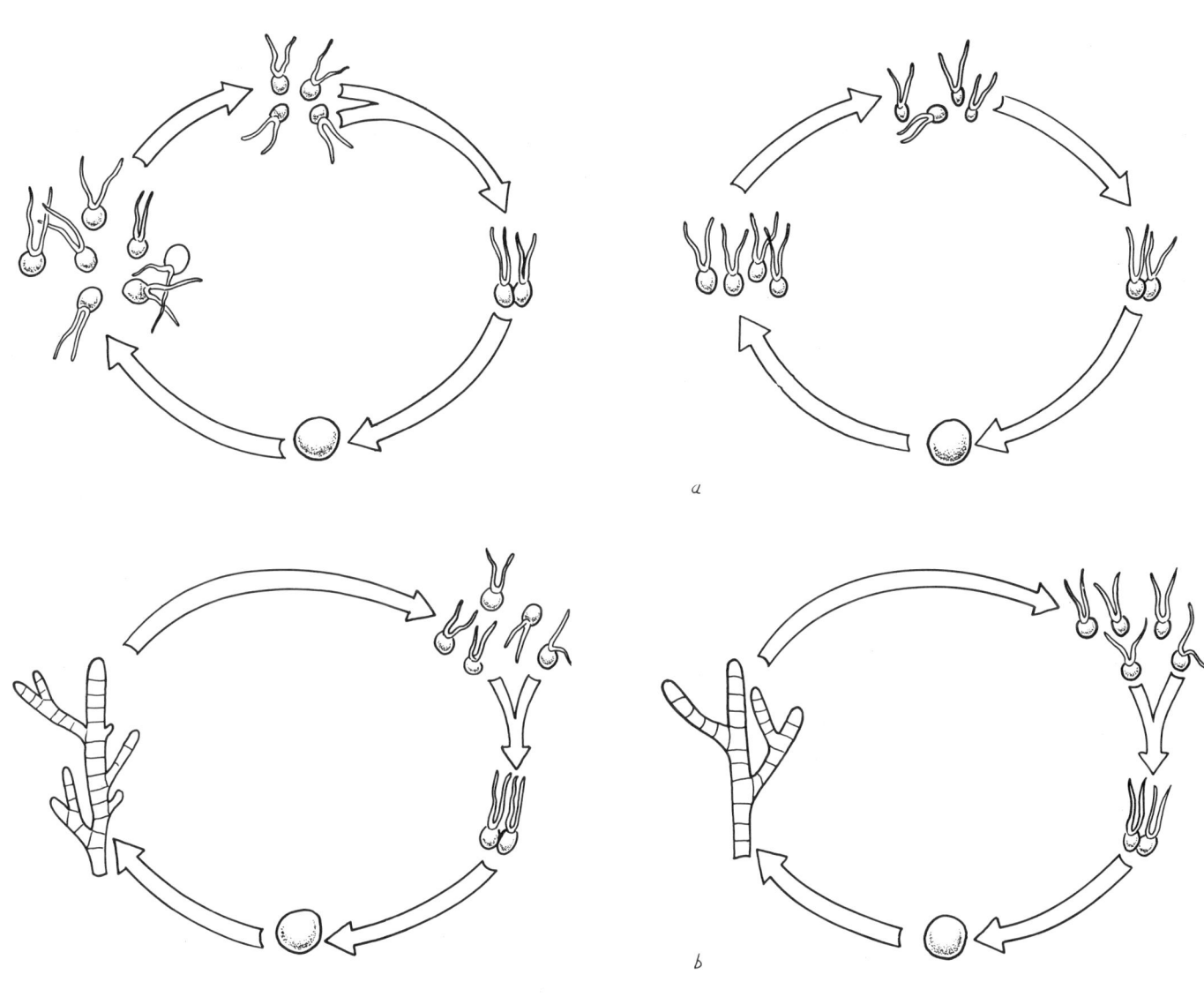

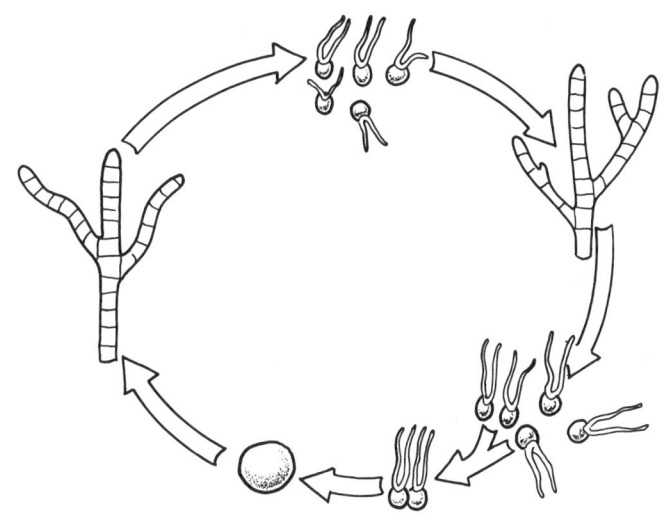

a

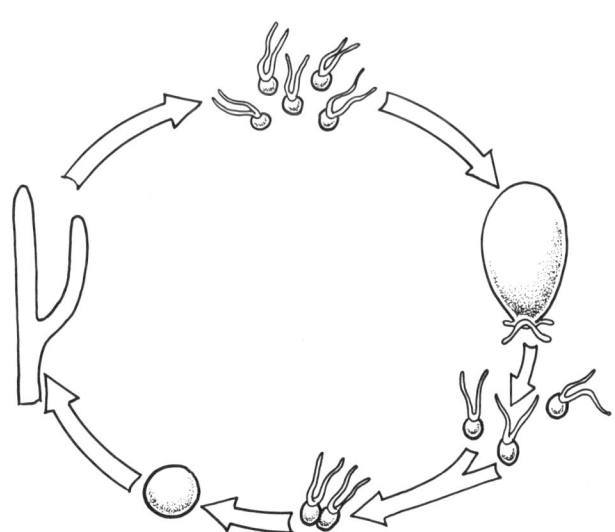

b

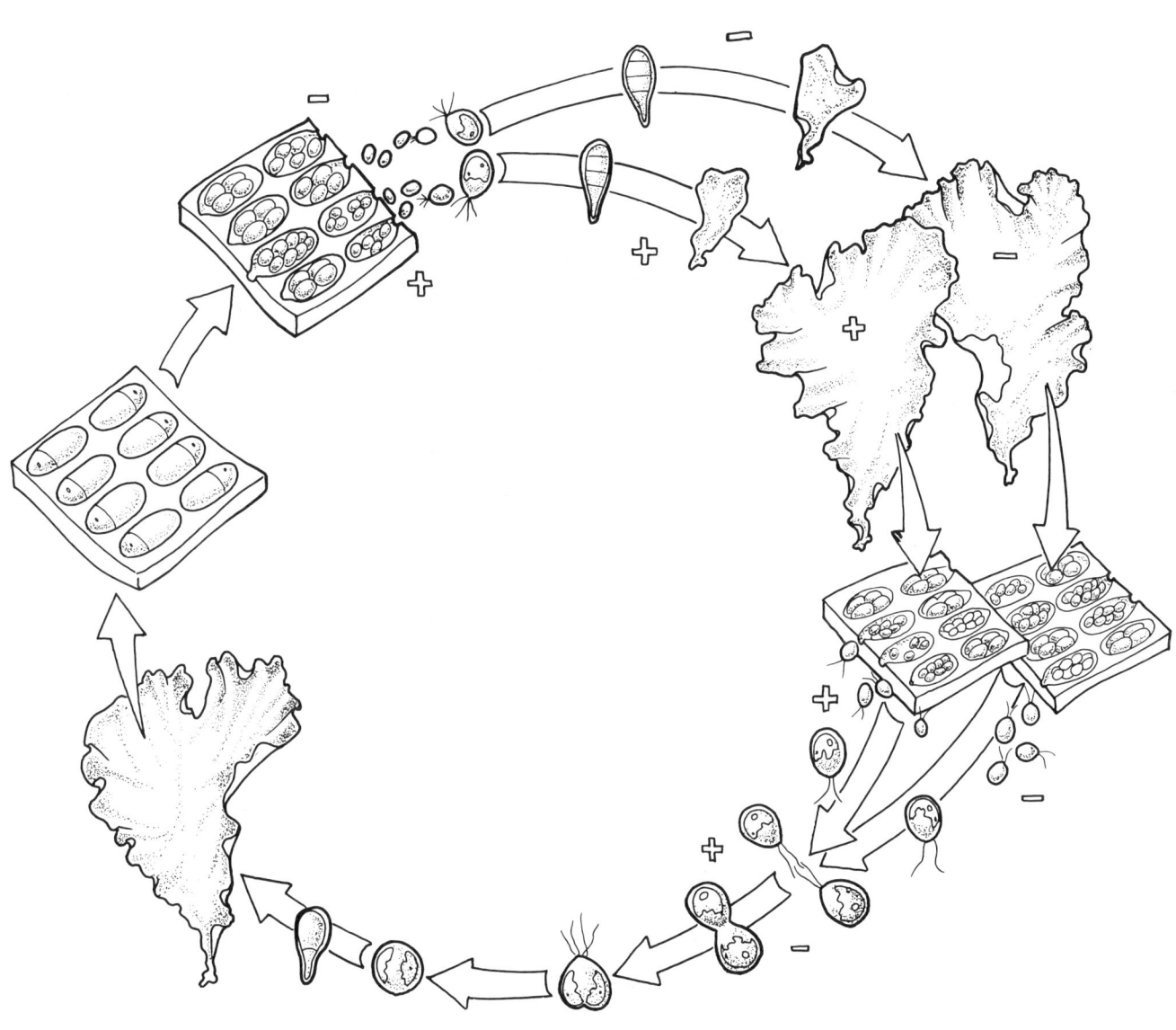

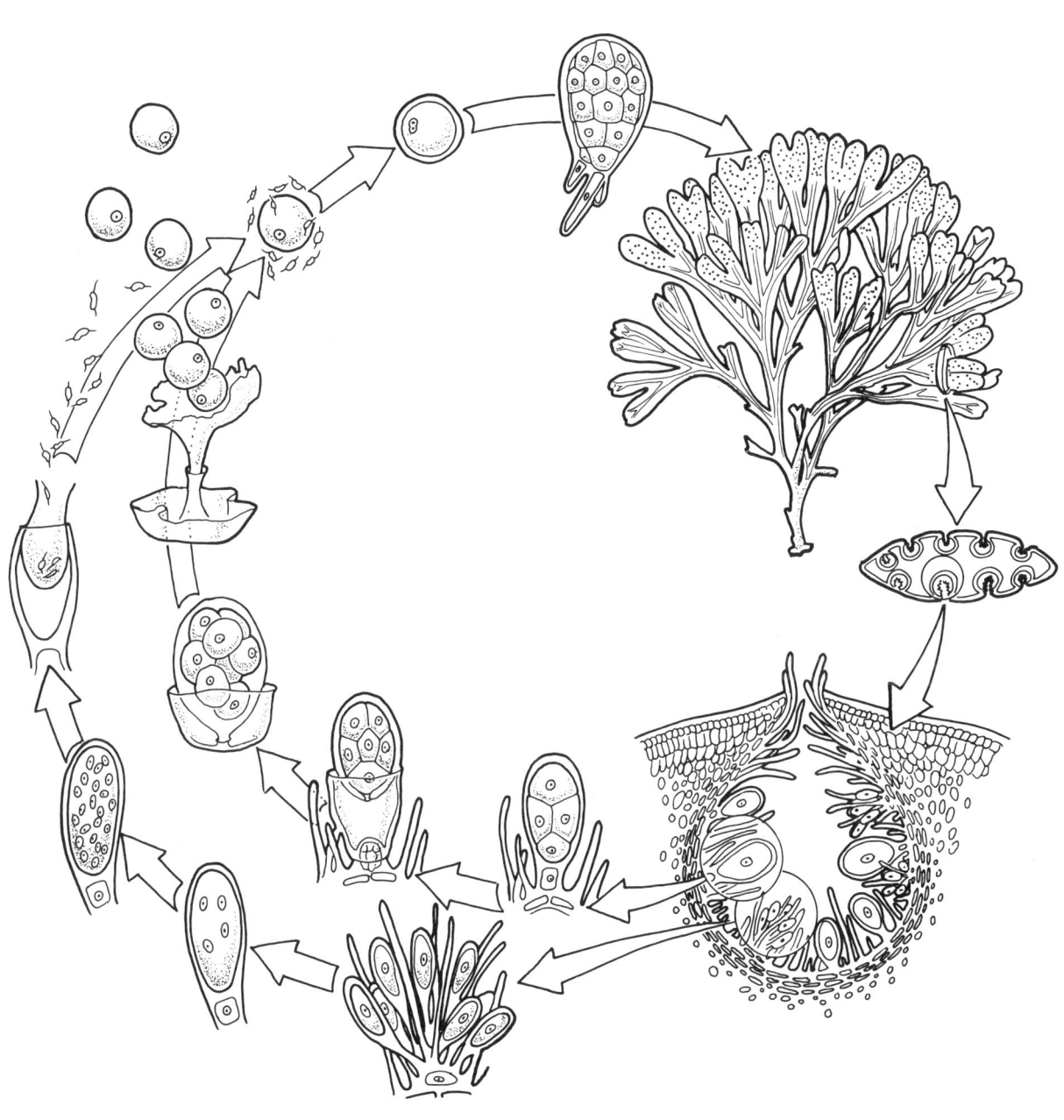

32

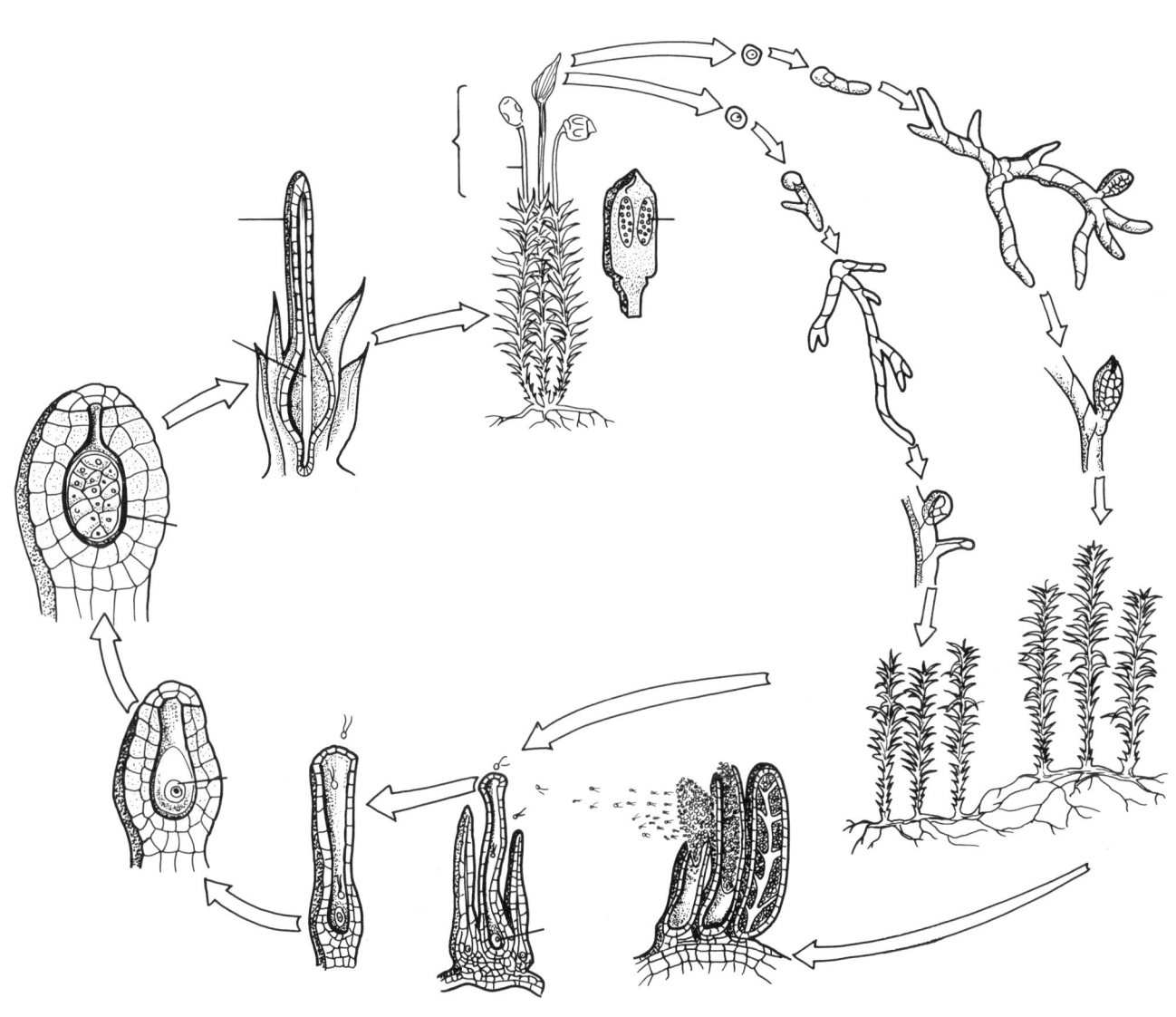

33

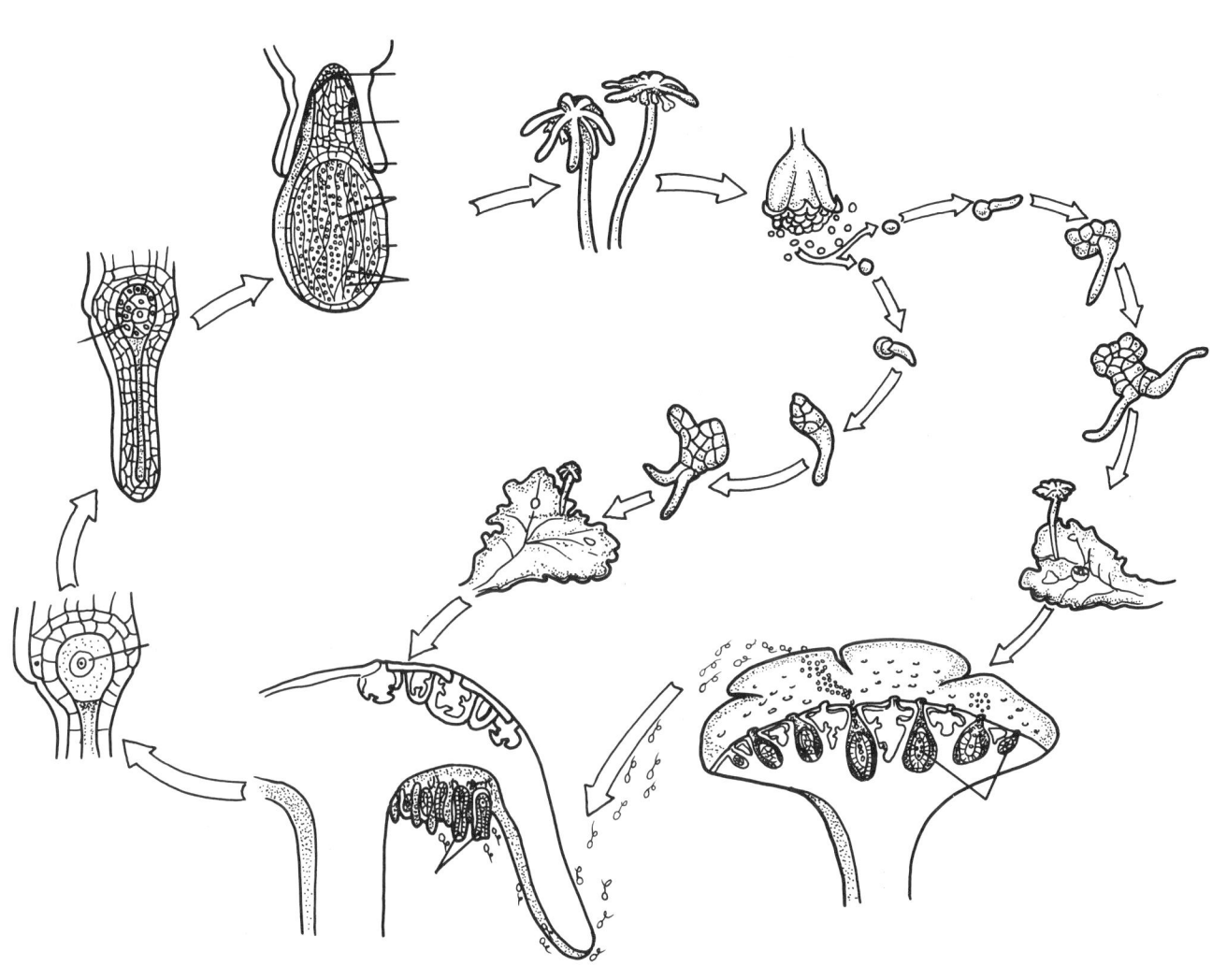

34

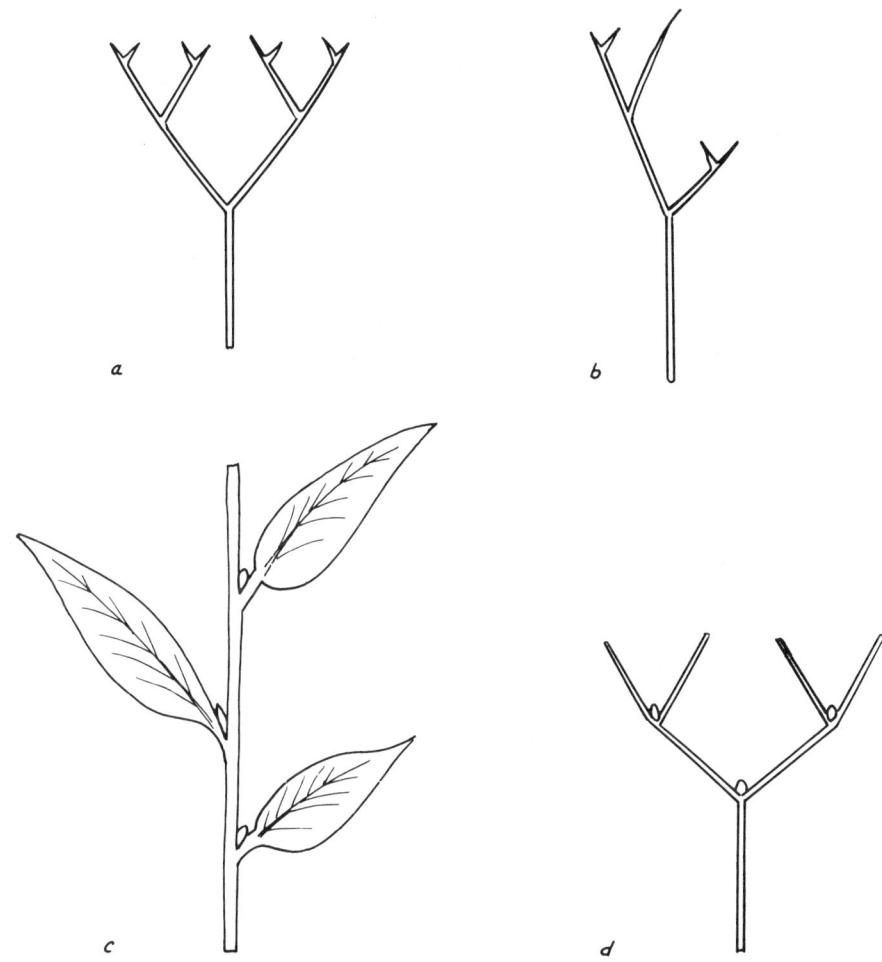

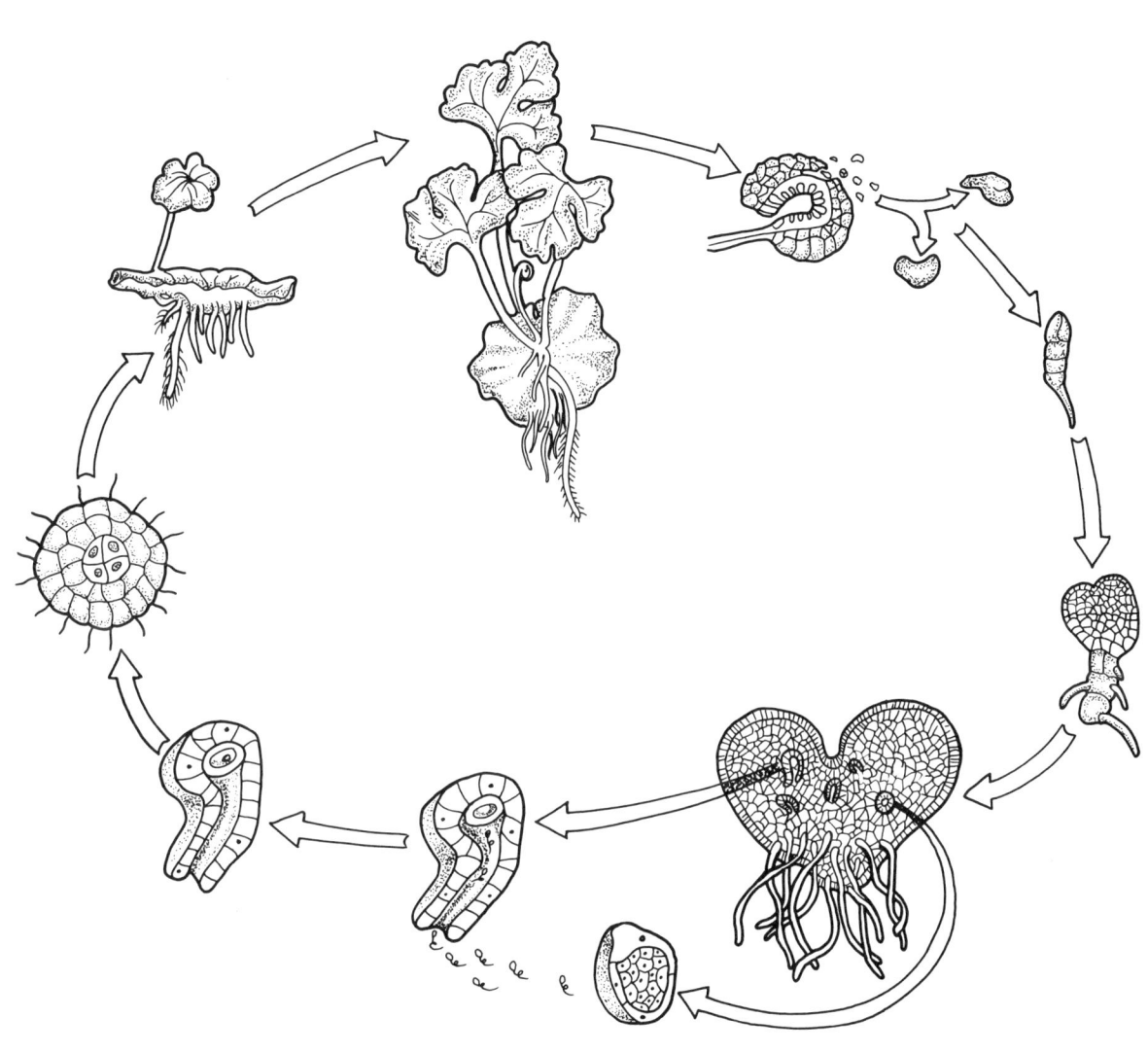

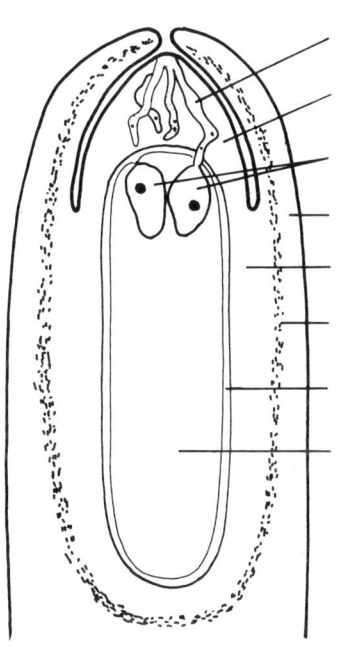